SEED INVIGORATION
TREATMENTS AND QUANTIFICATION OF VIGOUR

THE EDITORS

Professor R.N. Basu Ph.D.(I. A. R.I.), D.Sc.(Cal.) F.N.A.Sc. has been a teacher of the University of Calcutta since 1962. He retired as Khaira Professor of Agriculture after having served for over four decades in various capacities, including two four-year terms (1991-1999) as Vice-Chancellor in the University of Calcutta. In collaboration with his post-graduate students, pre-and post-doctoral research fellows, faculty members and guest researchers, Professor Basu has published around two hundred research papers in plant physiology in reputed national and international journals of which more than a hundred are in the field of seed physiology. This treatise is the outcome of extensive studies on seed invigoration carried out over a period of four decades, since 1974 at the University of Calcutta. The development of hydration-dehydration and dry seed treatments for maintenance of viability, vigour and improved field performance of crops provided new insights into the causes of seed deterioration and its control. Subsequently with the progressive development of concepts and ideas on seed vigour, Professor Basu proposed mathematical-statistical approaches towards quantification of seed vigour status.

Dr. Bina Punjabi is the Academic Advisor to Guru Nanak Vidyak Society, Mumbai, formerly Principal of Guru Nanak College of Arts, Science and Commerce, Mumbai. Dr. Punjabi obtained her Ph.D. from Calcutta University under the guidance of Professor R.N. Basu. She pursued her post-doctoral studies on the physiology of mid-storage hydration-dehydration seed invigoration treatments at the University of Calcutta on a Senior Research Fellowship of University Grants Commission, New Delhi. Thereafter, as a Senior Research Associate at the Bhabha Atomic Research Centre, Mumbai, she continued to work in the same field (1984-1989). Dr Punjabi has been associated with Professor Basu for nearly four decades and has published over two dozen research papers in plant and seed physiology and is presently working on quantification of seed vigour.

SEED INVIGORATION
TREATMENTS AND QUANTIFICATION OF VIGOUR

— Editors —

R. N. Basu

Bina Punjabi

2022

Daya Publishing House®

A Division of

Astral International Pvt. Ltd.

New Delhi – 110 002

ISBN 978-93-5222-065-6 (HB)

Published by : **Daya Publishing House®**
A Division of
Astral International Pvt. Ltd.
– ISO 9001:2015 Certified Company –
4736/23, Ansari Road, Darya Ganj
New Delhi-110 002
Ph. 011-43549197, 23278134
E-mail: info@astralint.com
Website: www.astralint.com

Digitally Printed at : **Replika Press Pvt. Ltd.**

Foreword

We have carried out research on seed vigour for over 40 years. Our emphasis has been the investigation of seed ageing and imbibition damage as the physiological bases of seed vigour and the application of these two phenomena in the practical evaluation of seed vigour. Over the last 20 years, as Chair (AAP) and member (SM) of the Vigour Committee of the International Seed Testing Association, we have introduced vigour tests into the International Rules for Seed Testing. A further aspect of our research has been the improvement of seed vigour using an aerated hydration treatment.

We first became aware of Professor Basu's work in the 1970's when AAP studied deteriorative changes in cell membrane composition during seed ageing as part of her PhD.In this context her attention was drawn to Professors Basu's work on physico-chemical control of seed deterioration. He illustrated that the accumulation of cytotoxic compounds and free radicals produced during lipid peroxidation can be counteracted by halogens such as iodine, and thereby help in maintenance of cellular membrane integrity. The importance of membrane integrity, lipid peroxidation and volatile aldehyde production in seed ageing and seed vigour can be found in the work described in Chapters 2 and 5 of this treatise. This work, and subsequent work on seed invigoration treatments, provided new ideas about the causes of seed ageing and its alleviation, which have been a stimulus to many seed scientists.By the time that we first met Professor Basu at an International Conference on Seed Science and Technology in New Delhi in 1990 his work in this area had developed further and was even more widely known and recognised internationally.

Professor Basu's work on the control of seed deterioration is significant because a major factor affecting seed quality is deterioration as a result of ageing during seed storage, which results in reduced vigour and eventually loss of the ability to germinate. His work on treatments that can be applied pre-storage or mid-storage to prevent or reduce the extent of deterioration has played an important role in the development of effective ways of maintaining and improving seed quality.

These have included use of non-aqueous organic solvents to apply chemicals to the seed, dry seed treatments and short periods of hydration. The latter were of particular interest to us in our own work on hydration seed treatments. To us, the most significant aspect of all Professor Basu's work in this area is that many of the treatments developed can be applied by farmers as well as seed companies, and they have relevance to developed, emerging, and developing economies. This is to be particularly commended as the practical application of scientific research is frequently overlooked in the race to publish the results of research. The reader is referred to the detail provided on this topic in Chapters 3 and Annexures

Identification of differences in seed vigour is of major interest throughout the seed sector, with many vigour tests identified, frequently based on the process of seed ageing. One such test is a bioassay developed by Professor Basu, based on the increase in the release of volatile lipid peroxidation products during seed ageing and the inhibitory effects of these gases on seedling growth of a high vigour control (see Chapter 5). This is a test that gives clear visual evidence of differences in seed vigour and has formed an effective evaluation of seed vigour and the impact of seed treatments in much of his work.

Despite the existence of many different vigour tests that have been developed in many laboratories worldwide, there has been no single numerical assessment of vigour that can combine the data from a number of tests and hence give an overall vigour classification that would be of practical value. This ambitious task has been a major outcome of Professor Basu's work which is described in Chapter 6. He has used a range of vigour-related physiological, biochemical, and morphological parameters to produce a simple arithmetic mean value which quantifies the vigour status of a seed lot. The reader is recommended to study his approach, which is described in detail in this chapter.

The research papers and reviews that contribute to this treatise are based on collaboration between Professor Basu and many others, including post- doctoral research scholars,post-graduate students and visiting researchers. They represent a lifetime of devotion to seeds and seed quality, which has resulted in a significant contribution towards providing a sound basis for successful crop establishment and food production. The publications have, of course, also been part of the education of past students and researchers and this treatise will provide valuable source material for the education of the seed scientists of the future.

Dr. Alison A Powell
Honorary Senior Lecturer, University of Aberdeen, UK
ISTA member
Chair of ISTA Vigour Committee and ISTA Seed Science Advisory Group

Dr. Stan Matthews
Honorary Senior Research Fellow,University of Aberdeen,UK
Member of ISTA Vigour Committee and ISTA Seed Science Advisory Group

Acknowledgements

The process of writing this monograph though an arduous task has given us immense satisfaction. We have endeavoured to consolidate studies on seed invigoration carried out in the College of Agriculture, University of Calcutta, spanning over four decades since 1974. The editors gratefully acknowledge the contributions of all faculty members, junior and senior pre-and post-doctoral research scholars, post-graduate students and visiting researchers who were associated with Prof. R.N. Basu in the studies on seed viability, seed vigour and seed invigoration during that period. The names of the authors of various publications resulting out of the researches on the subject (presented as abstracts in chapter 4) are given hereunder in alphabetical order of their surnames:

Dr. P.K. Bandopadhyay, Dr. P. Basu, Dr. A.K. Bhattacharya, Ms. Krishna Chattopadhyay, Dr. Nilanjana Choudhuri (*nee* Dhar), Dr. Malavika Dadlani (*nee* Dasgupta), Dr. B.K. De, Dr. C. Dharmalingam, (late) Dr. Gopa Dey (*nee* Pathak), Dr. B. Kapri, Dr. K. Krishnamoortthy, Ms. Pampa Kuchlan, Dr. C. Kundu, Dr. A.K. Lahiri, Dr. A.K. Mandal, Dr. R. Mitra, Ms. Anuradha Mukhopadhyay, Dr. Renuka Mandal (*nee* Saha), Dr. Prativa Pal, Dr. Bina Punjabi, Dr. Dolly Rudrapal (*nee* Pan), Dr. A.B. Rudrapal, Dr. F. Rahman, Ms. T. Ray, Mr. S. Ray and Dr A.K. Sengupta

Senior faculty members, scientists from several universities, institutes and organisations in West Bengal, as well as from across the country, have contributed by sharing their views during the course of our studies on seed invigoration, thus enriching this endeavour. We have been greatly inspired by eminent scientists in the relevant disciplines of seed science and technology at the international level who have highlighted our findings and views in their scientific publications and reviews on seed invigoration and thereby, in general, supporting and confirming them. We specifically acknowledge the very encouraging remarks by (late) Dr. Walter Heydecker of the University of Nottingham, U.K., one of the earliest scientists to have worked on the idea of seed invigoration for improving crop performance. He was a source of inspiration for us especially in the initial phases of our studies on mid-storage hydration-dehydration treatments for maintenance of seed viability, vigour and improved field performance. We thank (late) Prof. E.H. Roberts, University of Reading,U.K. who along with his team pioneered the concept of seed viability nomographs for recommending our humidification-drying treatments for long term preservation in seed banks which was highly motivating and indeed very encouraging . We appreciate Dr. S. Matthews and Dr. Alison Powell..... of the University of Aberdeen, U.K. for their keen interest in our studies and we acknowledge Dr M.B. McDonald, University of Ohio, U.S.A. for supporting our views on counteraction of lipid peroxidation as a major reason for the beneficial effects of seed invigoration treatments. The salient observations of these eminent scientists have been summarised and presented in Annexure No A-7.

The editors recall with gratitude the valuable suggestions, fruitful discussions, constant encouragement and support in various ways by (late) Prof. Arun Kumar Sharma, former Ghosh Professor of Botany, Calcutta University, (late) Dr. S.L. Basak former Director of Central Research Institute for Jute and Allied Fibres, Barrackpore, West Bengal and Prof. T.K. Bose, Ex-Director of Research, Bidhan Chandra Krishi Vishwavidyalaya, Kalyani, West Bengal.

(Late) Prof. T.M. Das (Editor, *Indian Biologist*) took keen interest in our studies on mathematical-statistical approaches towards seed vigour quantification and his persistence enabled the publication of the paper "Approaches to Quantification of Seed Vigour" in the *Indian Biologist* in 2004, and the second paper on "A Holistic Approach to Quantification of Seed Vigour through Integration of Morphological and Physiological Parameters" in 2007, in the same journal. We also thankfully recall the verbatim publication of this paper in *Seed Research* (2016) with his kind permission. The initiative taken by Dr. P.K. Agrawal, former editor of the journal for its publication is acknowledged.

Thanks are due to Dr. Malavika Dadlani, President of the Indian Society of Seed Technology and former Joint Director (Research) Indian Agricultural Research Institute, New Delhi, for her valuable inputs and for editing the final manuscript.

We thank Mr. V. Shinde (Administration), Tata Institute of Social Sciences, Chembur, for meticulously typing the manuscript, Mr. R. Savant for the art work and figures in the text, Mrs. Maria F. Fernandes for her assistance in the proof reading and Mr. V.K. Mishra and Mr. A. Vishwakarma of G.N.V.S. Institute of Management for secretarial assistance.

We take this opportunity to express our gratitude to the Management of Guru Nanak Vidyak Society for support.

Finally we appreciate our families for their patience and encouragement throughout our endeavour. We sincerely acknowledge our friends and well wishers and thank Astral Publications for their kind cooperation and keen interest in publishing this book.

R.N. Basu

Bina Punjabi

Editorial Submission

Seed invigoration implies a significant improvement in seed performance by any post-harvest physiological treatment that would result in better storability (post-storage germination and seedling vigour) and improved field performance and productivity of the resultant crop. Seed invigoration of orthodox seed with easily practicable 'wet' hydration-dehydration and 'dry' (non-aqueous via organic solvent, vapour phase or dry dressing with powdered materials) treatments has been duly standardized for the preservation of seed vigour and viability and improvement of field performance and productivity of a wide range of crop seeds. Extension of storage life of a few recalcitrant seeds has also been possible. The efficacy of the different invigoration treatments depends on the botanical (genetical, morphological and physiological) nature of the seeds with initial vigour level of the seed at the time of treatment being a very important consideration in choosing the appropriate treatment. In view of the biological differences between different kinds of seeds preliminary routine standardization with small samples should be taken up before large-scale application with big commercial seed lots. A critical perusal of the list of abstracts of selected publications in Chapter 4 may be advantageous for the purpose. The significant improvement of correlations between plant growth and yield performance attributes with the different physiological and biochemical parameters across genus and species barriers would reflect the benefits of seed invigoration treatments.

Subsequently, with the progressive development of ideas and concepts on quantification of seed vigour, it has been possible to mathematically express the extent of beneficial effects of seed invigoration treatments in quantitative terms in percentage values (with a theoretical maximum value for the seed vigour status of a seed lot not exceeding 100 per cent as in case of standard seed germination tests).

R.N. Basu

Bina Punjabi

Contents

Figures and Diagrams

Abbreviations

AA: Accelerated Ageing

AERS: Amylolytic Enzyme Resistance Score

ALRS: Amino Acid Leaching Resistance Score

AOSA: Association of Official Seed Analysts

ATP: Adenosine Triphosphate

BCKV: Bidhan Chandra Krishi Vishwavidyalaya

BHT: Butylated Hydroxytoluene

CCD: Charged Coupled Device/Close Circuit Television Camera

CGIAR: Consultative Group on International Agricultural Research

CMSA: Community Managed Sustainable Agriculture

CR: Conductivity Resistance

CRIJAF: Central Research Institute of Jute and Allied Fibres

CSVS: Correlated Seed Vigour Score

CT: Cold Tests

CU: Calcutta University

Cv: Cultivar

D-D: Dipping Drying

DEAS: Dehydrogenase Enzyme Activity Score

DNA: Deoxyribonucleic acid

EC: Electrical Conductance

ECRS: Electrical Conductance Resistance Score

EDTA-Na: Ethylenediaminetetracetic acid Sodium Salt

EPR: Electron Paramagnetic Resonance

ESR: Electron Spin Resonance

FE: Field Emergence

G%: Germination Per cent

GA: Gibberellic Acid

GF: Germination Factor

GHGs: Green House Gases

GHz: Gigahertz

GM: Genetically Modified

GMO: Genetically Modified Organism

GSF: Germination Cum Seedling Factor

HD: Hydration-dehydration

HMV: High Medium Vigour

HV: High Vigour

HYVs: High Yielding Varieties

IAA: Indole Acetic Acid

IAASTD: International Assessment of Agricultural Knowledge Science and Technology for Development

IARI: Indian Agricultural Research Institute

ICAR: Indian Council of Agricultural Research

ICRISAT: International Crop Research Institute for Semi-arid Tropics

IPCC: Intergovernmental Panel on Climate Change

IPR: Intellectual Property Rights

IRRN: International Rice Research Newsletter

IRs: Ionising Radiations

ISST: Indian Society of Seed Technology

ISTA: International Seed Testing Association

JSBS: Jute Seedling Bioassay Score

KR: Kilorads

LHPO: Lipid Hydroperoxides

LMV: Low Medium Vigour

LP%T: Lipid Peroxides per cent Transmission

LP: Lipid Peroxidation

LPO: Lipid Peroxidation

LP-OD: Lipid Peroxides Optical Density

LPRS: Lipid Peroxidation Resistance Score

LPRS-reci: Lipid Peroxidation Resistance Score Reciprocal

LPRS-rev: Lipid Peroxidation Resistance Score Revised

LV: Low Vigour

MBTH: 3-methyl-2- Benzothiozolinone Hydrazone

MED: Moisture Equilibration Drying

ME-SD: Moisture Equilibration Soaking Drying

MFs: Magnetic Frequencies

MSC-D: Moist Sand Conditioning Drying

MSC-S-D: Moist Sand Conditioning Soaking Drying

MV: Medium Vigour

MWs: Microwaves

NPM: Non-Pesticidal Management

NV%: Non-Viabillity Percentage

NVF: Non-Viability Factor

OD: Optical Density

PABA: *p*-Aminobenzoic Acid

PEG: Polyethylene Glycol

PGS: Participatory Guarantee Scheme

PM: Physiological Maturity

RH: Relative Humidity

RNA: Ribonucleic Acid

S: Score

S-D: Soaking Drying

SD: Standard Deviation

SE: Seedling Evaluation

SF: Seedling Factor

SLRS: Sugar Leaching Resistance Score

SMP: Solid Matrix Priming

SOD: Superoxide Dismutase

SRI: System of Rice Intensification

TBA: Thiobarbituric acid

TNAU: Tamil Nadu Agriculture University

TSL: Total Seedling Length

TZ: Topographical Testrazolium Test

U14C-Glycose:Uniformly Labelled Radiosensitive Glucose

UN: United Nations

UNCTAD: United Nations Conference on Trade and Development

UPBS: Unified Physiological-Biochemical Score

USA: United State of America

USSR: United Soviet Socialist Republic

UVA, UVB UVC: Ultraviolet Radiations

VE reci: Volatiles Emanation Reciprocal

VE: Volatiles Emanations

VERS: Volatiles Emanations Resistance Score

VF: Viability Factor

Chapter 1

Prefatory Chapter: Aim, Scope and Structure of the Book

In this prefatory annotation to the treatise, "SEED INVIGORATION: Treatments and Quantification of Vigour", we have attempted to cover the brief thematic and applied aspects of seed invigoration. The aim, scope and structure of the subject matter as covered in the different chapters and sections of the book are outlined herein.

Seed Vigour Concept *vis-à-vis* Viability

Vigour and viability are interrelated characters and generally loss of vigour precedes loss of viability. Seed viability is the capacity of a seed to germinate under favourable conditions provided that any dormancy in the seed is broken beforehand and the germination percentage in a standard germination test would give a quantitative measure of viability. Vigour which stands for the performing ability of the germinated seed, though understood well, is more abstract and, as such, more difficult to define and quantify. Further information on the development of the vigour concept is given in Chapter 2 and that on its quantification in Chapter 6.

Assessment of Seed Vigour

Various approaches have been made towards the development of an acceptable vigour test. The brick grit test, cold test, seed leachate study, controlled ageing test, *etc.*, while capable of broadly grading seed quality cannot answer categorically how vigorous a high vigour seed is or in short fails to give mathematical value to vigour as is obtainable for viability. Though indirect and a bit lengthy, an otherwise simple seed vigour bioassay method has been standardized in the College of Agriculture,

Calcutta University that would give a quantitative value in the percentage scale to a stock seed lot of unknown vigour status. In this bioassay method, high vigour jute (*Corchorus olitorius* L.) seed (bioassay material) is grown along with stock seeds of unknown vigour status in air-tight glass or plastic (transparent) containers. The gaseous emanations of stock seeds containing among others volatile lipid peroxidation products would inhibit the jute seedling growth (bioassay material) depending upon the concentrations of the inhibitory products in the emanations in a quantitative manner. The total seedling growth of the bioassay material, grown along with the stock seed, when expressed as a percentage value of the growth of the blank set (growth of the bioassay material alone in an exactly similar container, without stock material) would stand for the vigour status of the stock seed. The chemical assay of volatile lipid peroxidation products (aldehydes) consistently corroborated the bioassay findings.

Seed Invigoration Treatments

A relevant corollary of a discourse on seed vigour is invigoration of seed as a practical proposition. While discussing the genesis of vigour concept and the implications thereof, eminent seed scientist Walter Heydecker (1972) argued strongly in favour of seed invigoration and highlighted its scope and possibilities in future (see Chapter 2).

Seed invigoration implies an improvement in seed performance by any post-harvest treatment resulting in improved germinability and field performance. The qualitative change in the seed should persist after the treatment is removed and the invigoration treatments are basically physiological in nature. Broadly, the seed invigoration treatments include (i) presowing treatments for improved field performance, (ii) midstorage hydration-dehydration treatments for vigour, viability and productivity , and (iii) prestorage dry treatments of harvest-fresh high vigour seed for the maintenance of vigour, viability and field performance, specifically yield per unit area.

Though proved effective in many situations, lack of consistency has been a problem with pre-sowing treatments which were developed by Henckel (1946) in the USSR basically for plant hardening to tide over adverse field conditions. The treatments involved a few long duration soaking (24 h) drying cycles. Our modified pre-sowing short duration (3 h) soaking followed by drying treatments, however, gave highly consistent positive results.

In view of the great importance of seed vigour and viability in agriculture, particularly in developing countries, we have placed greater emphasis on the midstorage hydration-dehydration treatments and prestorage dry seed treatments for the maintenance of vigour and viability and improvement of crop performance in the field with focus on achieving yield enhancement with very little additional cost.

Midstorage Hydration-Dehydration (HD) Treatments

These treatments first developed in the College of Agriculture, Calcutta University in 1974 are very effective in high-medium, medium and low vigour seeds. They are, however, not suitable for harvest-fresh high vigour seeds. The principle behind the midstorage (also called midterm HD treatment) is to allow the stored seed to take up moisture to around 20-25 per cent (on wet weight basis) for some prophylactic (preventive) reactions in the seed, followed by drying back to the original weight for safe restorage. Besides soaking (full immersion in water, or in dilute solutions of a range of chemicals), seed hydration, depending on the initial vigour status, can be done by spraying the seed with a fixed quantity of water (spraying-drying), dipping the seed in water for 5-10 minutes and then keeping the surface-wetted seed covered for 2-6 h (dipping-drying), moisture equilibration with a water saturated atmosphere (100 per cent RH, also called humidification) for 24-48 h, incubation with pre-moistened sand (moisture content around 5 per cent, not sufficient for germination) for 16-20 h, *etc.* In all cases, unless the hydrated seed is dried back to safe limits for restorage (at least to the original weight), the benefits of the treatments will not be fully realized.

Soaking-drying, which is a very effective vigour and viability maintaining treatment for medium and low vigour orthodox non-leguminous seeds it cannot be given directly to most leguminous crop seeds because of the possibility of soaking injury. However, a brief period of soaking following humidification (moisture equilibration at 100 per cent RH) followed by drying back would effectively maintain vigour and viability during subsequent storage with due manifestations in field performance. Except in the humidification-drying treatment, in treatments using liquid water, a range of effective chemicals in dilute (aqueous) solutions have been shown to give additional benefits. However, compared to the large beneficial effect of water (hydration) alone, the chemical effects are rather small and often masked by the hydration effect. More details about the treatments, their manifestations and possible modes of action are given in subsequent relevant Chapters.

Prestorage 'Dry' Seed Invigoration Treatments

These treatments were also developed in the College of Agriculture, Calcutta University five years after the first midstorage hydration-dehydration (HD) treatment was done. As hydration-dehydration proved ineffective in high vigour harvest-fresh seed, avoidance of the use of water (hydration) was imperative; in a broad sense, 'dry' treatments would include vapour phase treatment (with gases), permeation of effective chemicals through nonaqueous organic solvents like acetone or dichloromethane (solvent infusion) and use of dry effective chemicals (with or without carriers), that may enter seeds in various ways. Based on several considerations, we selected iodine(a solid halogen which readily passes into vapour form) as a candidate for prestorage dry seed treatment and fortunately that proved to be a successful one. Treatment with iodine way back in 1979 very effectively retained the vigour and viability of high vigour (harvest-fresh) mustard seed and pioneered the use of various other effective substances in the dry state. For example, dry seed treatment with bleaching powder (active ingredient calcium oxychloride that slowly releases another halogen, chlorine) has proved very successful in all kinds of agricultural and horticultural seeds so far tested. Also, a wide range of pharmaceutical and botanical preparations have been found to be effective in maintaining seed vigour, viability and crop productivity. Technically, such dry treatments (with or without carriers) are very advantageous to the farmers as they can be given soon after harvest during bagging and storage. If required, depending on circumstances, the dry invigorated seed may be given further midterm hydration-dehydration treatment for reinvigoration.

The details of different kinds of seed invigoration treatments, their manifestations and possible modes of action have been presented in Chapter 3.

Besides presenting abstracts of selected papers published from the College of Agriculture, Calcutta University, on a range of crop seeds variously treated for vigour, viability and productivity (arranged chronologically in Chapter 4), selected theme papers highlighting major developments and observations have been covered in Chapter 5. Extracts of a few selected papers have also been appended at the end of chapter 4.

Quantification of Seed Vigour: Bioassay and Seed Invigoration Treatments

The development of bioassay of seed vigour, a very useful method of determination of vigour of a seed lot of unknown vigour status, as mentioned above in this chapter, has been of great advantage to us in our endeavor to quantify seed vigour in general and of the different seed invigoration treatments in particular. Bioassay of seed vigour along with its development as a useful method of quantification of seed vigour has been dealt with in sufficient details in the Annexure No A-1.

Quantification of Seed Vigour: Mathematical-Statistical Approaches

Vigour status of different seed lots and seed invigoration treatments has been quantified using a range of vigour-related morphological and physiological-biochemical parameters such as germinability and early seedling growth, the leaching of electrolytes, sugars and amino acids (as indicators of membrane permeability), activity of enzymes associated with viability and vigour, *etc.* Vigour rating data have been so calculated that the same would range between 0 and 100 as with germination per cent data (all with positive connotations). The incorporation through multiplication of the germination factor GF (germination per cent divided by 100, range 0 to 1) in the mathematical and statistical analyses of the data would help to eliminate the effect of non-viable seeds on vigour rating. Conversion of all negatively correlated parameters to positive values (many are originally positive and as such would need no change) in the same scale and range would allow taking average values of vigour from a large number of vigour parameters. A simple arithmetic mean value would then quantify the overall vigour status of a particular seed invigoration treatment and would enable selection of the most suitable ones from a range of treatments and seed lots. These issues dealing with vigour quantification of 'wet' and 'dry' seed invigoration treatments that would be of considerable practical value to our predominantly small and marginal farmers has been discussed in details in Chapter 6. For more details on the subject, refer to the paper by Basu R.N. 2007, on 'A Holistic Approach to Quantification of Seed Vigour through Integration of Morphological and Physiological–Biochemical Parameters', (Annexure No A-2).

In Chapter 7 on "Supplement to thematic and allied issues – some important observations and issues", the major findings have been highlighted. Nevertheless, it is ardently hoped that in the near future still more effective seed invigoration treatments would be developed with greater benefits to agriculture in general and the farming community in particular. Seed invigoration treatments would impart increased tolerance to crops to overcome to some extent, the challenges of climatic irregularities in an era of global climate change with eventualities such as submergence and flooding, intermittent droughts, soil salinity fluctuations, *etc.* Within certain limits, healthier plants raised from duly invigorated seeds would always be expected to surpass the performance of corresponding control plants from non-invigorated seeds and appreciably mitigate adverse climatological effects on crop performance.

A general review of literature on seed priming treatments for improved crop performance has been discussed in Chapter 8.

Several important methodologies, points and observations that need special mentioning and certain other matters requiring further clarification are described in details or details in the Annexures. Bibliography broadly covers the literature cited in the different chapters and sections.

Chapter 2

Conceptual Issues on Seed Vigour, Viability and Seed Invigoration

Seed vigour is a character of the living seed that stands for potential performance ability of the seed in the "seed to plant to seed" regeneration cycle. Seed vigour is intimately associated with seed viability (no viability would mean no vigour) and seed lots with the same viability may have different levels of vigour and ability to perform in the field. For example, separate seed lots of a cultivar (may be all with 100 per cent germination), may have different levels of vigour. While a viable seed is one which can germinate under favorable conditions provided that any dormancy that may be present is removed, vigour is tantamount to the property of the seed that promotes a satisfactory performance as opposed to weakness or unsatisfactory performance (Heydecker, 1972).

Efforts Towards Assessment of Seed Vigour

As early as 1876, F. Nobbe introduced the term seed vigour in his book '*Handbuck der Samenkunde* (*vide* Marcos Filho, 2015). Attempts to capture vigour have also been made by various researchers and one of the earliest being the *Ziegelgrus* (brick grit) method of Hiltner and Ihssen, 1911; the cold test (Isley, 1950) for screening cereal seedlings for survival under suboptimal temperatures, and exhaustion tests (Germ, 1960) of cereals that would give growth potential of the seedlings before they begin to photosynthesize.

Walter Heydecker, a leading researcher in the field of seedling vigour, published a paper in 1960 with the title "Can we measure seedling vigour?" According to him, the first documented experimental use of the German near equivalent of the term vigour as *Triebkraft* was coined by plant pathologist L. Hiltner as early as in 1911 (Hiltner and Ihssen, 1911) whereby the said term was used for seeds capable of penetrating a 30-40 mm layer *Ziegelgrus* (brick grit, 2-3 mm diameter size) under specified conditions. The percentage of successful seeds (*Ziegelgrus* value) obviously stood for more vigorous seeds.

Triebkraft (vigour), according to Lindenbein and Bulat, (1955), is a germination test under aggravated conditions, designed to discover and show the weakness of seeds. Heydecker (1960) defined vigour as the ability to germinate and produce a stand in a suboptimal environment but later the relevance of an adverse environment was no longer stressed (Heydecker,1972).

The electrical conductivity test of Matthews and Bradnock (1967) which is based on greater leaching of electrolytes from low vigour seeds (Hibbard and Miller, 1928) has been a widely accepted vigour test for a large number of seeds. The topographical tetrazolium (TZ) test based on detection of dehydrogenase enzyme activity has also been a successful biochemical test. Various other tests have also been developed but the International Seed Testing Association (ISTA) is still to incorporate any such tests in the official germination test certificate as an international recommendation because of technical and administrative reasons.

Based on the vigour concept by a number of investigators, Heydecker (1972) observed that over the years the International Seed Testing Association took up the issue of seed vigour very earnestly and almost every report of the vigour test committee of ISTA deliberated on the subject. Although there has not been any internationally accepted vigour test, through the efforts of ISTA the standard germination test itself has been acting as a mechanism of a realistic assessment of seed vigour in terms of the capacity of the seed to give rise to plants expected to be of good performance in the field; that being the ultimate objective of any vigour test.

Vigour has been a subject of critical appraisal over a pretty long period as is evident from the deliberations of the ISTA Vigour Test Committees. The vigour definition as accepted by ISTA, "seed vigour is the sum of those properties which determine the potential level of activity and performance of the seed lot during germination and seedling emergence" (Perry, 1978) will bear out the qualitative nature of the vigour concept. The definition of vigour proposed by the Association of Official Seed Analysts (AOSA) of the USA, "seed vigour comprises of those seed properties which determine the potential for rapid, uniform emergence and development of normal seedlings under a wide range of field conditions" (McDonald, 1980), though quite comprehensive is still difficult to quantify. To the producer and the customer of seed a quantitative approach would be most desirable.

According to Isley (1950,1957) two ideas namely, i) rapidity of growth, and ii) non-susceptibility to unfavourable growing conditions are basic to the concept of vigour; as such, he looked upon vigour as the sum total of all seed attributes which favour stand establishments under unfavourable field conditions.

Factors influencing vigour of seeds (Schoorel,1956, 1960 and Isley, 1957) include the following:

1. Weather during ripening and harvest;
2. postharvest treatment of seeds such as threshing, drying, cleaning, *etc.;*
3. conditions during storage and its duration;
4. activity of the seed borne insects and pests including microorganisms;
5. use of chemicals and pesticides, and
6. genetic characteristics of the seeds.

Heydecker who has indeed been an authority on the subject further noted that vigour results from a combination of all of the following: adequate maturity at harvest; absence of physiological ageing, *i.e,* physiological deterioration or presence of necrosis which are usually started by mechanical damages but increase in extent due to mechanical activity or autotoxic reactions or both; all making for a well-functioning enzyme synthesizing biological apparatus. He also noted that the rate of seed deterioration would be determined by genetic factors, species and cultivar and also by treatment and environment prevailing before and certainly during storage, accounting for differences between seed lots and even individual seeds within the same lot. "Loss of germination is clearly an important indication of loss of vigour but it is equally clearly the last relevant indication of the final catastrophe" (Heydecker, 1972).

Summarizing the causes of vigour differences among different seed lots, Heydecker (1972) outlined them as (i) genetical (cultivational differences, heterosis effect, *etc.*), (ii) physiological (mitochondrial activity, higher metabolic turnover, *etc.*), (iii) morphological (within a cultivar difference in vigour between big and small seeds), (iv) cytological (chromosomal aberrations, mutation-linked effects), (v) mechanical (gross breakages in seeds especially machine use damages, necroses, *etc.*), (vi) microbial (fungal and bacterial load, invasion through surface cracks, necroses, direct and indirect effects) and summed up four effects of seed vigour. Those are: (i) superior survival (a vigorous seed is one that remains vigorous), (ii) it resists its adversaries in the field, (iii) it is characterized by having right kind of reserves ensuring good performance in the heterotrophic phase, and (iv) continues to grow well in the autotrophic phase of growth.

Sequence of Events in Deterioration of Stored Seed

According to Delouche (1969) the sequence of seed deterioration under aging conditions would be as follows:

1. Cellular membrane degradation and loss of semi-permeability;
2. impairment of energy-yielding and biosynthetic mechanisms;
3. consequently with reduced respiration and biosynthesis;
4. slower germination and slower heterotrophic seedling growth;

5. reduced storage potential;
6. slower growth and development of the autotrophic plants;
7. less uniformity in growth and development amongst plants in the population;
8. increased susceptibility to environmental stresses (including microorganisms);
9. reduced stand-producing potential;
10. increased percentage of morphologically abnormal seedlings and lastly;
11. loss of germinability.

Membrane Integrity and Seed Vigour

Most phenomena associated with loss of vigour are believed to be caused by the weakening of the vital lipoprotein plasma membrane with built-in semi-permeability (selective permeability) which is a vital lifẹ-sustaining property of the living cell. The consequent leakage of electrolytes, sugars and amino acids would lead to the deterioration of the metabolic and transport efficiency besides encouraging colonization by microorganisms. Takayanagi and Murakami (1968) successfully used rapid urine glucose testing paper for detection of sugar leaching from seeds. However, according to Abdul-Baki and Anderson (1970) the same may not always be a good vigour index.

The electrical conductivity test of Matthews and Bradnock (1967, 1968) is, however, much more promising especially when applied to seed samples with high germination percentage to avoid excess contribution through leaching from dead seeds that would obscure the contribution of living seeds. The introduction of the germination factor (GF) by Basu (2004) would improve the situation to a great extent. (see section on quantification of seed vigour).

Oxidation of Lipids and Loss of Membrane Semipermeability

Koostra and Harrington (1969) showed a decline in polar lipids, chiefly owing to their oxidation as the main cause of leaking cell membranes of seeds. Studies at the College of Agriculture, University of Calcutta have also strongly supported this view.

Many researchers have suggested the involvement of free radical-induced lipid peroxidation in seed deterioration (Pammenter, Adamson and Berjak 1974; Basu, 1976; Rudrapal and Basu, 1982; Harman *et al.,* 1982; Buchvarov and Gantcheff, 1984; Wilson and McDonald, 1986b and others).

Increase in membrane permeability has been attributed to peroxidation of membrane lipids (Van Staden *et al.,* 1976) and decrease in membrane fluidity.

Auto-oxidative (non-enzymic), catalytic (non-enzymic) as catalysed by haeme proteins and enzymic (catalysed by lipoxygenase enzyme) lipid peroxidation may take place in various ways (see Figures 3.1, 3.2 and 3.3 in Chapter 3). Despite differences in the mode of oxidation the oxidative products of the same unsaturated lipid moiety would be broadly similar. Thus, oxidation of linoleic acid would give rise to linoleate hydroperoxide, which depending on the specific resonance hybrid forms would finally yield volatile aldehydes such as hexanal, pentanal, butanal, *etc.* (Herman, *et al.,* 1982; Wilson and McDonald, 1986b; Pal and Basu, 1989; Sur and Basu, 1990a). The presence of volatile aldehydes in the gaseous emanations of low vigour germinating stock seeds has been the major reason of inhibition of the growth of jute seedlings (bioassay material for quantification of seed vigour of different lots of stock seeds) in the study of Sur and Basu (1990a).

Despite some reservations in certain quarters on the involvement of free radical induced lipid peroxidation as many of the observations were indirect, the direct measurement of organic free radicals by Buchvarov and Gantcheff (1984) unequivocally implicated free radical induced lipid peroxidation in the senescence of soyabean seeds. Hailstones and Smith (1988) demonstrated the highly significant negative correlation between soyabean seed vigour and lipid peroxide levels coupled with identification of the volatile aldehydes, hexanal, pentanal and butanal as thermal breakdown products of the hydroperoxide thereby lending further support to this well entrenched concept of lipid peroxidation in seed deterioration. Using gas chromatographic studies Smith and Adamson (1989) also convincingly implicated lipid peroxidation in the deteriorative senescence of lettuce seeds.

Reduced Lipid Peroxidation and Volatile Aldehyde Production under Seed Invigoration Treatments

Midstorage hydration-dehydration seed invigoration treatments always showed significantly lower post-ageing lipid peroxidation than in the control (non-invigorated seeds). Reduced lipid peroxidation by viability and vigour maintaining treatments has been consistently demonstrated using different crop seeds (Rudrapal and Basu, 1979 in wheat; Basu *et al.,* 1985 in mustard; Dey and Basu, 1985 in mustard; Dey and Basu, 1982 in sunflower; Choudhuri and Basu, 1988 in onion; Pal and Basu, 1989 in rice; Saha *et al.,* 1990 in soybean; Ramamoorthy and Basu, 1996 in groundnut, *etc.*). In a number of crop seeds considerable reduction in volatile aldehyde production has been recorded with the hydration-dehydration invigoration treatments.

In the prestorage "dry" seed invigoration also, lipid peroxidation as well as volatile aldehyde emanation were significantly reduced by the viability maintaining seed treatments implying thereby a broad similarity in the mode of action of the two kinds of invigoration treatments. In most of the prestorage treatments volatile aldehyde production, which is a few steps ahead in the usual lipid peroxidation process and is considered a part of the same, was assayed using the volatile aldehyde absorbing chemical 3-methyl-2-benzo-thiozolinone hydrazone (MBTH as a 2 per cent w/v solution) followed by colour development and spectrophotometry, along with jute seedling bioassay developed by us (Sur and Basu, 1990, Bhattacharya and Basu, 1990; see Annexure No A-1 for details).

Basu and Rudrapal (1980) recorded that the maintenance of high viability of high vigour (HV) mustard seeds exposed for 48 hours to a partially saturated iodine vapor (a dry invigoration treatment) was associated with a considerable reduction of lipid peroxidation during the post accelerated ageing early germination phase. In the aforesaid study lipid peroxidation was studied using thiobarbituric acid (TBA) reagent and volatile aldehyde production was not assayed. It was suggested that iodination stabilized the unsaturated fatty acid component of lipoprotein cell membranes rendering them less susceptible to lipid peroxidation and free radical reactions. However, similar reduction in lipid peroxidation has been recorded with a wide range of diverse "dry" (non-aqueous) seed invigoration treatments. Reduced volatile aldehyde production during the early postageing germination phase of prestorage "dry" seed invigoration treatment in a number of crop seeds has been recorded in studies conducted in the University of Calcutta by Pal and Basu, 1989 in rice; Bhattacharya and Basu, 1990 in peas; Pal and Basu, 1993 in wheat; Pal *et al.*, 1996 in wheat and sunflower; Mukhopadhyay *et al.*, 1997 in French bean and soybean, *etc.*

Seed Invigoration as a Practical Proposition

On the question of seed invigoration as a practical proposition Heydecker (1972) observed, "No seed is perfect and invigoration (and not only reinvigoration) of seeds may be possible at some future time."

The presowing seed hardening treatment developed by Henckel (1946) in the former USSR for imparting resistance to drought should be looked upon as a seed invigoration treatment that involved 2-3 cycles of long duration (24 hours or so) wetting–drying (drought hardening) and would yield significant beneficial effects in terms of productivity. May *et al.* (1962) reviewed the voluminous Soviet literature on the subject which, in general, was very positive. The same, however, could not be consistently confirmed by researchers in Western countries. On the other hand, Dawson (1965) in India reported a 40 per cent increase in yield of finger millet with seed presoaking. Chinoy *et al.* (1968) obtained significant beneficial effect of seed hardening employing ascorbic acid (vitamin C) in the soak water in particular. Rajashekhar *et al.* (1970) recorded noticeable increase in yield of ragi (*Eleusine coracana* L.) at different levels of soil fertility using presowing seed hardening treatment.

Presowing osmotic treatments (osmoconditioning) using polyethylene glycol (PEG, molecular weight 6000-8000) have shown more consistent beneficial effects, in expensive vegetable and flower seeds in particular, may also be looked upon as useful seed invigoration treatments (also see Chapter 3).

At the time when Heydecker made the aforesaid statement (1972), only research work directly dealing with seed vigour and viability and showing positive manifestations of the vigour effect would come under the purview of seed invigoration (or reinvigoration). Nevertheless, in a most comprehensive review paper in 1977 on the world literature on seed treatment, the official journal of

International Seed Testing Association, Seed Science and Technology, Volume 5, pp. 353-425 recorded in Table 2 (p.361) item, "Hardening against drought, heat, salinity, frost," the following observation "... it is claimed that the drying back process is the means by which the resistance mechanisms are derepressed. If that is agreed upon, then "seed hardening" that involves long duration wetting-drying cycles should be considered as an act of invigoration. There are similar arguments for osmoconditioning with polyethylene glycol (PEG, molecular weight 6000-8000), with or without other chemicals of low concentrations, followed by drying back, that should also be considered as seed invigoration treatments.

Working for over four decades in the College of Agriculture, University of Calcutta, Basu and co-researchers have convincingly invigorated seeds of a wide range of agricultural and horticultural crops fulfilling such expectations. It is worthwhile to recall the views of Heydecker and Coolbear (1977) (Seed Science and Technology, vol.5 pp 353-425) who stated in their most comprehensive review of world literature on seed treatments with respect to our early researches on wheat, rice and jute as follows: *"Brief soaks against deterioration in storage*: We must now mention a completely different but most interesting discovery concerning a benefit from a wet seed treatment. This has been reported by Basu and Dasgupta (1974) and Dasgupta and Basu (1975). They found that three hours soaking (immersion) of wheat grains after six months' conventional storage, in solutions of a range of chemicals, but even in water, largely counterbalanced the deterioration that would have occurred otherwise by the end of a year's uninterrupted storage ......". Similar invigoration treatments for the maintenance of vigour and viability of seeds were also recorded at around the same time for rice and jute seeds (Basu, Chattopadhyay and Pal, 1974).

Our success story with prestorage "dry" seed invigoration treatments for high vigour harvest fresh seed of a wide range of agricultural field crops and vegetables would certainly have made Dr. Heydecker exceedingly happy but unfortunately, he suffered a fatal cerebral attack soon after his last aforesaid publication. He had been a visionary of the highest order in the field of seed vigour and invigoration and his sincere efforts during his last days to highlight our researches at the University of Calcutta in various ways through encouraging letters and extending invitation for presenting our views on seed invigoration were indeed highly encouraging.

Terminological Points Concerning Invigoration and Reinvigoration

Regarding the use of the terms "invigoration" and "reinvigoration", we would, however, like to point out certain reservations despite their widespread usage in the literature on seeds. The dictionary meaning of the term "invigorate" is to give vigour or strength (Oxford Quick Reference Dictionary) or to give life and energy (Webster's New Collegiate Dictionary) is more literal than scientific. The question is "whether one can increase vigour level or quantum of vigour of a truly high vigour seed?" Perhaps not, we cannot increase the vigour level from the initial highest vigour status but try to maintain the level or slow down the fall in vigour under adverse/stress situations. With our long experience (more so with our more recent success with quantification of vigour status of seed invigoration treatments) there has never been a consistent and significant increase in vigour level immediately after the treatment under normal situations; the observed "hardening effect" would be the consequence of overcoming stress, a situation similar to ageing faced under field conditions. However, despite reservations we have used "invigoration" in its popular meaning, specifically in comparison to corresponding untreated control seeds. Incidentally, we would consider hardening, osmoconditioning and allied treatments as invigoration treatments for reasons stated earlier in this discussion. We prefer to use the term "reinvigoration" for the second 'wet' treatment given to prestorage 'dry' invigorated (for high vigour harvest-fresh material) seed after a storage period (midterm, hydration-dehydration treatment). Similarly, a 'dry' seed invigoration treatment immediately after a "wet" (H-D) treatment would be a "reinvigoration" treatment.

A logical corollary of the aforesaid discourse on seed vigour is invigoration of seed as a practical proposition. Seed invigoration thus implies an improvement in seed performance by any postharvest treatment resulting in improved germinability, greater storability and better field performance than the corresponding untreated (control) seed. An invigoration treatment should bring about a qualitative improvement in the seed which should persist after the treatment is removed, and the treatments are basically physiological in nature.

Exclusion of preharvest treatment of mother plants *in situ* with specific chemicals from the purview of the present discussion has been deliberate with the sole purpose of narrowing the scope of a rather complex subject. Also, physical treatments employing radiation, magnetism, electricity, *etc.* have been excluded for the sake of brevity.

Chapter 3

Seed Invigoration Treatments

It would be convenient to discuss physiological seed invigoration treatments under the following headings:

1. Presowing treatments for improved field performance
2. Midstorge treatments for vigour, viability and productivity
3. Prestorage treatments for better storability and field performance

Presowing Treatment for Improved Field Performance

The presowing seed 'hardening' treatment, also known as presoaking treatment, involving 2 to 3 cycles of 24 h wetting-drying, was developed by Henckel (1946) in the former USSR for imparting resistance to drought conditions (drought-hardening). The voluminous Soviet literature on the subject (May, Milthorpe and Milthorpe, 1962; Ovcharov, 1977) would show significant beneficial effects of seed hardening. However, such effects could not be consistently confirmed by studies conducted in the Western countries.

In India, Chinoy and co-workers (Chinoy, Gurumurti, Shastri, Abraham, Dave, Shah, Pandya and Saxena, 1968) obtained beneficial effects of seed hardening, especially employing ascorbic acid. Earlier, Dawson (1965) working with finger millet (*Eleusine coracana* L.) reported a significant beneficial effect of seed presoaking on the growth of the root system and a 40 per cent increase in yield. Krisnasastry and co-workers (Rajasekhar, Rama Rao, Krisnasastry and Mallana, 1970; Sashidhar, Mekhri and Krisnasastry, 1977) induced stress tolerance in finger millet and groundnut (*Arachis hypogaea* L.) seeds with dilute calcium chloride and ascorbic acid solutions and correlated the stress resistance with accumulation of proline. Using the relatively short-term soaking in water or dilute solution of sodium phosphate (dibasic), Ray (1982) and Chatterjee and Singh (1983) obtained a significant increase in grain yield of wheat (*Triticum aestivum*) and barley (*Hordeum vulgare* L.). Beneficial effects of presowing seed treatments have been reported by a number of Indian researchers in a range of agricultural and horticultural crops (Basu and Sur, 1988).

Short-term hydration of seed prior to planting greatly benefits stand establishment but use of chemicals in soak water like potassium or sodium phosphate (dibasic) would give additional advantages. Invigoration of soybean (*Glycine max* L. Merr.) a leguminous seed, has been reported using ferrous sulphate which was accompanied by a significant reduction in lipid peroxidation (Hailstones and Smith, 1991). Our own results on rice (*Oryza sativa* L.) and wheat have confirmed the beneficial effect of presoaking with ferrous sulphate (10^{-3} to 10^{-2} M), GA (10^{-3} M) and MBTH (10^{-3} M); in addition, a significant improvement in poststorage germinability has been recorded with these chemicals (unpublished).

The lack of consistency in the effect of some presowing treatments is indeed a discouraging factor in their large-scale acceptance by farmers and the situation needs a critical appraisal. The following reasons may be put forward for the inconsistency of results of presowing seed treatments:

1. The kind of seed is an important factor. Leguminous seeds are less responsive to wetting-drying especially when long-term soaking durations are employed resulting in soaking injury (Armstrong and McDonald, 1992).
2. Vigour of seed at the time of treatment would greatly determine the efficacy of the treatment (Pan and Basu, 1985a; Hofmann, Aschermann-Koch and Steiner, 1992).
3. Prolonged soaking may show adverse effects because of anaerobic respiration leading to accumulation of toxic metabolites. Hydration by placing on moist media or by moisture equilibration with a water-saturated atmosphere, or through moist sand conditioning would be desirable.
4. Drying of seed following hydration should be slow and compatible with the degree of hydration. Often the embryo enlarges to such an extent that drying is injurious. Slow drying employing temperature not exceeding 35°C at relatively low humidity and moderate air velocity is recommended. In some cases, surface-drying or very light air-drying may be employed.
5. Chemical effects are sometimes claimed without keeping proper 'water' control (besides the 'untreated' control). Very often the chemicals do not give significant additional effects over that obtained with water alone.

Nevertheless, with proper care and judgement, especially with respect to hydration and its duration, presowing seed treatment should be readily acceptable as a practical proposition. The results of modified presowing treatments conducted in our laboratory (similar to midstorge treatments) are much more promising and consistent.

In this connection, the presowing osmotic treatments, particularly those employing polyethylene glycol (PEG, molecular weight 6000-8000), needs special mention. Use of PEG as a presowing treatment has been much more successful in terms of germinability, uniformity of stand and subsequent crop performance (Heydecker, 1977; Heydecker and Coolbear, 1977) than the simple water presowing treatment. According to our findings, PEG effectively regulates entry of water into the seed, without causing injury in leguminous seeds, and advances germination reactions to a more or less fixed level determined by the water potential. When not followed by air-drying, osmoconditioning improved seedling development of soybean as demonstrated by increased plumule and radical length. However, water soaked or osmoconditioned seed upon air-drying showed increased membrane damage and resulted in the development of abnormal seedlings (Armstrong and McDonald, 1992). On sowing, individual seeds of a lot subjected to osmoconditioning would start from near identical levels of metabolism and would therefore show rapid and uniform germinability, synchrony in growth and development and, in many cases, increased biological and agricultural yield (Heydecker, 1977; Knypl, 1981; Knypl and Khan, 1981; Leskovar and Sims, 1987). Osmoconditioned seeds generally do not store well. In sweet pepper (*Capsicum frutesence* L.), however, osmoconditioning in 0.4M mannitol solution for 4 days considerably delayed ageing and increased seed longevity (Georghiou, Thanos and Passam, 1987).

Further, for specific situations requiring use of hormonal and other chemicals, those may be supplied to the seed during osmoconditioning (Khan, 1977; Knypl, 1981). Priming or osmoconditioning has been mostly used in vegetable and flower seeds (Haigh, Barlow, Milthorpe and Sinclair, 1986; Barlow and Haigh, 1987; Argerich and Bradford, 1989) because of the technical advantages of handling relatively small seed lots. However, in field crops like soybean, beneficial effects on field stand and yield have been reported (Knypl, 1981). Contradictory reports where osmoconditioning has not proved successful in improving yield in soybean are also available (Helsel, Helsel and Minor, 1986; Khan, Miura, Prusinski and Ilyas, 1990).

The benefits obtained from osmoconditioning with PEG or salt solutions may be summarised as follows:

1. Faster emergence and more uniform field stand in normal as well as in stress situations. Although in many cases there has been no difference in final germination, there are several reports showing increased field stand due to priming, especially in nonleguminous seeds.
2. Besides uniformity and synchrony, significant yield increase in many vegetable crops in particular, has been reported.
3. Osmoconditioning would very effectively overcome the serious problem of soaking injury in many legumes.

4. Osmoconditioned seeds can be rinsed and carefully dried for restorage for short periods in a number of crops.

Osmoconditioning has, however, a number of limitations which need to be sorted out before large-scale commercial application. Some of them are:

1. Osmotic seed treatment is expensive, especially when osmotica such as PEG are used. As such, for the present, only low volume, high-value seed would be suitable for treatment.
2. The bathing solution needs continuous aeration, otherwise the seed will suffer from ill effects of anaerobic respiration.
3. As the treatment is done over a prolonged period, proper caution against microbial attack should be taken.
4. Vigour of the seed at the time of treatment may determine the efficacy of treatment.
5. Although some counteraction of ageing effect has been reported, usually storability of seeds may be adversely affected by osmoconditioning because of advancement of germination reactions to a level not compatible with drying back of the seeds required for restorage.
6. In certain materials, there is a possibility of induction of secondary dormancy, particularly under unfavourable temperature and light conditions.

To overcome the problem of anaerobic respiration of seeds during PEG treatment, we have developed a moist sand conditioning treatment incorporating PEG 6000 for the maintenance of viability of orthodox pea (*Pisum sativum* L.) and recalcitrant tea (*Camellia sinensis* L.) seeds with considerable success (Bhattacharya and Basu, 1990, and Bhattacharya, Rahman and Basu, 1994). The moist sand conditioning treatments broadly resemble the matriconditioning treatments with inert, water insoluble, porous inorganic carriers like Micro-Cel E and expanded vermiculite which have proved very effective in the establishment of a wide range of horticultural seeds (Khan *et al.,* 1990; Khan, 1992; Khan, Maguire, Abawi and Ilyas, 1992).

Notwithstanding the limitations, the osmoconditioning treatments deserve our serious attention and their efficacy and economic feasibility should be critically studied under tropical and sub-tropical conditions.

'Fluid drilling' of germinated seeds using a carrier like alginate gel, developed in the UK (Currah, Gray and Thomas, 1974) has advantages nearly similar to the osmoconditioning treatment. However, there is little possibility of adoption of such techniques by vegetable growers in the third world countries.

Low temperature presowing treatment (water imbibition at a temperature of 10°C for 3 weeks) would substantially enhance germination rates and has been found to be more effective at improving the uniformity of germination than the PEG treatment (Coolbear, Newell and Bryant, 1987). Treated tomato (*Lycopersicon esculentum* L.) seeds could be maintained at lag phase moisture content while certain aspects of germinative metabolism began. The low temperature requirement for arresting too much germination advancement would, however, limit the acceptance of this promising treatment by average third world farmers.

Provision of a specified amount of moisture through a sand medium (moist sand conditioning) will allow pregermination reactions to a certain extent, depending upon the seed moisture content and ambient temperature, under near aerobic conditions. Standardization of such treatments may be of considerable advantage to the resource-poor farmers in the third world countries.

An understanding of the mode of action of different physiological treatments is not merely of theoretical interest but also of great practical significance. A clear insight into the mechanism of action of water and chemicals would enable us to know the basic reasons for differences in the response of various materials to the treatments and the means to overcome the limitations. Further, if the mechanism is understood, it would be possible to devise even simpler and easier methodologies to invigorate seeds and improve their performance in the field. With these justifications, the possible modes of action of the different seed invigoration treatments are discussed.

Mode of Action of Presowing Treatments

According to Henckel (1964), presowing seed hardening treatments are responsible for activation of numerous subtle drought resistant mechanisms which otherwise remain latent. The treated seeds respond due to increased hydrophilicity associated with high osmotic pressure of bound water, increased phosphorylative activity of the mitochondria, higher DNA in growing points, less ribonuclease activity, and well-preserved cellular ultrastructure. Austin, Longden and Hutchinson (1969) attributed the improved germination of hardened carrot seeds, subjected to three 24 h imbibition-dehydration cycles, to germination advancement and enlargement of embryo within the seed. Hegarty (1970) also made similar observations and Sen and Osborne (1974) provided biochemical evidence in favour of germination advancement. However, if the effects were simply due to germination advancement and embryo enlargement, similar results would accrue by advancing germination dates. Many of the benefits of presowing treatments cannot perhaps be explained in this way implying alternative possibilities.

The osmoconditioning treatments, as well as the low temperature imbibition treatments, basically work on the same principle of presowing treatment but do so with greater perfection and consistency with the added advantage of overcoming the problem of soaking injury of legume seeds by osmoconditioning. Perhaps some of the important pre-and early germination reactions (up to a desired level) could take place in a favourable environment *vis-à-vis* the tough difficult-to-control field conditions so that the treated seed could make a good start upon sowing. This presupposes that it is the early reactions which are likely to suffer most under stress situations. The stresses suffered by the seed immediately after sowing could be similar to ageing stress (its intensity depends on edaphoclimatic conditions) with consequent physiological and biochemical manifestations. Presowing treatments may counteract deteriorative processes such as free radical-induced lipid peroxidation reactions leading to the production of toxic volatile compounds which favour pathological deterioration (Harman, Nedrow and Nash, 1978; Harman and Stasz, 1986). An effective enzymatic corrective and restorative action during a favourable hydration phase would also improve subsequent seed performance in the field.

The utility of drying back (or light air-drying), besides the technical convenience of sowing dry seed, however, needs critical physiological and biochemical scrutiny.

Modified Presowing Seed Treatments

Original presowing seed treatments as developed by Henckel in the erstwhile USSR around 1946 were undertaken primarily for 'hardening' of plants to tide over drought like situations in the field and involved two to three long-term soaking (24 h) drying cycles.

Although the Soviet literature on the subject has been extensive and generally positive, the same could not be consistently confirmed by a number of researchers in some western countries. Our preliminary studies suggested that shorter durations similar to our midstorage viability and vigour-maintaining seed treatments would give more consistent positive effects. However, hydrated seeds need not be dried back to original weight, light air-drying would do for the convenience of sowing in case of presowing treatments (which do not require restorage for a longer time as in case of midstorage treatments).

A comparative study on the modified presowing treatments and midstorage treatments on field performance and productivity of two vegetable crops, lettuce and carrot (Pan and Basu, 1985) was conducted as discussed hereunder to show the usefulness of the modified presowing seed treatments and the overall similarity in the efficacy of the modified pre-sowing treatments with midstorage treatments particularly when the treatments are given to medium vigour (MV) seeds.

Midstorage soaking-drying (S-D) treatment of high vigour (HV) lettuce seed was not effective for the maintenance of vigour and viability; while moisture-equilibration drying (ME-D) and dipping drying (D-D) treatment very effectively reduced the physiological deterioration in storage. Presowing (S-D) treatment of high vigour (HV) lettuce seeds did not show any beneficial effect on field performance; only a marginal non-significant improvement was noted with ME-D and oxalic acid soaking drying.

Medium vigour (MV) lettuce seeds given presowing treatments showed significant improvement in field performance and productivity than the untreated control seeds. Also midstorage treatment of MV lettuce seed significantly improved field performance with MED. D-D and S-D (particularly with oxalic acid and *p*-hydroxybenzoic acid) showing increased fresh weight of heads (Table 3.1a, Pan and Basu, 1985a). In case of both HV and MV carrot seeds all hydration-dehydration treatments showed significant retention of seed viability. However, MV seeds gave better response than HV seeds.

Table 3.1a: Effect of Mid-storage and Pre-sowing treatments of Medium-Vigour Lettuce Seeds on Weight of Heads (fresh weight/plant, g.) (*vide* Pan and Basu, 1985)

Treatments	*Time of Treatment*		*Mean*
	Mid-storage	*Pre-sowing*	
Control	80.3^{e}	79.0^{c}	79.6
Moisture equilibration-drying	101.3^{bc}	90.0^{ab}	95.6
Dipping-drying			
- Water	98.6^{cd}	84.6^{bc}	91.6
- Oxalic acid	110.0^{a}	90.6^{ab}	100.3
- *p*-Hydroybenzoic acid	108.3^{a}	88.6^{abc}	98.5
Soaking-drying			
- Water	93.0^{d}	89.3^{abc}	91.1
- Oxalic acid	103.0^{b}	99.0^{a}	101.0
- *p*-Hydroybenzoic acid	106.0^{ab}	98.3^{a}	102.1
Mean	100.0	89.9	

Note: Mid-storage treatments were given to 15-month-old seeds, re-stored after treatment for a further 3 months under ambient conditions (average RH 77.5 ± 3.8% and average temperature 27.6 ± 1.2ºC) until sowing. Pre-sowing treatments were given to 19-month-old seeds only 2 days before sowing. Within a time of treatment, values scored by the same letter, using Duncan's multiple range test, are not significantly different at $P = 0.05$.

L.S.D. (P = 0.05) for main effects and interaction: treatment, 8.7; time of treatment, 4.4; interaction, not significant.

In MV carrot seeds presowing as well as midstorage ME-D, D-D and S-D treatments showed greater weight of roots (fresh weight/root) over the respective controls with the improvements being more pronounced in the midstorage treatments than in presowing treatment.

In HV carrot seeds presowing ME-D treatment was more effective than S-D. Oxalic acid soaking-drying and water-dipping-drying proved better than water soaking-drying for seedling emergence, yield of roots/m^2 as well as fresh weight per root (Table 3.1b, Pan and Basu, 1985a).

As a modified presowing treatment, high medium vigour (HMV) wheat seed (8.5 month-old seed kept in rubber stoppered 2.5-litre capacity glass bottles) was given 2 h soaking in water and sodium phosphate dibasic, 10^{-3}M at room temperature followed by drying back, 24h moist sand conditioning-drying (incubation with dry sand moistened to 5 per cent prior to drying), moist sand conditioning-soaking-drying in both cases employing water and sodium phosphate as above. Seedling emergence, grain yield and 1000 grain weight were significantly higher in the presowing treatments as compared to the untreated control (Table 3.2, Mandal and Basu, 1987).

Table 3.1b: Effect of Mid-Storage and Pre-sowing Treatments of Medium-Vigour Carrot Seeds on Weight of Roots (fresh weight/root, g.) (*vide* Pan and Basu, 1985)

Treatments	*Time of Treatment*		*Mean*
	Mid-storage	*Pre-sowing*	
Control	48.3^{c}	52.0^{b}	50.3
Moisture equilibration-drying	67.0^{ab}	63.0^{a}	65.0
Dipping-drying			
- Water	64.0^{b}	59.6^{a}	61.8
- Oxalic acid	67.0^{ab}	60.3^{a}	63.6
- *p*-Hydroybenzoic acid	66.6^{ab}	60.0^{a}	63.3
Soaking-drying			
- Water	65.0^{b}	59.0^{a}	62.0
- Oxalic acid	69.0^{a}	60.0^{a}	64.5
- *p*-Hydroybenzoic acid	67.7^{ab}	61.6^{a}	64.3
Mean	64.3	59.4	

Note: Mid-storage treatments were given to 15-month-old seeds and re-stored for a further 3 months under ambient conditions (average RH 77.5 ± 3.8% and average temperature 27.6 ± 1.2ºC) until sowing. Pre-sowing treatments were given to 19-month-old seeds only 2 days before sowing. Within a time of treatment, values scored by the same letter, using Duncan's multiple range test, are not significantly different at P = 0.05.

L.S.D. (P = 0.05) for main effects and interaction: treatment, 6.0; time of treatment, 3.0; interaction, not significant.

The same treatments when given midway during storage to 5-month-old wheat seeds (midterm or midstorage treatment) and stored (after drying back to original weight) for another 3-5 months in rubber-stoppered glass bottles showed significantly better field performance and productivity. In fact, in terms of grain yield (g/m^2) midstorage treatment was better than the presowing treatments. Perhaps it would have been still more pronounced if the conventional storage practice of using moisture pervasive gunny bags would have been followed.

We suggest that any of the modified presowing treatments of well dried seeds in moisture impervious containers be carried out whereby improvement in field performance would be assured. If viability maintenance is a problem midstorage or prestorage dry treatments are recommended. Often prestorage 'dry' treatments may be followed by modified wet presowing treatments for invigoration of seeds for a still better field performance.

Conventional presowing hardening treatments as originally developed by Henckel (1946) to confer resistance to adverse edaphoclimatic conditions would involve 2-3 24 h soaking-drying cycles. In the midstorage H-D treatments we noted that under our conditions long soaking durations in wheat are not compatible with drying back, and even moderate air-drying may cause harm (Mandal and Basu, 1982). Our own studies as well as by others would suggest that shorter durations (often similar to midterm treatments) would serve the purpose. In barley, Chatterjee and Singh (1983) reported considerably improved field performance with presowing treatment employing 6 h soaking with sodium phosphate (dibasic). Mandal and Basu (1987) used very short soaking duration as recommended for midterm treatment in wheat (only 2 h). In fact, they compared midstorage treatment with presowing treatment employing similar H-D techniques, and besides soaking also used moist sand-conditioning with or without further soaking in water or sodium phosphate (dibasic, 10^{-3}M) for the alleviation of any possible soaking injury. The wheat seed lot used in the study was, in fact, a very well stored relatively high-vigour lot with some possibility of adverse effects upon full immersion.

Table 3.2: Effects of Presowing Hydration-Dehydration Treatments on Plant Population/m², Grain Yield/m² and 1000-grain Weight of Wheat cv. Sonalika (*vide* Mandal and Basu, 1987; data for presowing treatment only considered here)

Treatments	*Plant Population (No./m²)*	*Grain Yield (g/m²)*	*1000-grain Weight (g)*
Control	123[a]	238[b]	47.18[b]
Soaking			
Water	132[a] (7.3)	271[a] (13.9)	49.27[a] (4.4)
Sodium phosphate	129[a] (4.9)	278[a] (17.5)	49.46[a] (4.8)
Moist sand conditioning			
Water	132[a] (7.3)	293[a] (23.1)	50.02[a] (6.2)
Sodium phosphate	128[a] (4.1)	284[a] (19.3)	49.32[a] (4.5)
Moist sand conditioning-soaking			
Water	129[a] (4.9)	279[a] (17.2)	49.24[a] (4.4)
Sodium phosphate	128[a] (4.1)	282[a] (18.4)	49.38[a] (4.7)

Within a treatment, values scored by the same letter are not significantly different at = 0.05. (Values in parentheses denote per cent increases over respective controls)

Note: Presowing hydration-dehydration treatment was given to 8.5 month old wheat seed by soaking (full immersion) in water or dilute dibasic sodium phosphate solution, 10^{-3} M for 2 h or moist sand-conditioning for 24 h using clean dry sand moistened to 5 per cent with water or sodium phosphate, dibasic 10^{-3}M solution (using 3 kg dry sand per kg of seed), with or without further soaking (moist sand conditioning - soaking) followed in all cases by drying back to the original weight before sowing (see Mandal, A.K. and Basu, R.N. 1987. *Field Crops Research*, **15**, 259-265) for more data on the comparative efficacy of midterm and presowing hydration-dehydration treatments.

Presowing treatment showed a marginal improvement in plant population/ m^2 which was not statistically significant (Table 3.2). Grain yield per unit area was, however, considerably improved in all the presowing treatments with the extent of improvement ranging from 13.9 to 23.1 per cent over untreated control. Even a more stable parameter like 1000-grain weight showed a statistically significant improvement of 4.4 to 6.1 per cent. However, moist sand-conditioning did not offer any particular improvement because that particular lot of wheat seed obviously did not suffer from soaking injury as anticipated at the beginning.

Another point, which is worth pointing out in connection with the aforesaid comparative study between midterm and presowing treatments, is the fact that unlike many of our previous studies, in the particular study with relatively high vigour seeds used for sowing, the superiority of the midterm treatment over the presowing treatment was less magnified than our studies with medium vigour seeds (attributable to less stringent storage conditions). As such, Mandal and Basu (1987) concluded that presowing treatments should usually be given to seeds kept in better storage conditions. Further such presowing treatments would not involve drying back to safe limits for restorage; surface-drying for the convenience of sowing should be enough. But, when storage conditions are bad, as in wheat stored in gunny bags under ambient conditions in eastern India, midstorage (midterm) treatment should be given for the maintenance of vigour and viability. Prestorage 'dry' seed invigoration would also be a better proposition before giving the midterm treatment. Additionally, if the situation so demands, prior to sowing, a short-term presowing soaking treatment followed by light air-drying could be a much better solution to the problem of successful field establishment and crop performance of wheat in many parts of Eastern India and elsewhere where conditions of seed storage are not congenial.

Midstorage Hydration-Dehydration (H-D) Treatments for Vigour Viability and Productivity

The wet seed treatments are very effective in medium and low vigour seeds and are not usually recommended for high vigour harvest-fresh seeds. All orthodox nonleguminous seeds, *viz.*, cereals, oilseeds, summer and winter vegetables and flowers so far tested have responded very well to these treatments (Basu *et al.*, 1974, 1975; Basu, 1976, 1990; Basu and Rudrapal, 1982; Basu and Mandal, 1985; Basu and Mandal, 1986).

The principle behind the mid-term hydration-dehydration (H-D) treatments is to allow the stored seeds to take up moisture around 20 to 30 per cent (on seed wet-weight basis) for certain preventive, corrective or restorative actions, followed by thorough drying back for restorage. Depending on the initial vigour status of the seed, hydration can be accomplished by spraying the seed with a fixed quantity of water, dipping the seeds in water for 2 to 5 minutes and then keeping the surface-wetted seed covered for 2 to 6 h, moisture equilibration with a water-saturated atmosphere (humidification *vide* Rao, Roberts and Ellis, 1987) for 24 to 48 h, incubation with premoistened sand for 16 to 48 h (moist sand conditioning), stepwise hydration with small quantities of water or simply soaking (full immersion) in water for 2 to 6 h depending upon the material being treated. In all cases, the hydrated seed is thoroughly dried back (as different from light air-drying in presowing treatments) to safe limits for restorage.

Except in the moisture equilibration-drying treatment (seed hydration by humid moisture-saturated air), in place of water, dilute solution of a range of chemicals can be used in the hydration-dehydration treatments. The effective chemicals include sodium and potassium hydrogen phosphate (di- and monobasic), potassium iodide, *p*-hydroxybenzoic acid, *p*-aminobenzoic acid and several others in concentrations ranging from 10^{-5} to 10^{-3} M. However, compared to the large effect of water alone, the chemical effects are rather small and are often reflected in field performance only.

Soaking-drying, which is a very effective vigour and viability-maintaining treatment for medium and low vigour seeds, cannot be directly given to most of the leguminious orthodox seeds because of the possibility of soaking injury. In the humidication-drying treatment the seed takes up moisture in vapour form in a very slow and progressive manner and is then slowly dried back, thereby greatly alleviating soaking injury. However, a brief period of soaking immediately after humidification has been found to be of considerable advantage in maintaining vigour and viability of soybean (Saha and Basu, 1984; Saha, Mandal and Basu, 1990), groundnut (Ramamoorthy and Basu, 1984) and pea seeds (Bhattacharya and Basu, 1990). Similar benefits may be derived from moist sand conditioning and moist sand conditioning-soaking treatments (each followed by drying back) (Ramamoorthy and Basu, 1984).

In the moist sand conditioning treatments, the dry seed is mixed with premoistened sand (5 to 8 per cent moisture content; sand:seed ratio 3:1 (w/w) for most kind of seeds) and incubated in covered containers for 16 to 48 h depending upon the material for a slow and gradual rise of seed moisture to the desired levels (20 to 30 per cent). By such conditioning, the seed can be kept for an even longer period at a fixed moisture level as the sand will be virtually air-dry and there will be no difficulty of aeration if layering is not very thick. In several materials, instead of sand, dried straw chopped finely and thoroughly washed has been successfully used.

Manifestations of Midstorage (H-D) Treatment

Immediately after treatment (before subjecting the seed to accelerated or natural ageing), only marginal differences in germinability between H-D treatments and controls are noted. However, after ageing, the treated and control seeds show significant differences with the former giving higher values for germinability and field performance (Layek, *et al.*, 2006, 2012, 2015). The invigorated seeds showed tolerance not only to ageing stress but also to stresses imposed by heat, salinity and X- and γ-irradiation. In medium vigour seeds of several materials, both pre- and post-irradiation H-D treatments very effectively counteracted the adverse effect of radiation on germinability and chromosomal aberrations (Dasgupta, Chattopadhyay, Basak and Basu, 1977; Punjabi and Basu, 1982; Dey and Basu, 1985; Sur and Basu, 1986).

The improved field performance of stored H-D seeds was manifested in better field emergence and ultimate stand and final crop yield through improvement of the major yield attributes (Dey and Basu, 1982; Mandal and Basu, 1983, 1987a; Pal and Basu, 1982). Even when the plant population was kept uniform by cultural operations such as, high seedling rates for planting and thinning, the yield per unit area was on an average 10 to 25 per cent higher implying a distinct seed invigoration effect as separated from an effect on viability and resultant population per unit area (Dey and Basu, 1982). In crops such as barley, a remarkable compensation for low population through increased tillering of surviving plants was recorded (Abdalla and Roberts, 1969) and a significant yield decrease was noted only after viability fell below 50 per cent. In our studies, a difference of even 10 to 15 per cent improvement in viability due to H-D treatment was significantly reflected in yield of wheat (Mandal and Basu, 1987a, b) and rice (Pal and Basu, 1982, 1988), and several other crops.

Notwithstanding the highly significant beneficial effects of mid-term H-D treatments on vigour, viability and yield potential, the methodology suffers from a number of limitations which are enumerated below:

1. Harvest-fresh, high-vigour seeds of most crops are not responsive to H-D treatment. It would have been more convenient if the treatments could be given before bagging and storage. In case of certified seed, the treatment can be given prior to revalidation only.
2. Handling of bulk quantities of seeds poses technological problems in hydration and drying back. Unless the seed is thoroughly dried back, restorage may accelerate deterioration. In the absence of artificial drying facilities, one has to depend on sun-drying which is often difficult.
3. Even with humidification by moist sand conditioning, a number of legume seeds do not respond well to midstorage H-D treatments.
4. Effects of chemicals are often marginal or nonsignificant. However, to safeguard against storage microflora and pests, proper fungicidal and insecticidal formulations may be incorporated in the soak-water.

The midstorage H-D treatments would be of great help in maintaining vigour and viability of low volume high-value vegetable and flower seeds including costly hybrid seeds. These treatments would also enable farmers to preserve the germinability of small quantities of seeds produced by them for their own use.

Mode of Action of the Midstorage Treatments

So far as the mode of action of the midstorage treatments is concerned, we explored a number of possibilities which included (i) removal of inhibitors and toxic metabolic byproducts of stored seed during hydration, (ii) germination advancement, (iii) reduced moisture uptake by treated seeds, (iv) antimicrobial effect, (v) enzymatic repair of biochemical lesions, and (vi) counteraction of lipid peroxidation reactions.

Although it was earlier suggested that leaching of toxic metabolities from the stored seed into the soak-water would extend the storage life of soaked-dried rice and jute seeds (Basu, Chattopadhyay and Pal, 1974), it was subsequently shown that the moisture equilibration-drying treatment, in which seed hydration was unaccompanied by leaching (Basu, 1976), was highly effective in retaining vigour and viability in storage. As such, leaching of inhibitory substances is not a major factor in seed invigoration. Metabolic removal of toxic substances, through activation of cellular detoxifying enzymes during the hydration phase may, however, improve prestorage germinability.

If germination advancement and embryo enlargement were the prime reasons for the beneficial effect of H-D treatments, harvest-fresh, high-vigour seeds characterised by greater germination capacity, would have been responsive to hydration-dehydration. On the contrary, the longevity of harvest-fresh seeds was reduced by soaking-drying. Further, in dormant lettuce seeds, H-D treatments extended storability to a great extent (Basu *et al.*, 1979; Pan and Basu 1985a; Punjabi and Basu, 1989) without causing any germination advancement. Therefore, the benefits of midstorage H-D treatments cannot be attributed to germination advancement as postulated by Austin *et al.* (1969) for carrot seeds subjected to presowing wetting-drying cycles.

Although differential moisture absorption by stored seeds could be an important factor in seed deterioration, no difference in the moisture uptake pattern of control and treated materials during the storage period could be observed. Even in very dry ageing conditions or when the seed was subjected to a severe heat stress for a brief period, in which situations moisture uptake would not be a factor, significant difference in germinability of treated and untreated seeds was recorded.

The H-D treatments, especially soaking-drying, may reduce microbial invasion by simply washing out the spores, or allowing them to germinate and subsequently killing them by drying. Thus the treated seeds would do well during storage, particularly under conditions favouring growth of storage microflora. But even in storage conditions unfavourable for microbial attack, such as ageing at a very low humidity and rather high temperature, the midstorage H-D treatments would continue to elicit significant positive effects.

Dormant lettuce seeds when maintained at a fully imbibed condition retain better germinability than corresponding dry-stored seeds (Toole and Toole, 1953). According to Villiers and Edgcumbe (1975), this is due to repair of vital bioorganelles by an endogenous enzymic restorative system in the imbibed seed, a viewpoint supported by Roberts (1981) and Osborne (1982). Whether such a repair would take place in such a short-term hydration phase of the H-D treatments is still a matter of conjecture. We have noted that the beneficial effects of midstorage H-D treatments are primarily prophylactic in nature rather than curative. The treatments do not show significant improvement of germinability before ageing. The electrical conductance of seed-steep water, which is a very useful index of membrane integrity (Matthews and Bradnock, 1968), would be considerably reduced by moisture equilibration at 100 per cent RH (humidification, no drying), but after drying back, the conductance would revert to the original value. That would indicate a temporary and reversible nature of changes in the cell membranes during hydration by humidification. A repair of damage of biomembranes should bring about a permanent change resulting in a significant lowering of electrical conductance of seed leachate immediately after humidification-drying. Of course, the treatment would show much lower conductance than the corresponding control after accelerated or natural ageing, indicating the prophylactic nature of the midstorage treatment.

Even if repair of bioorganelles takes place during hydration, this may not require *de novo* protein synthesis because in the presence of cycloheximide, a potent cytoplasmic protein synthesis inhibitor, the moisture equilibration-drying treatment substantially extends lettuce seed longevity (Pan and Basu, 1985b). Thus any repair of biochemical lesions, requiring protein synthesis, would depend on preexistent enzyme proteins. Activation of such enzymes may, however, be useful in metabolic breakdown of toxic substances accumulated in the seed during the previous dry storage. For example, if enzymes like superoxide dismutase (SOD) are activated during hydration that may lead to effective removal of free radicals. However, we have so far been unable to detect any relation between SOD activity and seed invigoration. Similar observations were made elsewhere (McDonald, personal communication). The role of other enzymes, which would act as scavengers of toxic byproducts of anerobic respiration, lipid peroxidation, *etc.* accumulated in the dry-stored seed, during the short hydration phase was subsequently studied in greater detail.

There is a distinct possibility of involvement of free radical-induced lipid peroxidation reactions in seed deterioration (Pammenter, Adamson and Berjak, 1974; Basu, 1976; Berjak, 1978; Buchvarov and Gantcheff, 1984). Peroxidation of membrane lipids would increase permeability (Van Staden, Darvey and du Plessi, 1976) and decrease membrane fluidity (Dobrestov, Borschevskaya, Petrov and Valdimirov, 1977). Seed invigoration treatments may improve storability by controlling free radical reactions and consequent peroxidative damage to lipoprotein cell membranes.

Lipid peroxidation may take place in different ways, namely, autooxidative (non-enzymic), catalytic (non-enzymic, catalyzed by haeme proteins) and enzymic (catalyzed by lipoxygenase). Whatever may be the course of oxidation, the products would be broadly similar. Thus if linoleic acid is the substrate, oxidation will give rise to linoleate hydroperoxide, which depending on the specific resonance hybrid forms, will finally yield volatile aldehydes such as hexanal, pentanal, butanal, *etc.*

The presence of such volatile aldehydes can be readily demonstrated by trapping them with reagents like MBTH followed by colorimetric or spectrophotometric analysis (Harman, Nedrow, Clark and Mattick, 1982; Wilson and McDonald, 1986a; Pal and Basu, 1989; Sur and Basu, 1990a). Part of the inhibition of jute seedling growth in our bioassay for seed vigour (Basu *et al.,* 1990; Sur and Basu 1990a) is attributable to the presence of these volatile aldehydes in the gaseous emanations of deteriorating seed. Seed invigoration treatments always showed significantly lower post-ageing volatile aldehyde production than the control, thereby implying reduced lipid peroxidation by such treatments. In the earlier reports from this laboratory, reduced lipid perioxidation in the viability maintaining treatments was demonstrated by the thiobarbituric acid (TBA) reaction in wheat (Rudrapal and Basu, 1979), sunflower (Dey and Basu, 1982), mustard (Basu and Rudrapal, 1980), onion (Choudhuri and Basu, 1988) and in soyabean (Saha *et al.,* 1990).

Although the role of lipid peroxidation in seed ageing has been the subject of controversy (Koostra and Harrington, 1969; Spencer, Earle, Wolff and Tullent, 1973; Van Staden *et al.,* 1976; Priestley and Leopold, 1979, 1983; Pearce and Abdul-Samad, 1980; Flood and Sinclair, 1981; Stewart and Bewley, 1980), the direct measurement of organic free radicals has clearly implicated free radical-induced lipid peroxidation with the senescence of soybean seeds (Buchvarov and Gantcheff, 1984). The highly significant negative correlation between soybean seed vigour and hydroperoxide level and identification of hexanal, pentanal, butanal as thermal breakdown products of the hydroperoxides (Hailstones and Smith, 1988) would lend credence to the concept of lipid peroxidation as a major reason for seed ageing. Smith and Adamson (1989) employing gas chromatographic techniques have also implicated peroxidation in the deterioration of lettuce seeds.

The beneficial effect of chemicals dissolved in soak water on post-ageing vigour and viability could be due to their role as antioxidants or antioxidant-synergists (Basu, 1976). As synergists, some of the exogenously supplied chemicals may increase the activity of endogenous antioxidants such as tocopherols. In this connection, the suggestion that short-term hydration could serve to temporarily increase the lateral mobility of oxidizing membrane lipids allowing contacts with endogenous antioxidants (Wilson and Mcdonald, 1986 b) deserves experimental verification.

Although the loss of membrane integrity is most likely due to peroxidative attack on unsaturated fatty acid moieties of lipoprotein biomembranes, the damage to protein component of the membrane cannot be ruled out.

Nevertheless, we may conclude that a significant reduction in free radical-induced lipid peroxidation is instrumental, at least in part, for the beneficial effects of seed invigoration treatments. Further critical research on the subject particularly through direct study of free radicals is, however, urgently required for the confirmation of this hypothesis.

Prestorage Treatments for better Storability and Field Performance

Prestorage treatments of harvest-fresh seed are primarily aimed towards protection against deteriorative senescence during storage and a seed lot retaining high vigour and viability is also likely to give better field stand and yield.

Soaking-drying treatment of freshly harvested high-vigour seed would reduce storability (Basu, 1976; Basu and Pal, 1979, 1980). However, some relatively high vigour materials, such as carrot (*Daucus carota* L.) seeds would show improved storability after moisture equilibration-drying treatment (Pan and Basu, 1985a) but many other seeds would not respond well to such wet treatments. 'Dry' treatments do not require hydration and therefore the question of drying back before restorage does not arise providing great technical advantages. The solvent infusion method of dry treatment in which chemicals are introduced into the seed using organic solvents such as acetone, dichloromethane, *etc.* (Khan, 1977) requires removal of the solvent but that can be done just by keeping the seeds in the open below an electric fan for some time. Exposure of seeds to very low concentrations of volatile chemicals, direct addition of such chemicals to the seed, or dry-dressing of seeds with inert carriers containing small amounts of volatile chemicals, are the usual methods followed for 'dry' seed treatments.

Solvent Infusion or Dry Permeation

The feasibility of introducing chemicals into the seed using organic solvent was first demonstrated by Milborrow (1963). Khan (1977) successfully used the technique for introducing bioactive chemicals into the seed for overcoming dormancy. The storability of lettuce (*Lactuca sativa* L.) (Basu, Pan and Punjabi, 1979) and mustard (*Brassica juncea Coss.*) seeds (Dey and Basu, 1985, Dey and Mukherjee, 1988) could be significantly improved by infusion of chemicals like *p*-hydroxybenzoic acid, potassium iodide, and *p*-aminobenzoic acid employing this technique. The treatment is, however, expensive and often risky because of possible injury to the seed by the solvent, particularly in seeds with high initial moisture, and does not hold hope of commercial acceptance in the resource-poor third world countries.

Vapour Treatment

Seeds are exposed to halogen vapours such or iodine, chlorine, *etc.* in a closed container (Basu and Rudrapal, 1980; Rudrapal and Basu, 1980, 1981). A very easy and practical method of seed treatment with chlorine vapour is the use of the common bleaching powder (calcium oxychloride) which slowly releases chlorine in contact with residual seed moisture. About 2 to 3 g of bleaching powder per kg of seed very effectively controlled the loss of vigour and viability of wheat (Mandal and Basu, 1986), rice (Pal and Basu, 1988), pea (Bhattacharya and Basu, 1990), and eggplant (*Solanum melongena* L.) and radish (*Raphanus sativus* L.) (Rudrapal and Nakamura, 1988) seeds. The treatments should be given to high vigour seeds before bagging and storage. Very small amounts of alcohols such as isopropanol, methanol, *etc.* can be directly added to the seed kept in a closed container. These chemicals may also be mixed with inert carriers like calcium carbonate, talc, French chalk, charcoal, clay, brick powder, *etc.* and then used for dry-dressing of seed in the same way as antifungal, antibacterial and insecticidal formulations commonly used by seed producers. For a number of seeds, we have used this technique employing 2 to 3 g of a carrier like calcium carbonate per kg of seed, with the carrier fortified with effective chemicals (each separately, for 2 to 3 g of calcium carbonate or French chalk). The carrier, calcium carbonate, alone has been found to improve storability in several materials. The dry treatments are very suitable for high vigour orthodox seeds including the soaking-susceptible legumes (Rudrapal and Basu, 1980; Mandal and Basu, 1986; Pal and Basu, 1988; Bhattacharya and Basu, 1990) Dey *et al.*, 2004, Sengupta *et al.*, 2005, Mishra *et al.*, 2005, Layek *et al.*, 2006, 2012, Mandal *et al.*, 2008, Biswas *et al.*, 2010, Guha *et al.*, 2012, Datta *et al.*, 2014, Bhattacharya *et al.*, 2015, .

Technologically, the dry seed treatments are ideal for large-scale application since drying back is not required and treatment can be given at any time of the year irrespective of high relative humidity of the air, overcast sky or rains, or when sun-drying is difficult. Prior to bagging and storage, these physiological treatments can be combined with dry fungicidal and insecticidal formulations without involving additional labour costs. Further, some of the dry treatments themselves may offer protection against storage fungi and pests as secondary effects.

Improved post-ageing germinability in the laboratory, better field stand, and yield have been reported in rice (Pal and Basu, 1988) and pea (Bhattacharya and Basu, 1990). However, as the dry treatments are basically meant for high vigour seeds, their efficacy may be comparable to that of the mid-term treatments given to seeds which have already shown some loss of vigour.

Our knowledge regarding the mode of action of prestorage dry treatments is yet to be fully elucidated. In the first report of the efficacy of very low concentrations of iodine vapour on the storability of Indian mustard seeds, Basu and Rudrapal (1980) suggested the role of iodine in the stabilization of double bonds on unsaturated fatty acid moieties of lipoprotein biomembranes as a possible reason for viability extension besides the possibility of iodine acting as a free radical controlling agent (Pryor and Lasswell, 1975). The role of chlorine (as well as bleaching powder, a source of chlorine) and bromine would be more or less similar. The antiomicrobial effects of the halogens may be of significance in high humidity and high temperature storage but their beneficial effects in conditions unsuitable for growth of microflora would suggest the antimicrobial role to be a secondary one.

The protective role of chemicals dry permeated into the seed on storability was recorded by Basu *el al.* (1979) in lettuce and Dey (Pathak) and Basu (1985) in mustard (*Brassica juncea*). Woodstock, Maxon, Faust and Bass (1983) successfully minimized the deterioration of stored parsley (*Petroselinum crispum* L.) and onion (*Allium cepa* L.) seeds using the antioxidants vitamin E (α-tocopherol) and butylated hydroxytoluene (BHT) which were ineffective in pepper seeds indicating

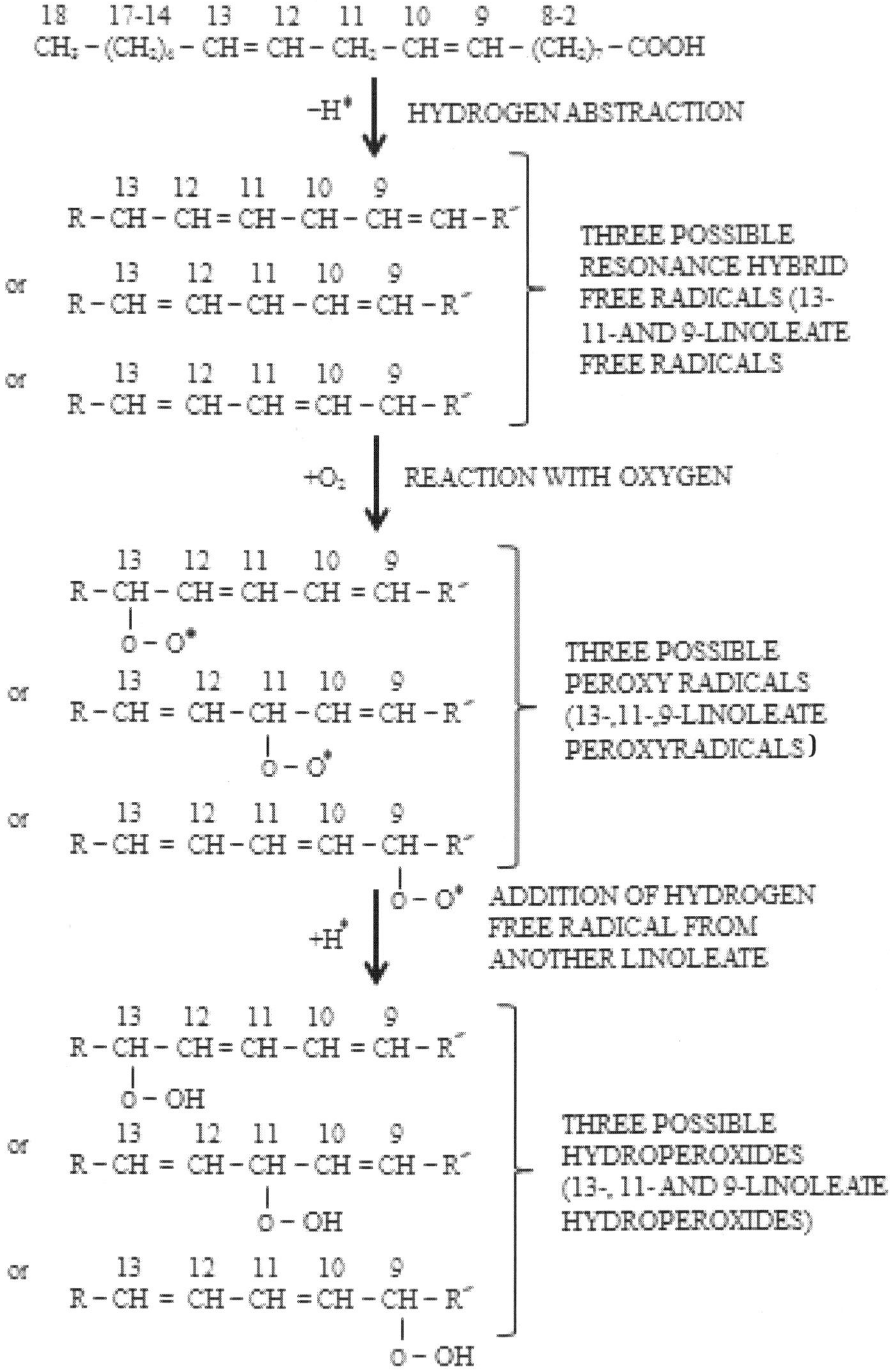

Figure 3.1: Suggested Pathway of Oxidation of Linoleic Acid to Linoleate Hydroperoxides (3 forms) by Free Radical-induced Lipid Peroxidation Reactions (*vide* Basu, 1993).

a differential response. Afterwards, Gorecki and Harman (1987) improved the storability of pea seeds by α-tocopherol and BHT and suggested that lipid peroxidation was probably reduced by their application prior to storage.

We have virtually no explanation for the beneficial effects of very low concentrations of the alcohols ethanol, methanol and isopropanol. Ethanol has been found to accumulate in ageing soybean seeds (Woodstock and Taylorson, 1981) and is a toxic metabolite. The toxicity of methanol prompted its use as a stress reagent (Musgrave, Priestley and Leopold, 1980) and isopropanol also has no apparent metabolic role in the dry seed. The effects of these chemicals on storage microflora and pests and their possible role as free radical controlling agents at very low concentrations require verification.

The carrier, calcium carbonate in particular, even when used alone (not mixed with active chemicals), has been found to improve storability of a number of seeds. Basu and Rudrapal (1982) showed better germinability of wheat seeds treated with a few calcium compounds including calcium carbonate but did not offer any explanation for its success. Bhattacharya and Basu (1990), using the jute seedling bioassay for vigour, demonstrated a substantial reduction in the emanation of

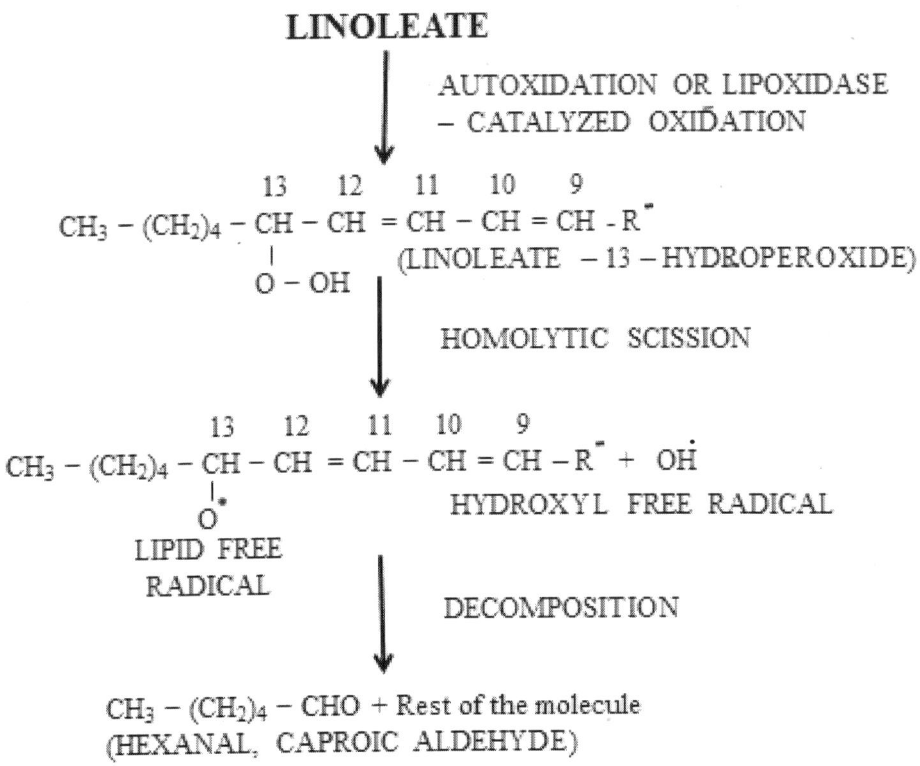

Figure 3.2: Possible Steps in the Breakdown of Linoleate-13-hydroperoxide to Hexanal (Caproic aldehyde, a volatile aldehyde) and an Unidentified Fragment (*vide* Basu, 1993).

A. Free radical formation and peroxidation

Radical initiation

(i) $LH \xrightarrow[\text{radiation or nonperoxide radical}]{\text{hydrogen abstraction by stray}} L^{\bullet} + H^{\bullet}$

(Unsaturated fatty acid) — (Lipid free radical) + (Hydrogen free radical)

(ii) $LH + O^{\bullet} - O^{\bullet} \xrightarrow[\text{Hydrogen abstraction}]{\text{Lipoxidase catalysed}} L^{\bullet} + O^{\bullet} - OH$

(Oxygen diradical) — (Peroxy free radical)

Propagation

$L^{\bullet} + O_2 \longrightarrow LO_2^{\bullet}$ (Lipid peroxy radical)

$LO_2^{\bullet} + LH \longrightarrow L^{\bullet} + LOOH$

(Another unsaturated fatty acid) — (Lipid free radical) + (Lipid hydroperoxide)

B. Possible control of free radical and lipid peroxidation reactions

(i) Reaction of lipid radical with antioxidant

$L^{\bullet} + AH \longrightarrow LH + A^{\bullet}$

(ii) Reaction of lipid free radical with peroxy radical

$L^{\bullet} + LOO^{\bullet} \longrightarrow LOOL$ (Polymeric form)

(iii) Recombination of free radicals in the presence of a medium like water

$R^{\bullet} + R^{\bullet} \longrightarrow 2\ R$ (Nonradical product)

(iv) Stabilization of polyunsaturated fatty acid (PUFA) by a halogen

$$R - \overset{13}{CH} = \overset{12}{CH} - \overset{11}{CH_2} - \overset{10}{CH} = \overset{9}{CH} - R^{\bullet} + I_2 \longrightarrow$$

(Linoleate) (Iodine)

$$R - \underset{I}{\overset{13}{CH}} - \underset{I}{\overset{12}{CH}} - \overset{11}{CH_2} - \underset{I}{\overset{10}{CH}} - \underset{I}{\overset{9}{CH}} - R^{\bullet} \text{ (Tetraiodostearate)}$$

Figure 3.3: Summary of Free Radical-Induced Lipid Peroxidation Reactions and the Possible Termination of Free Radical Reactions and Control Mechanisms of Peroxidative Breakdown of Lipids (*vide* Basu, 1993).

volatile toxic substances by stored pea seeds pretreated with calcium carbonate. Our results with Bengal gram (*Cicer arietinum* L.), besides confirming the beneficial effect of iodine, also showed a significant improvement of post-ageing germinability by the carrier calcium carbonate. The positive effects of the treatments, including the carrier alone, were consistently associated with reduced production of volatile aldehydes, which are presumably the products of lipid peroxidation (Wilson and McDonald, 1986 a).

The possibility of involvement of free radical and lipid peroxidation reactions in seed deterioration (Pammenter *et al.*, 1974; Buchvarov and Gantcheff, 1984) and seed invigoration is strengthened by the effectiveness of the free radical controlling agents (antioxidants) either in the wet or in the dry treatments. One of the earliest symptoms of ageing is loss of membrane permeability which most likely is caused by the breakdown of the lipoprotein membrane structure. The electrical conductance of the seed leachate is a good indicator of such damage (Matthews and Bradnock, 1968). The double bond in the lipid moiety of the biomembranes is especially susceptible for oxidative attack. Although the damage to the protein moiety of the membrane cannot be excluded (especially to the secondary and tertiary structures), the unsaturated lipids in the plasma membrane, the membranes of the inner lysosomic structures and the endoplasmic reticulum can readily undergo peroxidative changes.

Lipid peroxidation may be caused by (i) autoxidation (non-enzymic) due to exposure to oxygen; oxidation proceeds by free radical mechanism in which a chain reaction is initiated by removal of a hydrogen atom from an α-methylene group, (ii) chemical catalysis (non-enzymic) or counteraction of free radical and lipid peroxidation reactions in both wet and dry seed invigoration treatments by haeme compounds such as the cytochromes, and (iii) oxidation catalyzed by lipoxidase (lipooxygenase enzyme). The first two non-enzymatic processes are slow, the third one catalyzed by the lipoxidase enzyme, for which the substrates are unsaturated free fatty acids released from storage lipids by lipases or from membrane lipids by non- specific lipolytical hydrolases is rapid. Although the products of lipid peroxidation (hydroperoxides) are, similar in all the three cases; lipoxygenase-catalyzed reaction shows the following distinguishing features: (a) activation energy is smaller, and (b) the enzyme has a very specific substrate requirement in that the fatty acid must contain at least two *cis* double bonds interrupted by a α-methylene group and the double bond should be present 6 carbon atoms away from the methyl end of the chain (distance from the carboxyl end less critical). Linoleic acid (18 C atoms) present in storage lipid and lipoprotein biomembranes, having two double bonds, one between carbon atoms 9 and 10 and another between 12 and 13, provides an ideal substrate for peroxidative attack. The production of linoleate hydroperoxides from linoleic acid and the subsequent breakdown of linoleate-13-hydroperoxide to caproic aldehyde (hexanal, a volatile aldehyde) are shown in Figures 3.1 and 3.2.

Depending on the kind of resonance hybrid free radicals formed initially, different volatile aldehydes such as hexanal, pentanal, butanal, *etc.* may be produced. When linolenic acid is the substrate, acetaldehyde, crotonaldehyde and other unidentified fragments would be obtained.

Highly significant negative correlation between soybean seed vigour and hydroperoxide level has been reported by Hailstones and Smith (1989). Thermal breakdown of such peroxides produced volatiles which have been positively identified by gas chromatography as hexanal, pentanal and butanal. In lettuce also, volatile levels were closely related to hydroperoxide levels and seed viability (Smith and Adamson, 1989). As heating of solvent-extracted lipids of lettuce seeds yielded the same volatiles, these authors implicated lipid autoxidation in lettuce seed deterioration.

The bioassay system developed by us (Basu *et al.,* 1990; Sur and Basu, 1990 a) has clearly demonstrated the presence of growth-inhibitory gases in the emanations of low vigour seeds. A part of the gaseous emanation could be absorbed by the aldehyde-trapping reagent MBTH (3-methyl-2-benzo-thiazolinone hydrazone), suggesting the production of volatile aldehydes which are obviously products of lipid peroxidation. In the viability-maintaining seed invigoration treatments, the production of volatile toxic substances including aldehydes was less, implying reduced lipid peroxidation (Pal and Basu, 1989; Sur and Basu, 1990b). These results are in agreement with the reduced post-ageing lipid peroxide formation in wheat (Rudrapal and Basu, 1979), sunflower (Dey and Basu, 1982), mustard (Dey and Basu, 1985), onion (Choudhuri and Basu, 1988) and soybean (Saha *et al.,* 1990) following hydration-dehydration treatments.

If free radical-induced lipid peroxidation is accepted as one of the basic reasons of deteriorative senescence of the dry-stored seed (Wilson and McDonald, 1986b), the termination of the same should extend seed longevity. The possible termination reactions are summarized in Figure 3.3.

Basu *et al.* (2010) have documented a range of crop seeds which have shown significantly reduced lipid peroxidation and volatile aldehyde emanation in effective prestorage dry seed invigoration treatments.

Chapter 4

Abstracts of Selected Papers

1. Maintenance of Seed Viability in Rice (*Oryza sativa* L.) and Jute (*Corchorus capsularis* L. and *Corchorus olitorius* L.)

Basu, R.N., Chattopadhyay, K. and Pal, P. (1974). *Indian Agriculturist*, **18**(1), 75-79.

Soaking of rice and jute seeds in distilled water or dilute solutions of cinnamic acid (10^{-4}M), tannic acid (10^{-4}, 10^{-5}M) and sodium thiosulphate (10^{-5}M) and then drying to original weight in sun or in a current of air at 35°C significantly checked the loss of viability during subsequent storage under high humidity conditions. The soaking treatment was given before the seeds started to lose viability, for 6 hours in rice and 3 hours in jute. Treatment with IAA (10^{-3}M), hydrogen peroxide (3×10^{-2}M), Ethrel (10 ppm) and gallic acid (10^{-4}M) also proved beneficial for rice. In both rice and jute, the rate of germination and early seedling growth were greatly improved in the effective treatments. It is concluded that leaching out of autotoxic metabolites from the seeds is conducive to the maintenance of seed viability. The water-soaking and chemical treatments might check fungal attack during storage and also slow down the catabolic reactions in the senescing seed.

2. Control of Seed Deterioration in Wheat (*Triticum aestivum* L.)

Basu, R.N. and Dasgupta, M. (1974). *Indian Agriculturist.*, **18**(3), 285-288.

The loss of viability of wheat seeds subjected to accelerated ageing treatment (100 per cent RH and 45°C) or natural deterioration could be appreciably slowed down by soaking of stored seeds in dilute solutions (10^{-4}–10^{-3}M) of sodium chloride and sodium phosphate (di- and monobasic) or simply water, for 3 hours followed by drying back to original weight in a current of air at 35°C. The phenols, tannic acid (10^{-4}M) and gallic acid (10^{-4}M), and a combination of penicillin and streptomycin (each at 50-200 ppm) also proved beneficial. The speed of germination and vigour of seedlings raised from the treated seeds were much superior to the untreated control seeds. The beneficial effects of the treatments could be due to leaching out of inhibitory substances or to anticatabolic or antifungal properties of the water soaking and chemical treatments.

3. Prolonging the Viability of Wheat Seed

Dasgupta, M. and Basu, R.N. (1975). *14th All India Wheat Workers' Workshop*, BCKV, 1-4.

"Sonalika" wheat seeds were treated with dilute solutions of a range of chemicals including salts, organic acids, phenols and antibiotics for three hours and then dried back to original weights in a current of air at 35°C before storage. Among the different treatments, sodium phosphate (di- and monobasic, 10^{-4}M), sodium chloride (10^{-3}M), sodium thiosulphate (10^{-5}M), oxalic acid (10^{-5}M), gallic acid (10^{-4}M), tannic acid (10^{-5}M) and a combination of penicillin and streptomycin (50-200 ppm each) significantly slowed down the deterioration of seeds stored under accelerated ageing conditions (100 per cent RH and 45°C) and also under ordinary storage in cloth bags in the laboratory (natural deterioration). The speed of germination and vigour of seedlings, as measured by root and short length and dehydrogenase activity of germinating seeds, were significantly greater in the beneficial treatments. The importance of cell membrane functions in maintaining seed viability has been emphasized. Critical physico-chemical and pathological studies are, however, necessary to interpret the effects of the beneficial treatments.

4. Seed Treatment for the Maintenance of Vigour and Viability

Basu, R.N., Bose, T.K., Chattopadhyay, K., Dasgupta, M., Dhar, N., Kundu, C., Mitra, R., Pal, P. and Pathak, G. (1975). *Indian Agriculturist*, **19**, 91-96.

Treatment of stored seeds of rice (*Oryza sativa*), wheat (*Triticum aestivum*), jute (*Corchorus sp.*), sunflower (*Helianthus annuus*), gram (*Cicer arietinum*), lentil (*Lens culinaris*), mung (*Phaseolus aureus*), pigeon pea (*Cajanus cajan*), lathyrus (*Lathyrus sativus*), onion (*Allium cepa*), okra (*Abelmoschus esculentus*), carrot (*Daucus carota*), tomato (*Lycopersicon esculentum*), brinjal (*Solanum melongena*) and chilli (*Capsicum annum*) with water for short periods (usually 2-6 hours, depending on the material) followed by drying back, significantly extended their subsequent storage life under ambient conditions or accelerated ageing under different temperature and humidity regimes. A range of chemicals, including salts, phenols, vitamins and antipathogenic compounds, in low concentrations (10^{-5}–10^{-3}M), showed further improvement of germinability of seeds of the different materials. The easily available salts such as sodium chloride and sodium phosphate (mono and dibasic) have given promising results in majority of the seeds investigated.

5. Physico-chemical Control of Seed Deterioration

Basu, R.N. (1976). *Seed Research*, **4**, 15-23.

The loss of viability of seeds of wheat, rice, jute, sunflower and several pulses and vegetables, under ambient and accelerated ageing conditions, could be significantly slowed down by soaking the seeds halfway during storage, with water or dilute solutions (10^{-5}–10^{-2}M) of a range of chemicals for 2-6 hours, followed by drying back to original weight. Water soaking and drying itself has been found to be effective in majority of the seeds investigated. Instead of soaking in water, a light spray with water or solutions of the chemicals or even moisture equilibration with a saturated atmosphere, followed by drying back, prolonged the viability of stored wheat seeds. Soaking in double the volume of water, however, gave better results in majority of the nonleguminous materials. Treatment of fresh seeds or very old seeds was less effective than treatment of stored seeds (after 4-6 months of normal storage) of good germinability. A second treatment after several months of storage would further extend the viability. The soaking and drying treatments improved the immediate germinability of seeds but the beneficial effects were greatly magnified upon storage. Evidences have been presented to suggest that the treatments which extend seed viability possibly do so by reducing the free radical damage to the cellular components.

6. Seed Treatment for Vigour, Viability and Productivity of Wheat (*Triticum aestivum* L.).

Dasgupta, M., Basu, P. and Basu, R.N. (1976). *Indian Agriculturist*, **20**, 265-273.

Treatment of stored wheat (cv. Sonalika) seeds with water or dilute solutions of several chemicals for 3 hours, followed by quick drying back, significantly slowed down seed deterioration during the subsequent storage period. The viability of seed and vigour of seedlings were much superior in the different physico-chemical treatments compared to the untreated control seeds. The productivity of plants was greatly improved by such mid-storage seed treatment. In pot culture, highest increase in grain yield per plant was noted in plants from disodium phosphate treated seeds. Number of earheads per plant, weight of grains per earhead and 1000-grain weight were also significantly greater in plants raised from the treated seeds. The results were confirmed in a field experiment in which treatment of seed with disodium phosphate greatly increased grain yield per plot followed by those with sodium chloride and water. It was further noted that for a fuller expression of the yield potential of the treated seeds, optimum cultural conditions, especially an abundant supply of fertilizers, would be necessary.

7. Seed Treatment for Vigour, Viability and Productivity

Basu, R.N. (1977). *Indian Farming*, **27**, 27-28.

Studies in the College of Agriculture, Calcutta University have led to the development of a simple soaking-drying method for maintenance of germinability of a number of important field and vegetable crops. Stored seeds about mid-way during ordinary uncontrolled storage are soaked in water or dilute solutions of several chemicals for short periods (2-6 hrs) and then dried in the sun to the original weight and re-stored till planting in the field or nursery. Soaking in water itself is very effective in most seeds but further beneficial effects are noted with very dilute solutions of number of chemicals such as sodium phosphate or sodium chloride at 22-50 mg/l of water. Sodium phosphate has a significantly greater promoting effect on field performance and productivity of the crop raised from the treated seeds.

For maintenance of vigour and viability soaking-drying treatment of fresh seeds is ineffective. Maximum response is obtained with seeds treated after 4-8 months of ordinary uncontrolled storage. Very old seeds of poor germinability should not be used for the physico-chemical treatments.

8. Radioprotective Action of Seed Invigoration Treatments

Dasgupta, M., Chattopadhyay, K., Basak, S.L. and Basu, R.N. (1977). *Seed Research*, **5**, 105-110.

Treatment of stored wheat seeds with water or dilute solutions (10^{-5}–10^{-3}M) of sodium chloride, disodium phosphate, sodium thiosulphate, oxalic acid, cysteine, *p*-aminobenzoic acid, EDTA, potassium iodide, gallic acid and several other chemicals for 3 hours, followed by immediate drying back slowed down the loss of vigour and viability of seeds re-stored under natural and accelerated ageing conditions. Such seed invigoration treatments showed significant protection against X-irradiation damage. Pre-and post-irradiation soaking-drying treatments very effectively counteracted the deleterious effects of irradiation on the growth of the seedlings, activity of amylolytic and dehydrogenase enzymes, membrane permeability and chromosomal aberrations. Treatment with solutions of chemicals which increased the radiation damage, such as caffeine, sodium azide and 2, 4-dinitrophenol, showed an adverse effect on the germinability of seeds.

The beneficial effects of hydration could also be obtained by giving the seeds a light aqueous spray, or even equilibrating the seeds for 8-16 hours with a saturated atmosphere, followed in each case by rapid drying back. In jute, the effects of the chemicals were less pronounced; water soaking followed by drying of the stored seed slowed down the loss of vigour and viability and afforded significant protection against X-irradiation damage. These results indicate a parallelism between seed deterioration and radio-sensitivity and strongly suggest the involvement of free radical reactions in the process. It is concluded that the physico-chemical seed invigoration treatments minimize the loss of vigour and viability by counteracting the free radical pathology in the stored seed.

9. Seed Treatment to Maintain Viability, Vigour and Yield Potential of Stored Rice Seed

Basu, R.N. and Pal, P. (1978). *International Rice Research Newsletter* (*IRRN*), **3**(2), 5.

Soaking-drying treatments using water and sodium diphosphate (dibasic 10^{-4} M) were given to nine-month old seed of rice cv. Ratna stored under ambient conditions, was restored for four months before sowing in the nursery; 40-day-old seedlings were transplanted in the field and harvested after 95 days. Germination test in the laboratory at sowing showed significantly greater germination per cent and seedling vigour in terms of root and shoot growth. Yield of grains in the field was 6.1 and 6.7 tons per ha in the treatments against 4.6 tons/ha in the control.

10. Control of Seed Deterioration in Cotton (*Gossypium hirsutum* L.).

Dharmalingam, C. and Basu, R.N. (1978). *Current Science*, **47**, 484-487.

The deterioration of cotton seeds under warm-humid conditions can be controlled by hydration-dehydration pretreatment of stored (5-6-month-old) seeds. A soaking duration of 2-3 hours for the fuzzy seeds and about one hour for the delinted seeds, followed by drying back to the original weight, would greatly reduce the loss of viability of seeds during subsequent storage. The treated seeds not only maintained a high germination percentage but also showed a significantly greater vigour of seedlings in comparison to the untreated control seeds. Although hydration itself gave the major effect, dilute solutions (2×10^{-4}M) of disodium phosphate and *p*-hydroxybenzoic acid showed further improvement in germination percentage of seeds.

11. Maintenance of Vigour and Viability of Stored Jute Seeds.

Basu, R.N., Chattopadhyay, K., Bandopadhyay, P.K. and Basak, S.L. (1978). *Seed Research*, **6**(1), 1-13.

Under ambient temperature and humidity conditions of Calcutta jute seeds stored in cloth bags after harvest in November-December do not show significant deterioration till the time of next sowing in April-May. However, the left-over seeds which pass through the hot and humid monsoon season start to show signs of physiological deterioration and in *olitorius* jute, in particular, considerable deterioration takes place by the end of September. In both the species of jute, *Corchorus capsularis and Corchorus olitorius*, the carry-over seeds show a significant loss of vigour and viability by the sowing time in April-May.

The loss of vigour and viability of carry-over jute seeds could be greatly minimized by hydration-dehydration treatment of the stored seed towards the end of the rainy season or even during the winter months when sun-drying of hydrated seeds would be easier. The soaking treatment should be given for short periods, not exceeding 1-3 hours, followed by quick drying back to the original weight in the sun or in a drying cabinet at 35 ± 1°C. Although water-treatment alone has the major effect, some additional beneficial effects have been noted with a number of chemicals such as sodium phosphate (monobasic, 2×10^{-4}M), sodium chloride (2×10^{-4}M), sodium thiosulphate (2×10^{-4}M) and oxalic acid (10^{-3}M). Hydration could also be accomplished by moisture equilibration of the seeds with a saturated atmosphere for 12-24 hours. The beneficial effects of seed treatments have been noted under accelerated ageing conditions of 100 per cent RH and 40°C as well as under ambient warm-humid conditions.

Studies on the effects of seed treatment on productivity, employing carry-over JRC 321 seeds, have shown that the hydration-dehydration treatments improve the field performance of the crop raised from the treated seeds. The fibre yield per plant was significantly greater in the water-soaking-drying treatment than in the untreated control.

12. Control of Seed Deterioration by Free Radical Controlling Agents

Basu, R.N. and Dasgupta, M. (1978). *Indian Journal of Experimental Biology*, **16**, 1070-1073.

Treatment of stored wheat seeds with water and dilute solutions of several free radical controlling agents for 3 hr, followed by drying back, slowed down their loss of viability under accelerated ageing conditions. Hydration-dehydration showed the major effect in controlling seed deterioration. Cysteine, *p*-aminobenzoic acid, potassium iodide, disodium phosphate, oxalic acid and sodium chloride showed some improvement in germinability over water. Sodium azide and caffeine, on the other hand, counteracted the beneficial effect of water. The treatments which extended viability showed significant protection against X-irradiation damage, thereby indicating a relation between free radical pathology and seed deterioration.

13. Physiology of Hydration-Dehydration Treatments in the Maintenance of Seed Viability in Wheat (*Triticum aestivum* L.)

Rudrapal, A.B. and Basu, R.N. (1979). *Indian Journal of Experimental Biology*, **17**, 768-771.

Hydration of stored wheat seed by moisture equilibration with a saturated atmosphere (100 per cent RH at 23°C) for 12-48 hr followed by drying back significantly reduced the loss of vigour and viability under accelerated and natural ageing conditions. Beneficial effects of the treatments were associated with greater membrane integrity as indicated by lower electrical conductance of seed leachates and reduced leakage of sugars and amino acids from the seed, increased activity of amylolytic and dehydrogenase enzymes in the germinating seed and reduced lipid peroxidation. Hydration-dehydration treatments slowed down the loss of viability possibly by reducing peroxidative changes in the lipoprotein membranes and minimizing damage to vital biomolecules.

14. Seed Treatment for Maintaining Vigour, Viability and Productivity of Sugar Beet (*Beta vulgaris* L.).

Basu, R.N. and Dhar, S.N. (1979). *Seed Science and Technology*, **7**(2), 225-233.

The deterioration of sugar beet (*Beta vulgaris*) seeds under warm-humid conditions could be effectively controlled by soaking the stored clusters in five times the volume of water for six hours followed by drying back to the original weight before restorage. Soaking in dilute solutions of a range of chemicals such as sodium phosphate (monobasic, 10^{-4}M), sodium thiosulphate (10^{-5}M), tannic acid

(10^{-5}M) and several others did not show any significant improvement over water. The field emergence and productivity of plants raised from seeds, stored under ambient conditions for three months after treatment, were significantly greater than those from the untreated control seeds. The weight of roots from sodium phosphate (monobasic, 10^{-4}M) and EDTA (10^{-4}M) treatments even exceeded that of plants grown from control seeds stored during the corresponding period in a desiccator over fused calcium chloride.

15. Physico-chemical Control of Seed Deterioration in Rice (*Oryza sativa* L.)

Basu, R.N. and Pal, P. (1979). *Indian Journal of Agricultural Sciences*, **49**, 1-6.

Soaking stored seeds of rice (*Oryza sativa* Linn.) for 6 hr in water of volume twice the weight of the seeds, followed by drying in a current of air at 38°C, greatly slowed down the loss of seed vigour and viability during subsequent storage under ambient warm-humid and accelerated ageing conditions. Further advantages over hydration alone could be obtained with dilute solutions (10^{-5} to 10^{-3}M) of sodium phosphate, sodium chloride, sodium thiosulphate, oxalic acid and several other chemicals. Light aqueous spray or moisture equilibration with a saturated atmosphere for 24-48 hr followed by drying also prolonged the vigour and viability of stored seeds, indicating that physico-chemical events other than mere leaching out of inhibitors or toxic metabolites are involved in the process of seed invigoration for the maintenance of viability. The soaking and drying of harvest-fresh rice seeds overcame the problem of partial dormancy and non-uniform germination of seeds sown without further storage, but did not prolong seed viability. Treatment of very old deteriorating seeds was also less effective. The beneficial effects of the hydration-dehydration pre-treatment of the stored seed may be attributed to the counteraction of free radical damage to the cellular components.

16. Control of Lettuce Seed Deterioration

Basu, R.N., Pan, D. and Punjabi, B. (1979). *Indian Journal of Plant Physiology*, **22**, 247-253.

Hydration-dehydration pretreatment of one-year-old lettuce seeds, employing soaking-drying with or without chemicals or moisture equilibration with a saturated atmosphere followed by drying back, greatly slowed down the deterioration of seeds subjected to accelerated ageing (100 per cent RH and 40°C) after treatment. Introduction of chemicals like potassium iodide (10^{-4}M), *p*-hydroxybenzoic acid (10^{-4}M) and tannic acid (10^{-5}M) into the seed by the dry permeation technique, employing acetone as the solvent, effectively controlled seed deterioration. Germination advancement and leaching out of inhibitors from the seed are ruled out as the possible reasons of the beneficial effects of the treatments on the maintenance of vigour and viability. It is suggested that the physico-chemical treatments minimize lettuce seed deterioration by counteracting free radical damage to the stored seed.

17. Carbohydrate Utilization in Wheat Seed in Relation to Viability Maintenance by Soaking-Drying Treatments

Rudrapal, A.B. and Basu, R.N. (1979). *Seed Research*, **7**, 5-10.

Soaking of 5-month-old wheat seed in water and dilute solution of sodium phosphate (dibasic, 10^{-4}M) for 2.5 hours followed by drying back reduced the loss of vigour and viability during subsequent storage under accelerated and natural ageing conditions. The post-ageing amylolytic and dehydrogenese activities of seeds imbibed for 24 hours were considerably higher in the water and sodium phosphate treatments than in the control. The utilization of ($U^{-14}C$)-glucose, after imbibition for 4 hours, especially its incorporation into the ethanol-insoluble fraction of the embryo, was more in the pretreated seeds. The extent of ($U^{-14}C$)-glucose incorporation into the ethanol-insoluble fraction of the endosperm was less than that in the embryo but here also the pretreated seeds performed better than the control. [($U^{-14}C$)-glucose = uniformly labelled radiosensitive glucose]

18. Iodine Treatment of Seed for the Maintenance of Vigour and Viability

Basu, R.N. and Rudrapal, A.B. (1979). *Seed Research*, **7**, 80-82.

Seeds of wheat (*Triticum aestivum* cv. Sonalika), jute (*Corchorus capsularis* cv. JRC 321) and mustard (*Brassica juncea* cv. B85) of 6, 10 and 12 months of age respectively were iodinated in glass desiccators containing iodine crystals for 0, 5, 15, 24 hours and then placed in a current of air at 35°C for 24 hours. Significant beneficial effects of iodine treatment on both viability and seedling growth were noted in mustard seeds after accelerated ageing at 100 per cent RH and 42°C even under shorter durations. In wheat and jute, optimum duration of iodine was 15 hours above which there was a reduction in germinability. Compared to mustard seeds, however, the beneficial effect of iodine treatment on root and shoot growth was much less in wheat and jute seeds.

19. Seed Treatment for Vigour, Viability and Productivity of Tomato

Mitra, R. and Basu, R.N. (1979). *Scientia Horticulturae*, **11**, 365-369.

Soaking-drying treatment of stored tomato seeds greatly reduced their deterioration under accelerated and natural ageing conditions. Hydration by moisture equilibration at 100 per cent RH for 24 hours followed by drying-back was also very effective. Treatment of harvest-fresh seed was ineffective. The beneficial effects of the soaking-drying treatments were reflected in the productivity of the crop. Treatment with dilute solutions of sodium chloride (10^{-3}M) or sodium phosphate (*dibasic*, 10^{-4}M) proved better than water for field performance and productivity.

20. Control of Seed Deterioration in Sunflower (*Helianthus annuus* L.)

Pathak, G. and Basu, R.N. (1980). *Current Science*, **49**(2), 67-69.

Soaking-drying treatments were given to sunflower seeds which had been stored under ambient conditions for 5 months. Seeds were soaked in double the volume of water or solution of chemicals for 2 hr and then dried back in a current of air at 35°C to the original weight and the moisture content brought to 7.2 ± 0.2 per cent. The control seeds were dried without soaking.

The hydration-dehydration treatments effectively slowed down the deterioration of sunflower seeds stored for 4-6 months under ambient ageing conditions as well as under accelerated ageing conditions. However, no beneficial effect was noted immediately after treatment. Under accelerated ageing conditions of 100 per cent RH and 40°C none of the chemicals showed any significant effect over water in germination percentage and seedling growth. Under low humidity and high temperature ageing condition of 30 per cent RH and 45°C the chemicals *p*-hydroxybenzoic acid, oxalic acid, tannic acid and sodium phosphate, however, gave distinct advantage over hydration alone. According to Basu and co-workers, counteraction of free radical induced lipid peroxidation may be the primary reason of the beneficial effects of the physico-chemical treatments. The effective chemicals may control lipid peroxidation and free radical reactions directly or indirectly by synergizing antioxidants in the seed.

21. Iodine Treatment of Mung-Bean Seeds for the Maintenance of Vigour and Viability

Rudrapal, A.B. and Basu, R.N. (1980). *Current Science*, **49**(8), 319-320.

Eighteen month-old seeds of mungbean (*Vigna radiata* L. Wilczek cv. B^1) were exposed to a partially saturated iodine vapour in a desiccator for 8, 16, 24 and 48 hr. Germination tests conducted before subjecting seeds to accelerated ageing conditions showed no significant improvement in germination percentage and seedling growth. However, a highly beneficial effect of iodine treatment on germinability was noted after accelerated ageing for 20 days at 95 per cent RH and 40°C and also when seeds were subjected to a heat stress at 100°C for 45 minutes. In the accelerated ageing test, iodination for 8 h and 16 h was more effective but treatment for 24 h was found to be more effective in counteracting heat injury specially with respect to seedling growth. The stabilization of lipo-protein components of the cell membranes rendering them less susceptible to lipid peroxidation and free radical reaction could be the possible mode of action of beneficial effect of iodination on vigour and viability maintenance.

22. Iodination of Mustard Seed for the Maintenance of Vigour and Viability

Basu, R.N. and Rudrapal, A.B. (1980). *Indian Journal of Experimental Biology*, **18**, 492-494.

Exposure of mustard (*Brassica juncea* Coss cv. B.85) seeds to a partially saturated iodine vapour environment for 8-24 hr at 25°C significantly reduced the physiological deterioration of seeds during subsequent storage under different temperature and humidity regimes. The maintenance of viability of iodinated seeds was associated with greater activities of amylolytic and dehydrogenase enzymes and better integrity of cellular membranes and considerably reduced lipid peroxidation. It is suggested that iodine stabilizes the unsaturated fatty acid components of lipoprotein membranes rendering them less susceptible to lipid peroxidation and free radical reactions.

23. Control of Rice Seed Deterioration by Hydration-Dehydration Pre-treatments

Basu, R.N. and Pal, P. (1980). *Seed Science and Technology*, **8**, 151-160.

Hydration of 12-month-old rice seeds (*Oryza sativa* cultivar 'Ratna') by moisture equilibration with a saturated atmosphere (100 per cent RH at 25°C) for 24 to 72 hours or by soaking in water for six hours, followed in either case by drying back, greatly minimised their deterioration under accelerated and natural ageing conditions. Comparing the two methods of hydration, the germination percentage following ageing was slightly higher in the moisture equilibration-drying treatment. In one month-old (non-dormant) seeds, the moisture equilibration-drying treatment, followed by accelerated and natural ageing, did not show greater germination percentages than the control and contrary to the observation in the old seed, the soaking-drying treatment of fresh seeds significantly accelerated their deterioration on storage. It appears that the injury caused to the seed by rapid water uptake during soaking could be avoided by slow hydration accomplished by moisture equilibration with a saturated atmosphere.

24. Hydration-Dehydration Treatment of Stored Carrot Seed for the Maintenance of Vigour, Viability and Productivity

Kundu, C. and Basu, R.N. (1981). *Scientia Horticulturae*, **15**, 117-125.

Soaking of stored carrot seeds in water or in dilute solutions of chemicals such as sodium thiosulphate (10^{-5}M) and disodium phosphate (10^{-4}M) for 2 hours, followed by drying back to the original weight, greatly reduced physiological deterioration during subsequent storage. The treatment effects on storability were more pronounced in carry-over seeds than in seeds of the current year's harvest. Instead of soaking, even dipping of stored seeds for 2 minutes in water or solutions of chemicals, or hydration by moisture equilibration with a saturated atmosphere for 24-48 hours followed by the usual drying back, very effectively minimized the loss of seed viability. Hydration-dehydration alone proved highly beneficial, but further improvement over water was noted with some of the chemicals.

The beneficial effects of the mid-storage seed treatments were also noted on field performance and the roots of plants grown from treated seeds were bigger than those from control plants. As in the case of laboratory germination tests, the effects of the treatments on field emergence and productivity were more marked in carry-over seed than in current year's seed.

25. Role of Embryo and Endosperm in Rice Seed Deterioration

Mandal, K. and Basu, R.N. (1981). Proceedings of the *Indian National Science Academy*, **B47**, 109-114.

Embryos of rice seeds deteriorated to different degrees by storing under ambient conditions for 0-3 years were transplanted on to endosperms of freshly-harvested (high-vigour) seeds. Age-induced deterioration adversely affected the germinability of embryo and its subsequent growth; storage for three years under ambient conditions resulted in complete non-viability of the embryo. In the reciprocal transplantation experiment, high-vigour embryos were transplanted on to differentially aged endosperms. The condition of the endosperm considerably influenced the growth of the embryo. However, the endosperms of non-viable seeds could support the growth of high-vigour embryo to a certain extent. While the effect of ageing on the sprouting of the embryo is direct, that of the endosperm on seedling growth is indirect. Our results suggest that of the damage to the aleurone cells of the endosperm by ageing reduced the capacity of the endosperm to support the development of the sprouted embryo. Growth of high-vigour embryos transplanted on to endosperms of high-vigour seeds was proportional to the amount of aleurone tissue retained on the latter and the embryos completely failed to grow on aleuroneless endosperms. Root and shoot growth of transplants was inversely related to the induction of α-amylase in the endosperms. Ageing or artificial removal of aleurone cells reduced the growth of transplants because of failure of GA-induced synthesis of amylase and possibly other enzymes in the endosperm.

26. Maintenance of Soybean Seed Viability by Hydration-Dehydration Treatments

Saha, R. and Basu, R.N. (1981). *Indian Agriculturist,***25**, 275-278.

Moisture equilibration of stored (1-year old) soybean (*Glycine max*(L.) Merr. CV. Soyamax) seeds with a water-saturated atmosphere for 24 hr followed by soaking in water for 2 hr and then drying back to the original weight very significantly reduced the loss of germinability of seeds during subsequent storage. Moisture equilibration-drying alone was not effective and only soaking-drying treatment of stored seeds without preconditioning by moisture equilibration, accelerated the loss of viability. The moisture equilibration effect could be partly simulated by covering seeds with a double-layered thick moist cotton cloth. Use of the osmoticum, polyethylene glycol (PEG, mol. wt. 6000, 10 per cent w/v solution) was good for immediate germinability but the beneficial effect of PEG treatment was not obtained on drying back and storage.

27. A Note on the Involvement of Protein Synthesis in Hydration-Dehydration Treatment of Lettuce Seed

Pan, D., Punjabi, B. and Basu, R.N. (1981). *Seed Research*, **9**(2), 202-205.

Hydration of stored lettuce seed by dipping for 2 minutes or soaking for 30 minutes to several hours or moisture equilibration with a water saturated atmosphere (100 per cent RH, 28°C) for 24-48 hours, followed in each case by drying back greatly extended vigour and viability (Basu, *et al.*, 1999).

Several possibilities on the mode of action of such treatments *viz.* advancement of germination, leaching of toxic metabolites or differential moisture absorption by treated seeds have been discounted for by Basu, 1976. According to Villiers and Edgcumbe the damage caused to macromolecular structure during storage would be continually repaired by repair enzymes requiring *de novo* protein synthesis. Lettuce seeds soaked in the presence of protein-synthesis inhibitor, cycloheximide followed by drying back significantly reduced loss of vigour and viability during storage. This is indicative of the absence of any repair mechanism as a mode of action of the hydration-dehydration treatments. However, this does not preclude the possibility of repairs to biorganelles by preexisting enzyme proteins for which *de novo* protein synthesis is not required. Respiratory ATP may also not be necessary because 2,4-Dinitrophenol pretreatment within limits could not counteract the beneficial effect of the moisture-equilibration treatment.

28. Use of Chlorine and Bromine in Controlling Mustard Seed Deterioration

Rudrapal, A.B. and Basu, R.N. (1981). *Seed Research*, **9**(2), 188-191.

The deterioration of seeds of mustard (*Brassica juncea* Coss cv. B85) could be significantly controlled by treatment of harvest-fresh as well as one-year-old (high medium vigour) seeds with chlorine and bromine in the vapour phase confirming the earlier observations in the present laboratory on the great efficacy of halogenation with iodine for the maintenance of seed vigour and viability. The treatments with chlorine and bromine also imparted great resistance to the deteriorative senescence of seeds kept under accelerated ageing at 100 per cent RH and 42°C for 18 days, natural ageing under ambient conditions (44°C ± 1.5°C) and 65 per cent RH for 6 months and also to heat stress at 60°C for 30 minutes. Because of basic similarities in the reactivities of the halogens chlorine and bromine would be expected to mimic the responses of iodine with respect to the deteriorative senescence of the seed as well. In case of iodine the stabilization of the unsaturated fatty acid component of the lipoprotein membranes was suggested to be beneficial on the maintenance of seed viability and vigour.

29. Maintenance of Vigour, Viability and Yield Potential of Oilseeds

Dey, G. and Basu, R.N. (1982). In: *Pulses and Oilseeds in West Bengal*, A Government of West Bengal Publication, Volume **III**, pp. 217-223.

In West Bengal oilseeds stored under the ambient hot and humid conditions rapidly lose their vigour and viability. This is especially true for seeds of sunflower and safflower. The situation is less serious for current year's mustard and sesamum seeds but in carry-over seed lots of these materials, poor germinability at sowing is frequently met with. The loss of vigour and viability of oilseeds in ordinary uncontrolled storage could be greatly reduced by soaking-drying treatment with or without a range of chemicals. In sunflower, treatment of 3-4 month-old seed in July-August by soaking for 2 hours in water or solutions of sodium chloride (10^{-3}M), *p*-hydroxybenzoic acid (10^{-5}M) or *p*-aminobenzoic acid (10^{-3}M), followed by drying back, would carry them safely till next sowing. Not only the seed vigour and viability but also the productivity of the resultant crop would be significantly improved. In mustard, *p*-aminobenzoic acid (10^{-3}M), *p*-hydroxybenzoic (10^{-4}M) and sodium phosphate mono-basic (10^{-3}M), gave additional beneficial effects over hydration alone. Water alone was very effective in safflower and the chemicals did not give any significant additional effect. In mustard and sesamum, treatment of 12-18-month-old seed was effective but adverse effects were noted when relatively fresh seeds were treated. Drying back of the seed was most essential, as otherwise re-storage of inadequately dried seeds would lead to faster deterioration with consequent poor field performance and productivity.

30. Control of Rice Seed Deterioration by Physico-chemical Treatments

Pal, P. and Basu, R.N. (1982). In: *Rice in West Bengal.* A Govt. of West Bengal Publication, Volume III, 159-166.

The deterioration of rice seeds under the ambient hot and humid conditions of West Bengal could be greatly reduced by midstorage hydration-dehydration treatment of stored seeds. Seed hydration was accomplished by soaking in water for 2-6 hours, dipping the seeds in water for 5 minutes and then keeping the wet seed covered for 4-6 hours or even by moisture equilibration of the seed with a water saturated atmosphere for 48 hours followed in all cases by drying back to the original weight. In the soaking-drying and dipping-drying treatments, use of dilute solutions of chemicals such as sodium phosphate (mono- and dibasic, 10^{-4}M), sodium chloride (10^{-4}M) and sodium thiosulphate (10^{-4}M), showed some small improvements over hydration alone. The major effect was, however, due to hydration (followed by dehydration). Soaking-drying treatment of harvest-fresh seed adversely affected storability. The seed treatments, besides maintaining germinability, exhibited significantly better field performance and productivity than the control. Physiological studies have shown greater amylase and dehydrogenase activities in the treated seed. The leaching of electrolytes and sugars from the seed was less in the physico-chemical treatments. The greater loss of membrane permeability in control seeds resulting in increased leakage of electrolytes and sugars was possibly due to more damage to lipoprotein cell membranes. It is suggested that the physico-chemical treatments minimized the loss of seed vigour and viability by counter-acting lipid peroxidation and free radical reactions in cell membranes and other vital bioorganelles.

31. Preconditioning Soybean Seed for Viability Maintenance by Soaking-Drying Treatment

Saha, R. and Basu, R.N. (1982). *Seed Research*, **10**, 193-197.

Stored seeds of soybean (*Glycine max* L. cv. Soyamax) were preconditioned by moisture equilibration with a water-saturated atmosphere, soaking in polyethylene glycol (PEG 6000, 40 per cent w/v solution) for 24 hr and keeping in contact with moist sand (fine sand moistened to 10 per cent on air-dry basis) for 16 hr prior to soaking-drying treatment. The preconditioned seeds could be soaked for 2 hr in water, without encountering soaking injury, and then dried back for successful maintenance of vigour and viability during subsequent storage.

32. Dip-Dry Treatment for the Maintenance of Vigour and Viability of Stored Wheat Seed

Mandal, A.K. and Basu, R.N. (1982). *Indian Agriculturist*, **26**, 289-295.

Short-term dipping of stored (5-month-old) wheat (*Triticum aestivum* L. cv. Sonalika) seed for 2-10 minutes in water followed by moist placement (seeds kept covered in moist condition) for 2-4 hours and then drying back to the original weight significantly reduced the loss of vigour and viability. Dipping for 2-5 minutes followed by 2-3 hours moist placement showed greater beneficial effects on germinability. Moist placement after dipping (prior to drying) was important, otherwise drying of seeds immediately after dipping showed less effect. The dip-dry treatment exhibited greater effect in high and medium vigour seeds; in the later category of seeds, full immersion for 2 hours followed by drying (soak-dry) very successfully maintained vigour and viability.

33. Lipid Peroxidation and Membrane Damage in Deteriorating Wheat and Mustard Seeds

Rudrapal, A.B. and Basu, R.N. (1982). *Indian Journal of Experimental Biology*, **20**, 465-470.

In deteriorating wheat and mustard seeds, highly significant negative correlations were noted between lipid peroxidation and germinability and between electrical conductance of seed-steep water and germinability. Lipid peroxidation in seed and electrical conductivity of the leachate were positively correlated. The extent of different correlations were influenced to a certain extent by the ageing conditions but, in general, a broad similarity in this respect was noted between different ageing stresses. Hydration-dehydration treatments of stored seed, which slowed down physiological deterioration, exhibited lower lipid peroxidation and reduced the leaching of electrolytes and metabolites from the seed. These observations clearly implicate lipid peroxidation in the deterioration of wheat and mustard seed.

34. Alleviation of Salt Stress during Germination and Early Seedling Growth of Wheat by Hydration-Dehydration Pre-treatments

Mandal, A,K. and Basu, R.N. (1982). *Indian Agriculturist*, **26,** 121-128.

Midstorage soaking-drying (2 hr soaking followed by drying) treatment of stored (medium-vigour) wheat seeds greatly reduced subsequent physiological deterioration and showed significant alleviation of salt stress (induced by sodium chloride) as measured by germination and early seedling growth in a modified glass plate blotter technique. Pregermination soaking-drying treatments employing short (4 hr), medium (12 hr) and long (24 hr) soaking durations and a range of chemicals also showed an immediate beneficial effect of short and medium soaking durations. But when the treated seeds were stored in cloth bags under warm-humid ambient conditions for 75 days, only the beneficial effect of the short duration soaking-drying treatment was evident on germinability in a 0.2 M sodium chloride solution; prolonged soaking proved positively harmful. The midstorage treatment using short soaking durations is recommended because of its dual beneficial effects on storability and partial alleviation of salt stress.

35. Seed Viability

Basu, R.N. (1982). *Science and Culture*, **48,** 129-134.

The author has briefly covered the subject of viability of both orthodox seed and recalcitrant seeds with respect to the factors that affect seed longevity. The physiological and biochemical changes which occur in seeds (orthodox) during storage associated with loss of vigour and viability have been discussed.

The beneficial effects of the mid-storage hydration-dehydration physico-chemical seed treatments developed in the University of Calcutta for controlling the loss of germinability have been highlighted. The possible modes of action of such treatments have been put forward and the predominant role of the counteraction of free radicals and lipid peroxidation reactions (at least partly) in controlling loss of germinability during storage has been suggested.

The need for further studies on the mode of action and more information on the physiology of the recalcitrant seeds with respect to their behavioural responses *vis-à-vis* imbibed (dormant) orthodox seeds like lettuce has been emphasized.

36. Studies on the Maintenance of Seed Viability of Sunflower (*Helianthus annuus* L.).

Dey, G. and Basu, R.N. (1982). *Indian Journal of Plant Physiology*, **25**(1), 87-97.

Hydration-dehydration treatment of stored sunflower seeds, with or without chemicals, has been found to be very effective in controlling the loss of vigour and viability under different ageing conditions. The beneficial effects of the treatments were associated with greater dehydrogenase activity and reduced lipid peroxidation in the seed. It is suggested that in the effective treatments the damage to lipoprotein membranes by lipid peroxidation was considerably reduced and thus the membrane permeability was less affected than in the untreated seeds. The involvement of free radical reactions in the deterioration of sunflower seed and its control by physico-chemical treatments have been discussed.

37. Maintenance of Lettuce Seed Viability and Counteraction of Radiation Damage by Moisture Equilibration-Drying Treatment

Punjabi, B. and Basu, R.N. (1982). *Indian Biologist*, **14**, 9-12.

Lettuce seeds (thermodormant) could be maintained at a high level of germinability if stored in fully imbibed conditions. Also, storage in fully imbibed condition is known to significantly counteract irradiation damage. The results of the present investigation show that simply equilibrating the seeds with a saturated atmosphere for 24 hr followed by drying back to original weight very effectively counteracted both pre-and post- X- and γ-irradiation induced seed deterioration in lettuce seeds. Quenching of age or irradiation induced free radicals may explain the beneficial effect of moisture equilibration-drying treatments; post-irradiation H-D treatments are well documented but that of pre-irradiation (ME-D + irradiation) is not well understood. Perhaps the quenching of endogenous radicals or a biophysical stabilization of bio-organelles by the moisture equilibration-drying treatment, besides extending longevity would make the seeds more radio-resistant.

38. Control of Age– and Irradiation– Induced Seed Deterioration in Lettuce (*Lactuca sativa* L.) by Hydration-Dehydration Treatments

Punjabi, B. and Basu, R.N. (1982). In: *Proceedings of the Indian National Science Academy*, **B48**(2), 242-250.

Hydration-dehydration treatment of stored lettuce seed (1-year-old, medium-vigour), greatly slowed down their deterioration during subsequent storage under accelerated and natural ageing conditions. Hydration-dehydration of seeds, before or soon after X- and γ-irradiation, considerably minimized the adverse effect of irradiation on the development of biological after-effects responsible for the fall in germinability, especially the large reduction of root growth of seedlings. Pre-and post-irradiation treatments gave broadly similar effects. The results have been discussed in terms of a possible involvement of a cellular (biochemical) repair mechanism in the hydration phase and also on the basis of a physico-chemical control of free radical pathology in the aged and irradiated seed.

39. Testing Germination and Seedling Growth by an Inclined Glass Plate Blotter Method

Punjabi, B., and Basu, R.N. (1982). *Indian Journal of Plant Physiology*, **26**, 289-298.

A method of seed germination which gives highly uniform and reproducible results and enables easier recording of root and shoot growth of seedlings is described. In this method seeds are germinated on moist blotters on glass plates which are kept in polythene packets containing 20-30 ml of water and provided with two rows of holes on the back side. The plates are arranged at 66° angles on plastic dish-stands which can be very conveniently placed in germinators or in the room for required number of days. Day-to-day observation of the progress of germination can be made without disturbing the germinating seeds. The method has been very conveniently employed for testing germination and seedling growth of seeds of our common field crops, vegetables and flowers. (for details see extracts of the paper given in Chapter 5)

40. Maintenance of Vigour, Viability and Productivity of Stored Barley Seed

Punjabi, B., Mandal, A.K. and Basu, R.N. (1982). *Seed Research*, **10**(1), 69-71.

Five-month-old seeds of Barley (*Hordeum vulgare* L. cv. IB 65) were soaked in double the volume of water and dilute solution of sodium phosphate dibasic (10^{-4}M) for 2 h and then dried back to the original weight in a current of air at 35°C ± 1°C. Control seeds were not soaked but dried along with the treated seeds. Treated seeds maintained a high level of germinability as compared to the untreated control after storage under accelerated ageing conditions for 8 days at 100 per cent RH as well as under ambient storage conditions for 3 months. Sodium phosphate was more effective than water alone. Increase in yield of the crop raised from the treated seeds was noted which is in conformity with earlier reports from this laboratory. The improved field performance of the treated seeds may be looked upon as an indirect effect of reduced physiological deterioration of seeds during storage under natural ageing conditions.

41. Post-harvest Seed Physiology and Seed Invigoration Treatments

Basu, R.N. and Rudrapal, A.B. (1982). In: Frontiers of Research in Agriculture, Proceedings of the Indian Statistical Institute Golden Jubilee International Conference, Calcutta (Ed. S.K. Roy), pp. 374-397.

The authors have discussed the post-harvest physiology of orthodox seeds and physiological aspects of seed ageing in storage. The development of a highly beneficial seed invigoration hydration-dehydration treatment for the maintenance of vigour and viability and the possible mode of action of such treatments have been put forward. The counteraction of free radical and lipid peroxidation reactions have been suggested as a possible interpretation of the treatment effects.

42. Soaking and Drying of Stored Sunflower Seeds for Maintaining Viability and Vigour of Seedlings and Increasing the Yield Potential

Basu, R.N. and Dey, G. (1983). *Indian Journal of Agricultural Sciences*, **53**(7), 563-569.

Mid-storage soaking in water and drying of stored seeds of sunflower (*Helianthus annuus* Linn.) very effectively controlled the loss of viability and vigour of seeds during subsequent storage under the hot and humid conditions of eastern India. The germinability of the treated seeds was good and the field performance and productivity of the crop were significantly better. The increase in yield was attributable not merely to increase in plant population per unit area, because, even when the population was kept constant by cultural adjustments, the yield per unit area was more in the crops raised from the treated seeds. Seed treatment with water containing sodium phosphate (monobasic, 10^{-4}M), *p*-hydroxybenzoic acid (10^{-5}M) and *p*-aminobenzoic acid (10^{-3}M) and drying proved better than soaking with water alone and drying.

43. Maintenance of Vigour, Viability and Yield Potential of Stored Wheat Seed

Mandal, A.K. and Basu, R.N. (1983). *Indian Journal of Agricultural Science*, **53**(10), 905-912.

Soaking of 5-month-stored seeds of wheat (*Triticum aestivum* Linn. emend. Thell.) in water or dilute solution of sodium phosphate (dibasic, 10^{-4}molar) for 2 hr or dipping them for 2 min and keeping them covered for 3 hr, followed in either case by drying back to the original weight, could effectively prevent seed deterioration during subsequent storage in ambient conditions. Soaking-drying was more effective than dipping-drying. Dry-dressing of hydrated-dehydrated seed with ceresan, captan, dithane M-45 and thiram or spraying with malathion did not show any major improvement over hydration-dehydration. The protectants, however, showed no adverse effect on post-storage germinability. The field performance and productivity of the crop raised from the hydrated-dehydrated seed were consistently superior to the control in 1977-78, 1978-79 and 1979-80.

44. Preservation of Onion Seed

Basu, R.N. and Nilanjana Choudhuri (1983). *Seeds and Farms*, **9**(4 and 5), 17-19.

Onion seeds (*Allium cepa* L. cv. Sutton's Globe) stored under ambient conditions for 3-12 months were soaked for 2-6 hr followed by drying back to the original weight in a current of air at 35 ± 1°C. This hydration-dehydration treatment very significantly controlled the loss of germinability of onion seeds under all conditions of storage. These treatments were ineffective in harvest fresh high vigour seeds. The beneficial effect of the treatments was very significant in medium and low vigour seeds. However, in high vigour seeds, dipping-drying, moisture-equilibration drying and moist-sand conditioning treatments are more effective. A slow absorbtion of water by the high vigour seeds possibly eliminates the soaking injury to seeds. Seeds exposed to halogens like chlorine, bromine and iodine in vapour form for 24 hr or mixing the seeds with inert carriers like fresh chalk, calcium carbonate or kaolinite impregnated with the halogens would reduce physiological deterioration of high vigour onion seeds, but the magnitude of the beneficial effects of the hydration-dehydration treatments was much higher.

It is, however, not clear how these treatments reduce physiological deterioration of seeds in storage. Studies in this laboratory with other several crop seeds have indicated the counteraction of lipid peroxidation and free radical reactions as the possible mode of action.

45. Control of Seed Deterioration and Improvement of Field Performance of Okra by Hydration-Dehydration Treatment of Stored Seed

Kundu, C. and Basu, R.N. (1983). *Vegetable Science*, **10**(2), 84-93.

Physiological deterioration of okra (*Abelmoschus esculentus* L. Moench) cv. Pusa Sawani seeds under natural and accelerated ageing conditions was greatly reduced by midstorage hydration-dehydration (soaking-drying, dipping-drying and moisture equilibration-drying) treatment of stored seeds. The field performance and productivity of the crop grown from treated seed were significantly better and upto 16 and 39 per cent increase in fruit yield were recorded in the winter and summer crop, respectively. Treatment of harvest-fresh seed was not effective in the maintenance of vigour and viability.

46. Preservation of Wheat Seed by Hydration-Dehydration Treatment with Seed Protectants

Mandal, A.K. and Basu, R.N. (1984). *Pestology*, **8**(1), 20-23.

Hydration-dehydration treatment of stored (5-month-old) Sonalika wheat seeds effectively controlled seed deterioration during subsequent accelerated and natural ageing conditions. Seed hydration was done by soaking in water or fungicide (Dithane M-45) at 0.05 to 10 per cent solutions or insecticide (Satisfar 50 EC, an experimental organo-phosphorus insecticide formulation (etrimphos) of Sandoz (India) Ltd.) at 20 ppm concentration for 2 hours or dipping the seeds in solutions in the above concentrations of fungicide and insecticide for 5 minutes and then keeping the seed covered for 3 hours followed by drying back to the original weight. Soaking or dipping in low concentration of the fungicide and insecticide did not depress the beneficial effect of hydration-dehydration on seed germinability. Seeds treated with higher concentrations of Dithane M-45 and Satisfar 50 EC (at concentrations much higher than usually recommended) showed adverse effects on germination and seedling growth. The results, however, show compatibility of the protectants with the hydration-dehydration treatment.

47. Control of Seed Deterioration in Groundnut (*Arachis hypogaea* L.) by Hydration-Dehydration Treatment

Ramamoorthy, K. and Basu, R.N. (1984). *Plant Physiology and Biochemistry*, **11**(2), 148-152.

The ageing damage of 11-month-old groundnut (*Arachis hypogaea* L.) seeds could be prevented by soaking whole pods for 6-12 h followed by drying back. The viability of the kernels from pods of the same age could be maintained by moisture equilibration with a water-saturated atmosphere for 72 h or keeping in contact with moist sand (dry sand moistened to 5 per cent on air-dry basis) for

48 hr, with or without subsequent soaking in water (30 min), followed by drying back. In hydrated-dehydrated seeds the leakage of electrolytes and sugars and lipid peroxidation were much lower than in untreated control.

48. Control of Physiological and Pathological Deterioration of Stored Wheat Seed

Saha, R. and Basu, R.N. (1984). *Indian Journal of Mycological Research,* **22**(2), 111-116.

Hydration-dehydration treatment of stored wheat seed very effectively reduced physiological deterioration of seeds in storage. Dry-dressing of hydrated-dehydrated seed with the fungicides captan, ceresan and dithane M-45 showed only minor improvement in storability over hydration-dehydration alone. Surface sterilization of the seed with mercuric chloride before placing for germination improved the germinability of stored seeds by reducing fungal attack. When the treated and stored seeds were soaked in water for 2 hr, with or without surface sterilization with mercuric chloride solution before the germination test, there was no difference between water soaking-drying treatment and soaking-drying employing solutions of different fungicides in dry placement (no pre-soaking before germination test); however, soaking-drying with solutions of fungicides (treatment given prior to storage) proved better than water-soaking-drying. Pre-germination soaking in water, besides physiologically improving germinability, reduced fungal invasion during germination by washing off the fungal spores from seed surface.

49. Invigoration of Soybean Seed for Alleviating Soaking Injury and Ageing Damage on Germinability

Saha, R. and Basu, R.N. (1984). *Seed Science and Technology,* **12**, 613-622.

Soaking of 12-month-old soybean ('Soyamax') seeds in water for two hours at 28°C caused severe injury to the seeds. Moisture equilibration for 24 hours with a water saturated atmosphere (100 per cent relative humidity at 28°C) largely overcame the soaking injury. Drying back of moisture equilibrated and soaked (two hours) seed retained the beneficial effect and the moisture equilibration-soak-dry treatment proved very effective in counteracting soaking injury and physiological deterioration of the seeds. Germinability of seeds following the different hydration-dehydration treatments was significantly influenced by the initial vigour status of the seed. In six-month old seed, soaking-drying alone was effective but the alleviation of soaking and ageing damage was much more pronounced in the moisture equilibrated-soaked-dried seed. In one-month-old seed, no soaking injury was evident when seeds were kept fully immersed in water for two hours; the soaked-dried seeds also did not show any harmful effect on immediate germinability but following accelerated ageing or prolonged storage under ambient conditions, the latent deleterious effect of presoaking was clearly discernible. The physiology of alleviation of soaking injury and ageing damage and the interaction of treatments with the vigour status of seeds are discussed.

50. Response of different Wheat Cultivars to Hydration-Dehydration Treatment for the Maintenance of Vigour, Viability and Productivity

Mandal, A.K. and Basu, R.N. (1984). *Seeds and Farms*, **10**(9-10), 39-42.

The beneficial effects of the mid-term hydration-dehydration treatments were reflected on viability of seeds, vigour of seedlings and field emergence and productivity of four wheat cultivars Sonalika, Janak, UP 262 and WL 711. Hydration-dehydration alone (using water only) was effective in improving grain yield but disodium phosphate (10^{-4}M) showed some additional beneficial effect over water. On an average, the increase in yield over control ranged from 15-25 per cent in the four cultivars and the treatment effects were independent of varietal differences. We may mention here that seed hydration by the dip-dry technique would be easier than the soak-dry treatment especially under less favourable drying conditions. But it is necessary to keep the dipped seeds in moist conditions for several hours before drying back. If the technological problems in the hydration-dehydration treatment of large seed stocks are overcome, the present method of seed treatment would be of great advantage to our seed merchants and cultivators.

51. Influence of Hydration-Dehydration Treatments of Rice Seed on the Growth of Embryo-endosperm Transplants

Sur (Mandal), K. and Basu, R.N. (1985). *Current Science*, **54**(17), 862-864.

The present study clearly showed that the hydration-dehydration treatment besides benefiting the embryo, effectively prevented a decline in the functional activity of the aleurone cells of the endosperm during ageing and it resulted in a greater α-amylase production by the endosperm and consequent solubilization of the starch reserves and improved nourishment of the embryo transplanted on it.

52. In Search of Low Cost Technologies to Improve Seed Performance

Basu, R.N. and Mandal, A.K. (1985). In: Transferable Technology for Rural Development (ed. D.K. Dasgupta). Associated Publishing Company, New Delhi, 33-52.

Under high humidity and high temperature conditions, as prevailing in eastern India and the coastal belts of the country, it is very difficult to maintain vigour and viability of stored seeds under ambient conditions. The facilities of humidity and temperature-controlled storage are not available to the average seed producers. For small and marginal farmers an easily practicable storage technology would be welcome in such situations.

The midstorage hydration-dehydration treatments of stored cereals and other seeds developed in the College of Agriculture would help the farmers to overcome the problem to a great extent. The methodology of the treatments and the results so far obtained in the authors laboratory is discussed.

53. Effect of Irradiation on the Deteriorative Senescence of Embryo and Endosperm of Rice (*Oryza sativa* L.) Seed

Sur (Mandal), K. and Basu, R.N. (1985). *Indian Journal of Experimental Biology,* **28**(4), 303-309.

Employing embryo-endosperm reciprocal transplantation technique, it has been shown that γ-irradiated rice endosperm adversely affects the growth of non-irradiated embryos transplanted on them. One of the reasons of the poor growth of the non-irradiated embryo-irradiated endosperm transplant appears to be the fall in the gibberellin-induced α-amylase production in the aleurone cells of the endosperm. However, exogenous supply of gibberellic acid (GA_3) partly overcame the adverse effect of irradiation implying damage to other biomolecules and bioorganelles participating in the reactions leading to germination and early seedling growth.

54. Physico-chemical Control of Radiation and Ageing Damage in Mustard Seed

Dey (Pathak), G. and Basu, R.N. (1985). *Indian Journal of Experimental Biology,* **23**, 167-171.

Pre-and post-irradiation soaking-drying treatment of stored mustard seed significantly reduced the adverse effect of γ-irradiation on germinability. Incorporation of radioprotective chemicals, potassium iodide, *p*-aminobenzoic acid and hydroquinone into the seed by dry permeation technique, employing acetone as the solvent, showed significant radioresistance while caffeine reduced post-ageing germinability of non-irradiated and irradiated seeds. The beneficial effects of the physico-chemical treatments were associated with better membrane permeability and reduced lipid peroxidation. Besides the possible operation of a cellular enzymatic repair system in soaking-drying, the efficacy of the non-aqueous treatments suggests a physico-chemical mechanism in the counteraction of age- and irradiation-induced seed deterioration in mustard.

55. Absence of *de novo* Protein Synthesis in Moisture Equilibration-Drying Treatment for Maintenance of Lettuce (*Lactuca sativa* L.) Seed Viability

Pan, D. and Basu, R.N. (1985b). *Indian Journal of Experimental Biology,* **23**, 375-379.

Dry permeation of the cytoplasmic protein synthesis inhibitor, cycloheximide, into lettuce seed employing acetone as the solvent, reduced storability at high cycloheximide concentrations. Moisture equilibration-drying treatment of cycloheximide pretreated seed, however, showed better storability than corresponding non-equilibrated seed. Even when cycloheximide in aqueous solution was introduced into the seed by the soaking-drying treatment, similar beneficial effect of moisture equilibration on vigour and viability of stored lettuce seed was noted.

The results imply that *de novo* protein synthesis in the cytoplasm may not be the major reason of viability maintenance of lettuce seed by the moisture equilibration-drying treatment.

56. Mid-Storage and Presowing Seed Treatments for Lettuce and Carrot

Pan, D. and Basu, R.N. (1985a). *Scientia Horticulturae*, **25**, 11-19.

Mid-storage hydration-dehydration treatments with medium-vigour lettuce seed very effectively reduced physiological deterioration in storage and ultimately showed better field emergence and final yield. Soaking-drying treatment of high-vigour lettuce seed was not effective. Pre-sowing treatment of medium-vigour lettuce seed, though less effective than mid-storage treatment, gave better field performance, but high-vigour seed did not benefit from the treatment. For carrot, both mid-storage and pre-sowing seed treatments proved beneficial, with the former showing better performance than the latter. Unlike lettuce, well-preserved carrot seed responded positively to the pre-sowing treatment.

57. Physico-chemical Control of Mustard Seed Deterioration

Basu, R.N., Rudrapal, A.B. and Pathak, G. (1985). *Seed Research.*, **13**(2), 39-46.

The deterioration of mustard (*Brassica juncea Coss*) seed could be significantly slowed down by hydration-dehydration treatment of the stored seed with or without several chemicals and also by dry permeation of certain chemicals by the organic solvent infusion method. The hydration-dehydration treatments, which reduced the physiological deterioration of stored mustard seeds, very effectively counteracted the damaging effect of X-irradiation on germination and seedling vigour. Post-ageing bio-chemical studies showed that the effective treatments maintained better membrane integrity, higher amylase and dehydrogenase activities and reduced lipid peroxidation. It is suggested that the physico-chemical treatments may also slow down mustard seed deterioration by reducing lipid peroxidation and free-radical reactions which would otherwise cause damage to vital bioorganelles.

58. Effect of Culture Filtrate of *Aspergillus flavus* on the Vigour and Viability of Stored Wheat Seed

Saha, R. and Basu, R.N. (1985). *Indian Journal of Mycological Research*, **23**(1), 19-26.

Wheat seeds placed for germination in contact with high concentrations of crude culture filtrate of the storage fungus, *Aspergillus flavus*, showed reduction in germinability but lower concentrations of culture filtrate did not affect germination. The undiluted culture filtrate when used in soaking-drying treatment for viability maintenance, adversely affected germinability of wheat seed stored under high humidity and high temperature conditions. The adverse effect of the soaking-drying treatment with culture filtrate on the storability of seeds was associated with significant increase in membrane permeability of the treated seeds and substantial reduction in dehydrogenase activity.

59. Vigour and Viability of Wheat Seed Treated with Bleaching Powder

Mandal, A.K. and Basu, R.N. (1986). *Seeds and Farms*, **12**(10), 46-47.

Dry dressing treatment of wheat (*Triticum aestivum* L. cv. Sonalika) seed with calcium oxychloride (common bleaching powder active ingredient calcium oxychloride around 35 per cent) at the rate of 2–5 g/kg significantly reduced physiological as well as pathological deterioration during storage under accelerated and natural ageing conditions. Besides, germination percentage, seed vigour as measured by shoot and root growth of seedlings, was more in the seeds treated with Bleaching powder (2 g/kg of seed) than in untreated control.

60. Control of Age- and Irradiation-Induced Seed Deterioration in Rice (*Oryza sativa* L.) by Hydration-Dehydration Treatments

Sur (Mandal), K. and Basu, R.N. (1986). *Seed Research*, **14**(2), 197-205.

Soaking-drying (SD) treatment counteracted age-and γ-irradiation-induced deterioration of medium and low vigour rice seed. Pre-irradiation SD treatment of high vigour seed showed deleterious effect on germinability but post-irradiation soaking-drying slowed down the physiological deterioration of seed during storage. Moisture equilibration-drying (MED) treatment was more effective in maintaining the vigour and viability of high vigour seed than for medium and low vigour lots in which soaking-drying proved better. Germination percentage (sprouting) was less affected by irradiation but seedling growth, specially root growth was noticeably reduced. Besides germinability, the hydration-dehydration treatments also checked the decline in activity of the enzymes dehydrogenase, α-amylase and polyphenol oxidase by ageing and irradiation.

61. Maintenance of Vigour, Viability and Yield Potential of Cauliflower Seed

Mitra, R. and Basu, R.N. (1986). *Vegetable Science*, **13**(1), 34-40.

Soaking-drying treatment of stored cauliflower seeds reduced the loss of vigour and viability under accelerated and natural ageing conditions. Hydration by moisture equilibration with a water-saturated atmosphere (100 per cent RH at 23°C) for 8-24 hours, followed by drying back, also slowed down seed deterioration. The productivity of the crop was greatly improved by such mid-storage seed treatment. Treatment of harvest-fresh seed was, however, ineffective. Dilute solutions of potassium phosphate (monobasic, 10^{-4}M), sodium phosphate (diabasic 10^{-4}M) and sodium chloride (10^{-3}M) proved better than water in field performance and productivity. The curd weight in the different hydration-dehydration treatments showed an increase of 10-20 per cent over the control.

62. Effect of Sowing Water-Soaked and Dried Seeds on the Yield of Wheat with or without Fertilizer Application

Mandal, A.K. and Basu, R.N. (1987). *Indian Journal of Agricultural Sciences,* **57**(7), 517-519.

The results clearly show that the beneficial effect of mid-term seed treatment could be obtained with or without fertilizer application. The effects of fertilizer and seed treatment are just additive in nature as is evident from the statistical non-significance of their interaction effect.

63. Improvement of Germinability of some Forest Tree Seeds by Acid Scarification and Hydration-Dehydration Treatments

Sur, K., Lahiri, A.K. and Basu, R.N. (1987). *Indian Journal of Agricultural Science,* **31**(2), 115-122.

Exogenous coat-imposed dormancy of seeds of three common forest trees *viz.*, sona-jhuri (*Acacia auriculiformis*), babla (*Acacia nilotica*) and subabool (*Leucaena leucocephala*) could be removed by scarification for 30 minutes with sulphuric acid. Soaking-drying treatment of stored medium-vigour seeds given in conjunction with scarification showed a substantial protection against loss of vigour and viability of scarified seeds subjected to short-term accelerated ageing at 100 per cent RH and 40°C or to natural ageing under ambient conditions for 6 months. For long-term storage, however, the seed should be thoroughly air-dried and kept as such without scarification.

64. Midterm and Presowing Hydration-Dehydration Treatments for Improved Field Performance of Wheat

Mandal, A.K. and Basu, R.N. (1987a). *Field Crops Research,* **15**, 259-265.

Hydration-dehydration treatment of stored wheat seed (high-medium vigour) midway during the usual storage period, or just before sowing showed improved seedling emergence, grain yield and 1000-grain weight. Seed was hydrated by soaking in water or in a dilute solution of disodium hydrogen phosphate (10^{-4}M) or preconditioned with moist sand (dry sand moistened to 5 per cent with water or a solution of disodium phosphate and 1 kg of seed incubated with 3 kg of sand for 24 h) to avoid soaking injury and to obtain a slow and progressive rise in seed moisture content. The hydrated seeds were thoroughly dried back before re-storage in the midstorage treatment or lightly air-dried in the presowing treatment. As the control seed was stored in moisture-barrier containers, loss of vigour and viability was not very conspicuous; nevertheless, both midterm and presowing treatments showed improved field performance.

65. Hydration-Dehydration Treatments for Improved Seed Storability of Bamboo (*Dendrocalamus strictus* L.).

Sur, K., Lahiri, A.K. and Basu, R.N. (1988). *Indian Forester,* **14**(9), 560-563.

Soaking of bamboo seeds (*Dendrocalamus strictus* L.) in water or a dilute solution of disodium hydrogen phosphate (10^{-4}M, for 6 hr, followed in either case by drying back, greatly reduced their deterioration under accelerated ageing at 92 per cent RH and 40°C for 5 days. Hydration by moisture equilibration with a water saturated atmosphere for 48 hr followed by drying also slowed down the loss of vigour and viability of seeds. Soaking-drying treatment was, however, more effective than the moisture equilibration-drying treatment. Although the soaking-drying treatment with low concentration of disodium hydrogen phosphate (10^{-4}M) proved better than water (SD) for maintaining vigour and viability, GA treatment showed better seedling growth (not germination percentage) than the other treatments. Kinetin and ethephon adversely affected post-storage germinability.

66. Maintenance of Seed Vigour and Viability of Onion (*Allium cepa* L.)

Choudhuri, N. and Basu, R.N. (1988). *Seed Science and Technology*, **16**, 51-61.

Onion (*Allium cepa*) seeds stored for up to 15 months under ambient laboratory conditions were subjected to hydration-dehydration treatments by (i) soaking in water for six hours, (ii) dipping for five minutes then keeping covered for two hours, and (iii) moisture equilibration with a water-saturated atmosphere for 48 hours, followed in all cases by slow drying back to the original weight (7.8 per cent moisture content). Seeds were examined for loss of vigour and viability during subsequent storage under accelerated or natural ageing conditions. Hydration-dehydration treatments effectively slowed down the physiological deterioration; however, the effect depended upon seed vigour at the time of treatment. Use of sodium phosphate (monobasic, 10^{-4}M) and sodium sulphate (10^{-4}M) in the soak-water showed only minor additional advantages. The lower electrical conductance of seed-steep water and reduced leaching of sugars and amino acids of moisture equilibrated-dried seed following ageing indicated better membrane integrity maintenance than the corresponding untreated seeds. The improved storability of hydrated-dehydrated seed was also associated with greater dehydrogenase activity and appreciably lower lipid peroxide formation in cells.

67. Treatment of Rice Seed with Iodine and Chlorine for the Maintenance of Vigour, Viability and Productivity

Pal, P. and Basu, R.N. (1988). *Indian Agriculturist*, **32**(1), 71-75.

Exposure of harvest-fresh (high-vigour) rice seeds (*Oryza sativa*L. cv. Ratna) to very low concentrations of iodine vapours (50 mg of solid iodine crystal volatilized in a 5.5-litre-capacity glass desiccator for 100 g seed) and chlorine vapour (10 g bleaching powder kept in a 5.5-litre desiccator for 100 g of seed) for 8-72 hours or dry-dressing with iodinated french chalk (1 g iodinated french chalk containing 50 mg iodine for 1 kg of seed) and bleaching powder (2 g/kg of seed) significantly slowed down the deterioration of seeds stored under ambient conditions.

The productivity of the crop (grain yield/ha) raised from the iodinated french chalk (1 g/kg) and bleaching powder (2 g/kg) also showed significant improvement.

68. Seed Germination and Viability

Basu, R.N. and Sur K. (1988). In: Plant Physiological Research in India (ed. S.P. Sen), 117-137.

The authors have briefly reviewed the recent Indian works on various physiological aspects of seeds germination and viability such as water uptake, osmotic and salt stress in germination, temperature, light and radiation effects; metabolic changes during germination, the effects of chemicals, dormancy and dormancy-breaking treatments. Seed viability and seed invigoration treatments for maintenance of vigour and viability have been highlighted and the possible modes of action of such treatments discussed.

69. Storage of Rice Seed Samples for more than a Decade under Ambient Temperature and Low Humidity Regimes

Pal, P. and Basu, R.N. (1989). *Indian Agriculturist*, **33**(3), 181-183.

Soaking-drying treatment of 7-month-old Ratna rice seed samples followed by storage for 12 years under ambient temperature conditions in a glass desiccator over fused calcium chloride retained 85 per cent germination against 12 per cent germination of the untreated control seeds stored in the same desiccator. The methodology is suitable for storage of rice genetic stocks which after treatment can be stored over fused calcium chloride for a fortnight and then transferred to tightly corked rubber-stoppered glass bottles (sealed with molten paraffin wax) for decade-long storage under ambient conditions.

70. Effect of Sugarbeet Fruit Inhibitors on Seed Germinability

Choudhuri, N. and Basu, R.N. (1989). *Seed Research*, **17**(1), 16-22.

Water extracts of non-viable sugarbeet fruits at relatively high concentrations reduced the seedling growth of medium-vigour sugarbeet, rice, wheat, jute and mustard seeds placed on extracts-soaked blotters in petridishes during germination. There was a substantial recovery in germinability when the extract-treated seeds were transferred to water-moistened blotters. Soaking-drying treatments of medium-vigour sugarbeet seed employing sugarbeet fruit extract and water showed that storability (as measured by post-ageing germinability) of sugarbeet fruits was improved by the extract. With prolonged soaking duration (8-24 h soaking prior to drying) in particular the extract proved better than water due to nonadvancement of germination and compatibility with the drying back process. Soaking-drying treatment of stored rice seed employing different concentrations of viable sugarbeet fruit extracts showed that storability of rice was adversely affected by high concentration of non-viable fruit extract but the viable fruit extract was as effective as water.

71. Relation between Dormancy and Viability of Dry-Stored Lettuce (*Lactuca sativa* L. cv. Sutton's A1) Seed

Punjabi, B. and Basu, R.N. (1989). *Indian Journal of Experimental Biology,* **27**, 739-741.

The deterioration of dormant lettuce seeds in dry storage was similar to that of non-dormant (dormancy broken by low temperature and kinetin treatment) seeds. The results imply that while in imbibed storage dormancy is an essential prerequisite for the maintenance of viability, in dry storage endogenous dormancy does not impart any added longevity to the stored seed. Seed invigoration is independent of the status (dormant or nondormant) of the seed at the time of the treatment.

72. Volatile Aldehyde Production in Relation to Seed Vigour of Rice (*Oryza sativa* L.)

Pal, P. and Basu, R.N. (1989). *Indian Agriculturist,* **33**(4), 255-258.

Volatile aldehyde production during the initial phase of germination of rice seed (cv. Ratna) was significantly lower in high vigour than in the medium and low vigour seeds. Viability-maintaining soaking-drying and dipping-drying treatments showed substantially lower post-ageing volatile aldehyde production than the untreated control seeds.

73. Physiology of Seed Invigoration Treatments in Soybean (*Glycine max* L.)

Saha, R., Mandal, A.K. and Basu, R.N. (1990). *Seed Science and Technology,* **18**, 269-276.

Moisture equilibration-soaking-drying (ME-S-D) and moist sand conditioning-soaking-drying (MSC-S-D) treatments of stored seeds of soybean cultivar 'Soyamax' greatly reduced the loss of vigour and viability under accelerated as well a natural ageing conditions. Post-ageing physiological and biochemical studies have shown increased amylase and dehydrogenase activities and greater membrane integrity coupled with significantly lower lipid peroxidation in the treated seeds than in the control. It is suggested that the beneficial hydration-dehydration treatments, ME-S-D and MSC-S-D, improved seed vigour and viability of soybean, at least in part, by counteracting lipid peroxidation reactions.

74. Bioassay of Seed Vigour

Basu, R.N., Kuchlan, P. and Sur, K. (1990). *Indian Journal of Experimental Biology,* **28**, 166-170.

A simple and inexpensive method for the bioassay of seed vigour of wheat (*Triticum aestivum* L.) employing high-vigour seeds of jute (*Corchorus olitorius* L.) is described. In this method, in an air-tight gas jar, germinating jute seeds (bioassay material) were exposed to the gaseous emanations of germinating wheat seeds (stock material) of different vigour levels. Highly significant positive correlations between growth of wheat seedlings and growth of jute seedlings were obtained irrespective of whether the wheat seed was aged naturally, artificially or invigorated by hydration-dehydration treatments for obtaining stock materials of different vigour levels.

75. Vigour Rating of Wheat Seeds

Sur, K. and Basu, R.N. (1990b). *Seed Science and Technology*, **18**, 661-671.

The seed vigour of wheat (*Triticum aestivum* L. cultivar 'Sonalika') of different vigour levels obtained by ageing or hydration-dehydration treatments, was tested by their comparative germinability, bioassay of gaseous emanations employing high-vigour jute (*Corchorus olitorius* L. cultivar 'JRO 632') seeds and chemical assay of volatile aldehydes in the gaseous emanations using the aldehyde trapping reagent 3-methyl-2-benzothiazolinone hydrazone (MBTH, 0.2 per cent solution). The vigour of the stock seeds was positively correlated with seedling length but negatively correlated with the coefficient of variation (standard deviation of seedling length expressed as a percentage of the mean). The growth of jute seedlings (bioassay material)exposed to the gaseous emanations of germinating high-vigour wheat seeds was higher than with corresponding medium and low-vigour seeds. Hydration-dehydration treatments of stock seeds, such as dipping-drying, soaking-drying and moisture equilibration-drying, showed greater post-ageing germinability, coupled with greater seedling growth of jute grown along with wheat demonstrating high positive correlations between the stock seed vigour and bioassay results. Significant negative correlations were obtained between wheat seed vigour and volatile aldehyde production. The results have been discussed in-terms of lipid peroxidation and volatile aldehyde production in the stock seeds *vis-a-vis* inhibition of seedling growth of the bioassay material.

76. Retention of Vigour and Viability of Stored Pea (*Pisum sativum* L.) Seed

Bhattacharya, A.K. and Basu, R.N. (1990). *Indian Agriculturist*, **34**, 187-193.

The loss of vigour and viability of two garden pea cultivars, 'Arkel' and 'Bonneville' could be effectively controlled by moisture equilibration-drying and moisture equilibration-soaking-drying treatments. Dry-dressing of 3-month-old seeds with bleaching powder (calcium oxychloride at 3 g/kg of seed) successfully retained vigour and viability in ordinary storage for 9 months. Low concentrations of the alcohols, methanol and isopropanol, applied as dry-dressing employing calcium carbonate as a carrier (1 ml mixed with 3 g carrier for 1 kg of seed) also proved very effective, especially in 'Bonneville' seeds. The carrier (calcium carbonate, 3 g/kg of seed) itself significantly improved storability of seeds of both the pea cultivars.

77. Seed Invigoration of Red Beet (*Beta vulgaris* L.) for Improved Storability and Field Performance

Choudhuri, N. and Basu, R.N. (1990). *Advances in Horticulture and Forestry,* **1**, 125-133.

Hydration-dehydration treatment of stored red beet (*Beta vulgaris* L. cv. Sutton's Red Globe) fruits (seeds) employing soaking-drying and moisture equilibration-drying treatments significantly slowed down the physiological deterioration of seeds during subsequent storage under ambient warm-humid conditions. Soaking-drying with a range of chemicals showed a marginal improvement over soaking-drying with water alone. Weight of beetroots was significantly more in plants raised from sodium chloride (10^{-4}M), sodium phosphate (monobasic, 10^{-4}M) and sodium thiosulphate (10^{-5}M) soaked-dried seeds. Moisture equilibration-drying treatment, which was very effective in reducing seed deterioration in storage showed lower post-storage membrane permeability than corresponding control seeds indicating a close association between seed invigoration by hydration-dehydration and retention of membrane integrity.

78. Reduced Volatile Aldehyde Production in Wheat by Seed Invigoration Treatments

Sur, K. and Basu, R.N. (1990a). *Current Science,* **59**(16), 799-800.

Highly significant negative correlations were noted between germinability of wheat (*Triticum aestivum* L. cv. Sonalika) seeds of different vigour status and volatile aldehyde production. Mid-term hydration-dehydration seed invigoration treatments, which effectively maintained vigour and viability of wheat seeds, showed a substantial reduction in post-ageing volatile aldehyde production.

79. Influence of Ageing, Irradiation and Heat Stress of the Endosperm on the Growth of the Embryo

Sur, K. and Basu, R.N. (1990). In: *Proceedings of the International Conference on Plant Physiology,* Vol.**2** (Eds. S.K. Sinha, P.V. Sane, S.C. Bhargava and P.K. Agarwal). Published by the Society of Plant Physiology and Biochemistry, New Delhi, pp. 1279-1284.

Embryos of freshly-harvested (high vigour) rice seeds were transplanted on to the rice endosperms deteriorated by direct stresses, such as natural ageing, γ-irradiation (0.2 – 1,0 kg. and dry heat treatment (80°C). Aged irradiated heat-treated rice endosperms adversely affected the growth of high vigour embryo transplanted on them. Although the condition of the endosperm considerably influenced the growth of the embryo, even the endosperms of non-viable, irradiated and strongly treated seeds could support the growth of the embryos to a certain extent. The endosperm did not affect the germination but the subsequent growth of the germinated embryo was determined by the condition of the endosperm. The results showed that the damage to the aleurone cells made them less capable of supporting the growth of the germinated embryo. It was noted that the growth of the high vigour embryo was proportional to the amount of functional aleurone retained on the endosperm. One of the reasons for the poor growth of high vigour embryo appeared to be a decline in germination-induced α-amylase synthesis in the aleurone cells of the endosperm. The adverse effect of the various stresses on the endosperm could be substantially alleviated by hydration-dehydration treatments.

80. Improvement of Germinability of *Eucalyptus* Species by Pregermination Treatments

Bhattachyarya, A.K., Lahiri, A.K. and Basu, R.N. (1991). *Indian Forester*, **117**(8), 661-663.

The rate of seedling emergence of *Eucalyptus* species in nurseries is erratic as is the case with many other forest tree seeds which may be attributed to low seed viability due to rapid loss of moisture owing to its recalcitrant nature and exogenous coat imposed dormancy.

In the present laboratory the effect of gibberellic acid (GA_3) at 100 ppm, thiourea at 250 ppm and potassium nitrate at 500 ppm on poor and non-uniform germination of a number of *Eucalyptus* species was studied.

Seeds of each type were soaked for 4 hr in GA_3, thiourea and potasium nitrate. Control seeds were soaked in distilled water. Germination percentage root and shoot length were recorded after 14 days at 27 ± 1°C following ISTA (1976) rules. Pregermination treatments given to FRI-4 and FRI-5 with 100 ppm GA_3 gave maximum beneficial effect on germination and seedling vigour followed by 500 ppm potassium nitrate as compared to control. Thiourea at 250 ppm was ineffective.

81. Retention of Vigour and Viability of Jackfruit (*Artocarpus heterophyllus* L.) Seed

Bhattachyarya, A.K. and Basu, R.N. (1992). *Indian Agriculturist*, **36**(2), 65-74.

Storage of a typical recalcitrant seed like jackfruit is difficult because under ambient dry storage conditions with gradual loss of moisture the seed loses vigour and viability. Incubation of fresh seeds with sand containing 10 per cent Jalshakti (a modified starch cellulose preparation) or a 0.3 M solution of potassium dihydrogen phosphate (seed sand ratio 1:3 or 1:4) and keeping the same in loosely bound polythene bag could maintain seed vigour and viability (80-90 per cent) for 9 months under ambient storage conditions in Calcutta. The beneficial effects of the moist sand incubation treatments were associated with better membrane functions of treated seeds, increased dehydrogenase activity and reduced production of toxic gaseous metabolites.

82. Effect of Powdered Red Chilli, Turmeric and Neem Leaf on Pre- and Post-storage Germinability of Wheat Seed

Pal, P. and Basu, R.N. (1993). *Indian Agriculturist*, **37**(4), 267-271.

Treatment of high-vigour (harvest-fresh) wheat seed (*Triticum aestivum* L. cv. Sonalika) with 500 mg of red chilli powder, 1 g turmeric powder and 2 g neem leaf powder per kg of seed significantly slowed down the deterioration of seeds in storage. The beneficial effect of the above mentioned natural products were associated with reduced production of volatile aldehydes by the treated seeds.

83. Seed Invigoration for Extended Storability

Basu, R.N. (1993). *Seed Research* **Special Volume**, 217-230.

Seed invigoration for the maintenance of vigour and viability includes a range of 'wet' and 'dry' mid-storage and pre-storage treatments. Mid-storage hydration of stored seed by soaking for 2-6 hr depending on the material; dipping for 2-5 min and then keeping the surface-wetted seed covered for 2-6 hr moisture equilibration with a water-saturated atmosphere for 24-48 hr or incubating with premoistened sand for 16-36 hr followed in all cases by thorough drying back would greatly extend the storage life of seeds of majority of the orthodox, nonleguminous, agricultural and horticultural seeds. In leguminous seeds, many of which exhibit soaking injury, soaking-drying treatment of high or medium vigour seeds, in particular is detrimental to both immediate and post-storage germinability and treatments such as moisture equilibration or moist sand conditioning each followed by soaking for an hour and then drying would significantly extend longevity under accelerated as well as natural ageing conditions. Pre-storage dry treatments including acetone permeation of antiageing chemicals, use of low concentrations of halogens iodine, chlorine and bromine; and the alcohols, methanol, ethanol and isopropanol, would also improve the storability of several dry-stored orthodox seeds. However, except in soaking injury-prone leguminous seeds, the wet treatments are superior to the dry treatments. The mode of action of the seed invigoration treatments is still unclear but prevention of damaging oxidative reactions, especially free radical-induced lipid peroxidation reactions, involving unsaturated lipid moieties of lipoprotein biomembranes appear to be one of the primary reasons of invigoration.

84. Extension of Wheat Seed Viability with Plant Leaf Powders and an Aspirin-Containing Pharmaceutical Preparation

Pal, P. and Basu, R.N. (1994). *Indian Agriculturist*, **38**(2), 145-149.

Viability of wheat seed could be considerably extended by dry-dressing of the seed with powdered plant leaves (green leaves dried at 70°C) of bael (*Aegle marmelos* Corr), mango (*Mangifera indica* L.), tamarind (*Tamarindus indica* L.), nisinda (*Vitex negundo* L.) and kalke (*Thevetia peruviana* L.) at 1-2 g of leaf powder per kg of seed. These preparations proved as effective as neem leaf powder.

Acetylsalicylic acid (aspirin), a plant derivative and a widely used pharmaceutical formulation (ASPRO), proved very effective even at a much lower concentration (20-200 mg per kg of seed). The emergence of seedlings in soil and early seedling growth in seeds of some selected treatments after natural ageing for three months, were significantly better than in the untreated control.

85. An Appraisal of Research on Wet and Dry Physiological Seed Treatment and their Applicability with Special Reference to Tropical and Sub-tropical Countries

Basu, R.N. (1994). *Seed Science and Technology*, **22**, 107-126.

Physiological seed treatments for improved performance include presowing, prestorage and midstorage wet and dry treatments. As a presowing treatment, osmoconditioning of seeds, especially for vegetables, using polyethylene glycol has shown more consistent positive effects on seed performance than the conventional 2 to 3 cycles of 24 h wetting-drying treatments. Similar benefits are obtained with low-temperature holding of partially imbibed seeds. Single short-duration soaking for 4–8 h, followed by light air-drying also improves the overall germinability, rapidity and synchrony of germination of nonleguminous seeds. For leguminous seeds, instead of soaking-drying, moisture equilibration at 100 per cent relative humidity for 24 to 48 h, or preincubation with moist sand (5 per cent moisture content; 3 kg sand per kg of seed) for 16 to 24 h, or matriconditioning with highly porous, chemically inert, water insoluble, inorganic carriers like Micro-Cel E, are recommended. Prestorage dry-dressing of freshly harvested seed using the common bleaching powder active ingredient calcium chloride at the rate of 3 g per kg of seed, or mixing at the same rate with an inert carrier like calcium carbonate containing very low concentrations of iodine or alcohol, would improve storability of many dry-stored agricultural and horticultural seeds. Midstorage hydration-dehydration treatments which include short-term soaking-drying for low and medium vigour seeds and dipping-drying, moisture equilibration-drying and moist sand conditioning-drying for medium and high-medium vigour seeds, are very effective in extending storability and subsequent field performance. The seed following hydration must be thoroughly dried back before restorage. Use of very low concentrations of chemicals such as potassium or sodium phosphate (mono- and dibasic, 10^{-4}M) gives some additional advantage over water. The beneficial effects of the treatments on germinability are also reflected upon field emergence, crop growth and final agricultural productivity. More recently, beneficial effects of chemicals like ferrous sulfate (2×10^{-3}M) and 3-methyl-2-benzo-thiazolinone hydrazone (MBTH, 10^{-4}M) on pre- and post-storage germinability have been recorded. The mode of action of the different dry and wet seed treatments is yet to be clearly understood, but evidence so far indicates that retention of better membrane functions and a reduced production of toxic metabolites including volatile aldehydes, presumed to be products of free radical and lipid peroxidation reactions, are consistently associated with improved seed performance'. In this review, the practical implications of the physiological seed treatments have been discussed with special reference to tropical and sub-tropical countries.

86. Seed Invigoration Studies in the University of Calcutta: Glimpses of the Past and Present

Basu, R.N. (1994). Special Lecture (Felicitation Lecture), December 7-9, 1994, published by The Indian Society of Seed Technology (ISST) in: *Seed Tech News*, pp. 1-14, New Delhi on the occasion of the All India Seed Seminar at the Tamil Nadu Agricultural University, Coimbatore.

The author has traced the genesis of the seed invigoration studies in the University of Calcutta since 1974. The development of 'wet' and 'dry' seed treatments (vapour and powdered form) for maintenance of vigour, viability and productivity of various seeds like cereals, legumes, oil seeds, vegetables and flower seeds have been discussed. The mode of action of 'wet' (hydration-dehydration treatments) was based on sufficient evidences (through indirect) that lipid peroxidation reactions (catalysed by free radicals) involving unsaturated lipid moiety of the lipoprotein membranes is possibly the primary reason of deteriorative senescence of stored seeds.

Free radical reactions affect the structural integrity of proteins resulting in reduced activity of many vital enzymes. Besides other possibilities, like anhilation of free radicals in the dry-stored seeds by short-term hydration-dehydration could be a reasonable speculation. The effectiveness of the chemicals used in the soak water could be attributed to their role as antioxidants or antioxidant synergists. The dry treatment (in vapour form) and as powdered plant preparation possibly act through still unknown ways but ultimately resulting in lower lipid peroxidation and reduced production of volatile aldehydes. Iodination or chlorination has been suggested to remove, to a certain extent, the degree of unsaturation in the lipid moitey of lipoprotein membranes and render the same less susceptible to peroxidative attack. A systematic approach towards further understanding of the positive role of the different seed invigoration treatments in the protection of lipoprotein cell membranes in particular could be a productive field of research.

87. Preservation of Tea (*Camellia sinensis* L.) Seed Viability

Bhattacharyya, A.K., Rahman, F. and Basu, R.N. (1994). *Seed Research*, **22**(2), 108-111.

Tea (*Camellia sinensis* L.) seed exhibits recalcitrant behavior and is short-lived. The effect of various moist sand (water/chemical) preconditioning treatments was studied to extend viability of tea seed. The best treatment was moist sand incubation with PEG maintaining -5 bar water potential and containing 0.05 per cent mercuric chloride; closely followed by water or PEG with 0.25 per cent of Dithane M-45. The vigour of surviving seedlings as measured by root and shoot growth was also better in these treatments than in the untreated control. The percentage of cracked and sprouted seeds could be reduced substantially with the proper maintenance of requisite moisture level of sand media in storage.

88. Seed Viability

Basu, R.N. (1995). In: Seed Quality, Basic Mechanisms and Agricultural Implications (Ed. A.S. Basra), Chapter I, The Haworth Press Inc., NY, USA, 1-44.

An extensive coverage of research on seed viability with a broad review of research on seed invigoration treatments in the University of Calcutta.

89. Dry Physiological Seed Treatments for Improved Performance of Wheat (*Triticum aestivum*) and Sunflower (*Helianthus annuus*)

Pal, P., Ray, I., Mitra, R. and Basu, R.N. (1996). *Indian Agriculturist*, **40**(4), 219-232.

The loss of post-storage germinability of high vigour (harvest-fresh) wheat and sunflower seeds could be effectively controlled by pre-storage dry treatments with fine powder of turmeric (*Curcuma longa*) rhizome, powdered leaf of neem (*Azadirachta indica*), nisinda (*Vitex negundo*), brahmi (*Bacopa monnieri*) and spinach (*Spinacia oleracea*) at 1-2 g of crude powder per kilogram of seed.

Aspirin (*ortho*-acetylsalicylic acid, originally a chemical compound derived from Salix, now commercially synthesized), Aspro, a commonly used aspirin-containing pharmaceutical formulation and a vitamin C-containing formulation, Celin, showed beneficial effect on the germinability at a lower concentration, 200-500 mg per kg of seed. The commonly used bleaching powder (with 35 per cent as active ingredient) calcium hypochlorite, at 2-3 g per kg of seed, also showed significant improvement in germinability of wheat and sunflower stored under ambient storage conditions for 3 months. The beneficial effects of the treatments were also associated with reduced emanations of volatile gaseous products from treated seeds.

In wheat, emergence of seedlings, growth of plants and grain yield were better in the extent of improvements in the invigoration treatments than in the control both in the pot culture and field experiments. Similar improvements were noted in the field performance and yield of sunflower.

90. Seed Invigoration for Improved Vigour, Viability and Productivity of Peanut

Ramamoorthy, K. and Basu, R.N. (1996). *East African Agricultural and Forestry Journal*, **61**(3), 261-272.

To determine an effective midstorage seed treatment for maintenance of vigour and viability for peanut(*Arachis hypogaea* L.) cv. CO.l, three methods of hydration-soaking, moisture equilibration and moist sand conditioning each at various durations followed by drying, were evaluated. All the methods were effective, still seed invigoration of stored pods through 8 h water soaking or kernels from stored pods either by 72 h moisture equilibration or 48 h moist sand conditioning with subsequent soaking in phosphate for 30 min. followed by drying, greatly reduced the loss in vigour and viability during storage. Longer periods of hydration-dehydration were not very effective. Treated seeds showed reduced electrical conductivity, leakage of sugars, aminoacids and higher dehydrogenase activity. The lipid peroxidation was also minimum in hydrated dried seeds. Such beneficial effects could be attributed to maintenance of membrane integrity through counteraction of lipid peroxidation. Under field conditions, the invigorated stored seeds recorded higher dry matter production and yield. For maintenance of vigour, viability and productivity, depending upon the necessity and facilities available, one of the aforesaid methods can be practised with advantage.

91. Seed Invigoration Treatments for French Bean (*Phaseolus vulgaris* L.) and Soybean (*Glycine max* L.) for Improved Vigour and Viability after Storage

Mukhopadhay, A., Pal, P. and Basu, R.N. (1997). *Indian Agriculturist*, **41**(2), 79-85.

Treatment of high-vigour (harvest-fresh) French bean and soybean seeds with some powdered plant materials and pharmaceuticals significantly controlled the loss of vigour and viability of seeds stored under ambient conditions. Dry-dressing with 'haldi' – the powdered rhizome of turmeric (*Curcuma longa*), 'nisinda' (*Vitex negundo*) leaf powder, both at 2 g of crude powder per kilogram of seed, showed significant beneficial effect on germinability. Finely, powdered pharmaceuticals such as Aspro, an aspirin-containing formulation (at 200 and 100 mg per kg of seed for French bean and soybean, respectively), Celin, a vitamin C-containing formulation (500 mg per kg for French bean and 200 mg per kg for soybean) and *p*-aminobenzoic acid (PABA at 100 mg per kg for both French bean and soybean) showed beneficial effect on post-storage germinability of the two materials. The common bleaching powder (active ingredient, calcium hypochlorite) at 3 g per kg of seed also showed beneficial effect on the maintenance of vigour and viability of french bean and soybean stored in stoppered glass bottles for 24 months under ambient laboratory conditions in Calcutta.

The beneficial effects of the treatments on storability were associated with reduced post-storage emanation of volatile gaseous products from the treated seeds.

92. Effect of Dry Physiological Seed Treatments for Improved Vigour Viability and Productivity of Black Gram (*Phaseolus mungo*, Roxb.)

De, B.K., Mandal, A.K. and Basu, R.N. (1998). *Indian Agriculturist*, **42**(1), 13-20.

The loss of vigour and viability of high vigour (harvest-fresh) black gram (*Phaselous mungo* Roxb.) seed during storage could be effectively controlled by prestorage dry treatments with crude plant materials *viz.*, finely powdered catharanthus (*Catharanthus roseus* L.) leaf, hot chilli (*Capsicum frutescence* L.) fruit and turmeric (*Curcuma longa* L.) rhizome powder at the rate of 2 g, 1 g and 2 g per kg of seed respectively.

The pharmaceutical formulation, 'Aspro' (a commonly used aspirin containing formulation), vitamin C containing 'Celin' and 'Ibucon' (ibuprofen as the active ingredient an anti-inflammatory formulation) when used at a dose rate of 100 mg per kg of seed showed significant improvement in the germinability over untreated control under accelerated as well as natural ageing conditions. Common bleaching powder (active ingredient calcium hypochlorite) used at the rate of 2 g per kg of seed and commercial camphor (100 mg/kg of seed) also showed beneficial effects on the post-storage germinability of black gram seed. Seed treatments with Aspro, Ibucon, bleaching powder, turmeric (*haldi*) and hot chilli powder resulted in better field performance and productivity of the crop than the untreated control.

93. Dry-Seed Treatment for Improved Germinability and Productivity of Wheat (*Triticum aestivum*)

Mandal, A.K., De, B.K. and Basu, R.N. (1999). *Indian Journal of Agricultural Sciences*, **69**(9), 627-630.

Treatment of harvest-fresh (high-vigour) 'Sonalika' wheat seed (*Triticum aestivum* L. emend Fiori and Paol.) by dry-dressing with crude plant materials (dried and finely powdered ripe red chilli fruit and turmeric rhizome) @ 1-2 g/kg of seed and pharmaceutical formulations (aspirin-containing Aspro and ascorbic acid- containing Celin) @ 100 mg/kg of seed and commercial calcium hypochlorite, active ingredient of common bleaching powder @ 2 g/kg of seed greatly reduced the loss of vigour and viability under accelerated ageing at 100 per cent relative humidity and 40°C and natural ageing under ambient conditions. Mid-storage soaking-drying treatment with water (wet treatment), which was included for the comparison with dry seed treatment, was equally effective in slowing down seed deterioration. The crop raised from the treated and untreated seeds showed that water soaking-drying treatment as well as dry treatments, especially with Aspro and bleaching powder significantly improved field performance and productivity of the crop over untreated control. Between the wet (soaking-drying) and dry treatments only marginal differences were noted with respect to germinability and crop productivity.

94. Seed Invigoration Treatments for Improved Storability, Field Emergence and Productivity of Soybean (*Glycine max*. L. cv. Merril).

Mandal, A.K., De, B.K., Saha, R. and Basu, R.N. (2000). *Seed Science and Technology*, **28**, 349-355.

Dry-dressing of freshly harvested soybean (*Glycine max* [L.] Merrill) seed with finely powdered 'aspro' (an aspirin-containing pharmaceutical formulation; active ingredient, *ortho* acetyl salicylic acid 57 per cent) at 100 mg per kg of seed, ibucon (active ingredient, ibuprofen 28.57 per cent and paracetamol 71.42 per cent) at 100 mg per kg of seed and common bleaching powder (active ingredient, calcium hypochlorite, about 50 per cent) at 2 g per kg of seed showed significant beneficial effect on post-storage (ageing under ambient laboratory conditions; average 66.7 ± 3.4 per cent RH and 26.9 ± 1.3°C) germinability of soybean seed over the untreated control. Finely powdered dry red (hot) chilli (*Capsicum frutescens* L.) fruit (1 g powder per kg of seed) and catharanthus (*Catharanthus roseus* L.) leaf powder

(2 g per kg of seed) also improved germinability over the untreated control. A midstorage wet seed treatment, moist sand conditioning-soaking-drying (MSC-S-D), which was taken as a reference treatment on the basis of our earlier reports also showed higher germination percentage and seedling growth as measured by root and shoot length than the untreated control.

Field performance and productivity from the dry dressed seeds especially with 'aspro', red chilli powder and 'ibucon', were greater than that from the untreated control. MSC-S-D also showed a significant increase in yield over the control, but in this high-vigour seed lot, the dry treatment in general showed better field performance than the former. Dry-dressing treatment of freshly harvested soybean seed (high-vigour) with pharmaceuticals like 'aspro' at 100 mg per kg of seed is recommended for the improvement of seed storability and crop productivity of soybean because of the ready availability and technical ease of application.

95. Seed Invigoration Treatment in Wheat (*Triticum aestivum* L.) as Influenced by the Vigour Status of Seed at the Time of Treatment

Pal, P., Ray, I. and Basu, R.N. (2001). *Indian Agriculturist*, **45**(1), 1-9.

An assessment of the comparative efficacy of different seed invigoration treatments in wheat seed of high, medium and low vigour status was done by studying their respective germinability under ambient storage conditions in the laboratory and performance in the field. Seed invigoration treatment of high vigour wheat by dry-dressing with Aspro (active ingredient 56.5 per cent, *ortho*-acetylsalicylic acid or aspirin) at 0.1 g/kg of seed and bleaching powder(active ingredient chlorine-releasing calcium hypochlorite, 50 per cent) at 3 g/kg of seed effectively reduced physiological deterioration in storage and ultimately showed better field performance and final yield. The same dry seed invigoration treatment of medium vigour wheat showed marginal improvement with bleaching powder followed by Aspro but these dry treatments were not effective in low vigour wheat. Wet seed invigoration treatment employing water soaking-drying was not effective in high vigour but showed significant improvement in medium and low vigour wheat seed for storability as well as field performance. Field performance of' the crop raised from the high vigour seed treated with Aspro and bleaching powder showed 16 per cent and 13 per cent increase in yield, respectively. In case of medium vigour wheat seed, bleaching powder treatment increased grain yield by 13 per cent and treatment with Aspro showed a 10 per cent increase in yield. The dry treatments with Aspro and bleaching powder did not show any significant improvement of grain yield in low vigour seed lot. Water soaking-drying was not effective in high vigour seed but in medium and low vigour seed it showed 16 per cent and 8 per cent increase in yield respectively over the untreated control seed.

96. Pre-storage Seed Invigoration Treatment for Improved Performance of Mungbean (*Phaseolus aureus* Roxb.)

Mitra, R., Pal, P. and Basu, R.N. (2001). *Indian Biologist*, **33**(2), 21-25.

Seed invigoration treatment of harvest-fresh (high vigour) mungbean with powdered plant materials, pharmaceuticals and bleaching powder significantly controlled the loss of vigour and viability of seeds stored under ambient conditions. Dry-dressing with crude leaf powders of neem (*Azadirachta indica* L.), nisinda (*Vitex negundo* L.), spinach (*Spinacia oleracea* L.) and brahmi (*Bacopa monnieri* L.), powdered rhizome of turmeric (*Curcuma longa* L.) and red chilli (*Capsicum frutescens* L.) fruit at 2 g/kg of seed showed significant beneficial effect on germinability after storage. Aspro, a commonly used aspirin-containing pharmaceutical formulation at 0.2 g (containing 114 mg aspirin and 86 mg paracetomol) per kg of seed; aspirin (pure *ortho*-acetylsalicylic acid) at 0.1 g/kg seed; Celin, a vitamin C-containing formulation at 0.5 g/kg of seed also showed significant beneficial effect on post-storage germinability. The common bleaching powder (chlorine releasing active ingredient being calcium hypochlorite approx. 50 per cent) at 2 g/kg of seed also showed significant improvement of germinability of mungbean seed stored under ambient conditions for 6 months.

The beneficial effect of seed invigoration treatment was also reflected on field performance and productivity of the crop raised from the seed treated with Aspro, nisinda, and bleaching powder showing 23 per cent, 17 per cent and 9 per cent increase in seed yield respectively.

97. Seed Invigoration of Mustard (*Brassica juncea* L. Czern and Coss) as Influenced by Seed Vigour Status at the Time of Treatment

Rudrapal, D. and Basu, R.N. (2002). *Indian Agriculturist,* **46**(1 and 2), 1-9.

Mustard (*Brassica juncea* L. Czern and Coss) seeds of different vigour status were given seed invigoration treatment for the maintenance of vigour and viability. Pre-storage dry dressing of high vigour (HV) mustard seed, employing finely powdered tablets of aspirin containing 'Aspro' (0.2 g Aspro or 113 mg aspirin per kg) and vitamin C-containing 'Celin' (0.5 g/kg) and also common bleaching powder (active ingredient chlorine released from calcium hypochlorite) at 2 g/kg of seed, could retain better vigour and viability after natural ageing under ambient relatively hot and humid conditions for varying periods. Medium vigour (MV) seeds proved less responsive than HV seeds to such dry treatments, while in low vigour (LV) seeds the dry treatments brought about definite detrimental effect on vigour and viability. On the other hand, soaking-drying treatment was highly effective in medium and low vigour seed lots but not in high vigour seeds. It is suggested that the beneficial effects of the dry treatments may be linked to control of oxidation induced degradative reactions in mustard seeds.

98. Pre-storage Seed Invigoration Treatments for Improved Germinability and Field Performance of Okra (*Hibiscus esculentus*)

Kapri, B., Sengupta, A.K., De, B.K., Mandal, A.K. and Basu, R.N. (2003). *Indian Journal of Agricultural Sciences,* **73**(5), 276-279.

An experiment was conducted during 2000 and 2001 to study the efficacy of pre-storage dry treatment of freshly harvested (high-medium vigour) 'Pankaj' okra (*Hibiscus esculentus* L.) seed with finely powdered pharmaceutical formulations such as Aspro (active ingredient, *ortho*-acetylsalicylic acid 57 per cent) and ibucon (active ingredient ibuprofen 28.6 per cent and paracetamol 71.4 per cent) @100 mg/kg seed, Celin (vitamin C, ascorbic acid) @500 mg/kg seed and chemicals such as common bleaching powder (active ingredient, calcium hypochlorite, about 50 per cent) @2 g/kg seed. The results showed significant improvement on post-storage germinability over the untreated control. Finely powdered crude plant materials such as red chilli (*Capsicum frutescence* L.) fruit powder @1 g/kg seed, *Catharanthus* leaf powder @2 g/kg seed and *Trigonella* seed powder @1 g/kg seed also improved germinability over the untreated control. Pre-storage wet treatments (soaking-drying and dipping-drying) showed a marginal improvement of storability over the untreated control. Dry treatments showed better germinability than the wet treatments when both were given as a pre-storage treatment. Field performance and productivity from seeds of pre-storage dry treatments were greater than that from the untreated control. The invigoration effect of the dry treatments was especially noticeable with Ibucon, Celin, *Trigonella* seed powder and *Catharanthus* leaf powder. Pre-storage wet treatments of this relatively high vigour seed did not show any beneficial effect on field performance.

99. Seed Invigoration Treatments on different Seed Sizes of Wheat (*Triticum aestivum* L.) for Improved Storability and Field Performance

De, B.K., Mandal, A.K. and Basu, R.N. (2003). *Seed Science and Technology*,**31**, 379-388.

Pre-storage dry physiological treatments with a pharmaceutical formulation (aspro, active ingredient 56.5 per cent *ortho* accetylsalicylic acid, at 100 mg/kg of seed), chemicals (salicylic acid at 50 mg/kg of seed and common bleaching powder, active ingredient approximately 30 per cent calcium hypochlorite, at 2 g/kg of seed) and crude plant materials (finely powdered dry red chilli fruit and trigonella seed powder, each at 1 g/kg of seed) and a mid-storage wet treatment (soaking-drying) showed significantly improved post-storage germinability as well as field performance and productivity of the resultant crop over untreated control. The invigoration treatment effects were more or less similar in all the three seed sizes. Thus all good physiological treatments were equally effective in all seed sizes (large, medium and small) of the same seed stock. Use of large seed proved much better for germinability, field performance and productivity than the medium and small sized seed. Among the treatments, Aspro and trigonella seed powder, in particular, showed better results in improving storability, yield and other yield attributes. The interaction between treatment and seed size was statistically not significant implying thereby that the treatment effect was independent of seed size. The statistical non-significance of the interaction between seed size and invigoration treatment would indicate the quantitative nature of the effects of seed size. The results clearly demonstrate the suitability of large sized wheat seed as the planting material; as such, use of small seed which showed significantly poor performance in terms of storability and field performance should be avoided as far as possible.

100. The Role of Initial Seed Vigour Status in the Expression of Invigoration Effect of Wet and Dry Treatments in French Bean (*Phaseolus vulgaris* L.).

Rudrapal, D. and Basu, R.N. (2004). *Seed Science and Technology*, **32**, 205-212.

The deterioration of high vigour (HV) French bean (*Phaseolus vulgaris* L. cv. Pusa Parvati) seed during storage could be significantly slowed down by pre-treatment of the seed with powdered plant materials such as dried leaves of bitter gourd (*Momordica charantia* L.) and amrul (*Oxalis corniculata* L.) at the rate of 2 and 2.5 g per kg of seed, respectively, and pharmaceutical formulations namely, 'Aspro', a commonly used aspirin-containing formulation (56.5 per cent aspirin with 43.5 per cent paracetamol) and 'Celin' (vitamin C, an ascorbic acid containing formulation) at 0.2 and 0.5 g per kg of seed, respectively as well as by the common bleaching powder (active ingredient chlorine released from calcium hypochlorite) at 3 g powder per kg of seed. The vigour status of seed interacted with the different treatments to give differential response in respect of post-storage germinability and vigour. The high vigour (HV) seed lot proved more responsive to the dry treatments than the medium vigour (MV) seeds. On the other hand, in low vigour (LV) seeds, the dry treatments did not show any promotion of germinability and seedling vigour. The wet treatment, moisture equilibration-drying (humidification-drying treatment), was very effective in the medium vigour seed lot, but not in high and low vigour seeds. The dry treatments given to high vigour or the wet treatment to medium vigour seeds resulted in better post-ageing performance which would indicate that the beneficial treatments reduced degradative reactions responsible for the loss of seed vigour and viability.

101. Approaches to Quantification of Seed Vigour

Basu, R.N. (2004). *Indian Biologist*, **36**(2), 1-23.

Germinability is considered to be the most important physiological character of the seed. No viability means no vigour as such, there is usually a strong positive correlation between early seedling growth and germination as well as with many enzymes related to growth. Nevertheless, vigour parameters such as membrane integrity (which is positively related to seed vigour) is usually determined through leaching of electrolytes or metabolites and the data so obtained are negatively correlated with germination. Similarly many important physiological and biochemical parameters such as lipid peroxidation, volatile aldehyde emanation, *etc.* are negatively correlated with germination. The unification of data for overall vigour rating would therefore require transformation of such diverse data to uniform positive correlation mode as well as necessary conversion to uniform scale values. Besides such transformations it has been shown that multiplication of positively correlated vigour parameters by germination factor, GF (germination per cent ÷ 100 = GF) would effectively eliminate the relative contribution of non-viable seeds of a seed lot to the vigour score.

In the present paper these issues have been duly taken up with considerable rationalization of statistical correlations and predictive (regressional) improvements in analysis and evaluation of vigour scores of crop seeds. Basic data from different authoritative sources have been critically analyzed for the purpose of mathematical quantification of seed vigour score in percentage values (0-100) as in case of standard seed germination.

(See full paper in Chapter 6)

102. A Holistic Approach to Quantification of Seed Vigour through Integration of Morphological and Physiological-Biochemical Parameters

Basu, R.N. (2007). *Indian Biologist*, **39**(1), 1-44.

In the earlier paper on quantification of seed vigour (Basu 2004), the major emphasis was given to the germination factor (GF, germination per cent ÷ 100) of a seed lot in seed vigour rating. The role of GF upon integration through multiplication with positively correlated vigour parameters in eliminating the influence of non-viable seeds and significantly improving the correlations and regressional (predictive) efficiency of the transferred data was duly highlighted. In the present paper, a holistic approach to quantification has been made for seed vigour assessment by introducing another primary vigour factor namely, seedling growth factor (abbreviated as seedling factor, SF) obtained by dividing early (pre autotrophic) seedling evaluation per cent value by 100. A secondary vigour factor, germination-cum-seedling factor (GSF) has been derived through integration of the primary factors GF and SF by multiplication. Of course data transformation for the conversion of negative correlations to respective positive correlations, readjustment of scale values for uniformity and unification of score values, *etc.* would be the same as in the earlier paper.

The holistic nature of the present approach involving the three vigour factors (GF, SF and GSF) has been very clearly revealed in the highly significant positive statistical correlations of GSF- integrated morphological, physiological and biochemical vigour parameters or scores with the same GSF-integrated field emergence or other productivity related characters showing vast improvements compared to their corresponding nonintegrated counterparts. Such consistent observations in all the crops so far studied would certainly confirm the great utility of the functional togetherness of the vigour factors in the relevant cropping systems.

(See full paper in Annexure No A-2).

103. Seed Invigoration for Seed Saving by Small and Marginal Farmers and Ensuring Food Security

Basu, R.N., Mandal, A.K. and Pal, P. (2010). *Indian Biologist*, **Special Volume**, 153-167.

The present paper primarily deals with the results and prospects of prestorage dry treatments developed in the Calcutta University with the basic purpose of enabling smallholder farmers to save and use their own seeds (improved cultivars as well as locally adapted varieties) for the next sowing without depending on external commercial sources and thereby ensure seed security at the village level all over the country.

Though the treatments such as mid-storage and pre-storage ones have been developed primarily for the maintenance of vigour and viability, the seed treatments have been found to consistently improve field performance of all crops so far tested; the improvement in yield ranged from 5-20 per cent of the respective controls. The crops include cereals, pulses, oilseeds, fibre crops, vegetables and flowers.

The treatments can be very easily given by the farmers majority of whom in India, and West Bengal in particular, are small and marginal handling only small quantities of seeds if the procedural steps are explained (or demonstrated) in simple easy-to-understand local languages and vocabularies.

There is no point in saving seeds of plants grown from F_1 hybrid (mother crop) because such F_2 seeds (of subsequent generations) would not be true breeding. The situation is same for patented genetically modified (GM) seeds (presently only the commercially approved hybrid cotton seed; a wide range of under tribal GM seeds of many crops are in the pipeline, all of which are F_1 hybrids); they have been bred only as hybrids so that farmers are unable to save seeds and would be forced to go back to the company each year for seeds of major agricultural crops as well as agri-inputs tied to the seeds.

Under no circumstances, the genuine rights of farmers to save their own seeds should be compromised as that time tested traditional practice is essential for seed security which is integrally linked with food security as well as food sovereignty of the country.

104. Alleviation of the Adverse Effects of Climate Change by Seed Invigoration Treatments

Pal, P., Mandal, A.K. and Basu, R.N. (2017). In: Souvenir and Abstracts ISST – XIV National Seed Seminar, Food security through augmented seed supply under climate uncertainties, Jan., 28-30, 2017. ICAR-IARI, New Delhi.

According to the Intergovernmental Panel on Climate Change (IPCC), the irregularity of climate systems would be a major problem for the successful implementation of the existing cropping schedules. Besides, optimum temperature regimes, adequate rainfall and its timely distribution over the cropping season for a rainfed crop are the basic requirements for the success of arid zone agriculture. Inadequate rainfall causing drought and too much rainfall within a short period causing waterlogging would be undesirable. However, erraticism in the distribution of rainfall is increasing over the years and it is perhaps an indirect but expected outcome of global climate change.

Cultivators may choose less water requiring drought tolerant crop which may suffer from standing water as it might be inherently susceptible to waterlogging. Employing conventional breeding methods, development of cultivars equally tolerant to drought and waterlogging situations would be a very difficult proposition. The same would be true for simultaneous tolerance to opposite characters such as acidic and alkaline soil conditions, *etc.*

Preliminary studies in the faculty of agriculture in the University of Calcutta have shown that seed invigoration treatments would enable invigorated seedlings to tide over to a considerable extent, the adverse effect of partial drought, semisubmergence and medium levels of salinity under simulated laboratory conditions simply because the invigorated seedlings are morphologically and physiologically stronger and as such, capable of overcoming the short-term stress situations than the noninvigorated less vigorous counterparts.

In the present study nine local cultivars, grown in southern coastal region (Sunderbans) of West Bengal have been screened for their relative tolerance to several simulated stress conditions in the laboratory such as physiological drought (using polyethylene glycol, PEG 6000), submergence in water and salinity, and alleviation of the same through seed invigoration treatments by dry dressing with finely powdered pharmaceuticals such as aspirin (0.1g/kg), a botanical, neem leaf powder (2g/kg) and the common bleaching powder (2g/kg) and water soaking-drying as a reference treatment besides the untreated control.

The physiological and biochemical studies have been conducted only with a few selected cultivars. The seed invigoration treatments (given to the relatively high vigour seed lots) such as aspirin and neem leaf treatment as well as bleaching powder (active ingredient, calcium oxychloride) maintained post-ageing vigour and viability of seeds and consistently showed greater tolerance to the aforesaid simulated stress conditions. The indices of tolerance were significantly correlated to the quantified vigour ratings. The conjecture that the invigorated seed would always perform better than the non-invigorated seed after accelerated or natural ageing would justify the acceptance of such beneficial treatments the same was clearly achieved in the simulated laboratory findings. The post-aging physiological and biochemical studies revealed amongst others a considerably reduced lipid peroxidation a fully confirmed outcome of seed invigoration in general. The simulated laboratory studies, however, need to be corroborated by well designed field experiments in climatologically susceptible areas.

105. Quantification of Seed Vigour: Vigour Related Correlation-Regression Equations and Data Transformations

Basu, R.N. and Punjabi, B. (2017). In: Souvenir and Abstracts, Lead Paper – XIV National Seed Seminar on food security through augmented seed supply under climate uncertainties, ICAR-IARI, New Delhi, pp. 79-85.

In the earlier publications on seed vigour (Basu, 2004, 2007) the very important role of germination percentage, seedling evaluation per cent and their derivatives germination factor (GF) and seedling factor (SF) in vigour rating, an improvement of correlation-regression equations has been highlighted. The present discourse aims at more purposeful and efficient correlation-regression equations involving data transformation and integration of vigour related parameters for their use in practical and commercial agriculture.

Ten pairs of data sets arranged on the basis of (i) interchange of independent and dependent status of variates in bivariate correlation- regression analysis, (ii) appropriate data conversion and transformation to improve correlation-regression equations, and (iii) a rather non-conventional approach that involves estimation of germination factor (GF, which is simply germination per cent divided by 100) for subsequent integration with vigour parameters with a positive connotation. That improves the correlation through elimination of the numerical contribution of non-viable seeds in seed vigour rating.

Regarding data transformation the conversion of germination percentage to angles (arc-sin) would not show any noticeable improvement of correlations or associated statistical deravatives. No particular advantage of arc-sin transformations could be recorded with data of Ching and Schoolcraft (1968).

Thus GF integrated positive vigour parameters have been found to be most suitable for regressional predictivity.

Quantification of the vigour and viability status of seeds of agricultural and horticultural crop and assessment of their impact on field performance and productivity in mathematical/statistical terms would give a new dimension to the pre-eminent role of seed quality in commercial agriculture, especially so in view of the significant beneficial effects of seed invigoration treatments on seed performance under normal as well as abiotic stress situations in an era of global climate change and its projected adverse effect on agriculture in the developing countries in particular.

Review Papers on Seed Invigoration

In the review articles mentioned hereunder, the progressive methodological and conceptual developments of different seed invigoration treatments and their practical applications in agriculture and horticulture in the field, as followed by small and marginal farmers in particular, have been covered for bringing the same to the notice of researchers and students in the domain of seed science and technology.

The results of the seed invigoration treatments for the major cereals, leguminous pulses, oil seeds and horticultural field crops particularly vegetables and annual flower seeds, would very clearly show the general nature of the effectiveness of different treatments on seed vigour, viability, field performance and crop productivity.

As an understanding of the fundamental aspects involved in the *modus operandi* of seed invigoration treatments is considered to be of great importance; the review article on the same "Seed invigoration for extended storability", suggesting among others, the involvement of free radicals in lipid peroxidation and volatile aldehyde emanation, for critical attention of seed scientists and scholars especially in respect of their possible counteraction by the seed invigoration treatments. Also issues involving small and marginal farmers and the question of their seed and food security in a changing climatological and socio-political scenario have been highlighted in the Annexure No A-4. Many of the observations in the reviews have been duly covered in the text.

It needs mentioning that the reviews were not regular in timings but were undertaken mostly for presentation at important national and international conferences and at times upon invitation by editors of special subject-matter oriented publications in the relevant field of science.

The review articles, arranged mostly on the basis of their publication dates are as follows:

1. Seed Treatment for Vigour, Viability and Productivity of Rice (*Oryza sativa* L.). Basu, R.N. and Pal, P., 1978. In: Rice in West Bengal. A Government of West Bengal Publication, Volume **I,** pp. 163-170.
2. Maintenance of Vigour, Viability and Yield Potential of Oilseeds. Dey, G. and Basu, R.N., 1982. In: Pulses and Oilseeds in West Bengal. A Government of West Bengal Publication, pp. 217-223.
3. Control of Rice Seed Deterioration by Physico-chemical Treatments. Pal, P. and Basu, R.N., 1982. In: Rice in West Bengal. A Government of West Bengal Publication, Calcutta, Volume **III**, pp. 159-166.
4. Post-harvest Seed Physiology and Seed Invigoration Treatments. Basu, R.N. and Rudrapal, A.B., 1982. In: Frontiers of Research in Agriculture, Proceedings of the Indian Statistical Institute, Golden Jubilee International Conference, Calcutta (Ed. S.K. Roy), pp. 374-397.
5. In Search of Low Cost Technologies to Improve Seed Performance. Basu, R.N. and Mandal, A.K., 1985. In: Transferable Technology for Rural Development (Ed. D.K. Dasgupta), Associated Publishing Company, New Delhi, pp. 33-52.
6. Viability of Horticultural Seeds and Seed Invigoration Treatments for Vegetables and Flowers. Basu, R.N. and Mandal, K., 1986. In: Propagation

of Tropical and Subtropical Horticultural Crops (Eds.: T.K. Bose, S.K. Mitra and M.K. Sadhu), Chapter 8, Naya Prakash, Calcutta, pp. 572-596.

7. Seed Germination and Viability. Basu, R.N. and Sur, K., 1988. In: Plant Physiological Research in India (Ed. S.P. Sen), pp. 117-137.
8. Seed Invigoration for Extended Storability. Basu, R.N., 1990. In: Proceedings of the International Conference on Seed Science and Technology, 1990, New Delhi, Published in 1993, *Seed Research*, Special Volume **1**, pp. 217-230.
9. An Appraisal of Research on Wet and Dry Physiological Treatments and their Applicability with Special Reference to Tropical and Subtropical Countries. Basu, R.N., 1994. *Seed Science and Technology*, **22**, 117-126.
10. Seed Invigoration Studies in the University of Calcutta: Glimpses of the Past and Present. Basu, R.N., 1994. Special invited lecture at the Tamil Nadu Agricultural University, organised by ISST and published in *Seed Tech News*, pp. 1-12. (also see extracts of the paper in Chapter 5).
11. Seed Viability. Basu, R.N., 1995. In: Seed Quality, Basic Mechanisms and Agricultural Implications (Ed. A.S. Basra), Haworth Press Inc. New York, USA, Chapter 1, pp. 1-44.
12. Seed Invigoration for Seed Saving by Small and Marginal Farmers and Ensuring Seed and Food Security. Basu, R.N., Mandal, A.K. and Pal, P., 2010. *Indian Biologist* (Special Volume), pp. 153-167.

Chapter 5

Thematic Issues and Theme Papers

The thematic issues cover the important morphological, physiological and biochemical as well as agricultural production orientated aspects associated with seed invigoration. These include seed viability and vigour, field performance and productivity, and very importantly emphasising on the elucidation of the possible modes of protective and prophylactic action of the different seed invigoration treatments. Though positive effects are most likely through the counteraction of lipid peroxidation and free radical reactions we strongly believe that further clarification of the basic issues involved in invigoration would lead to even more effective seed invigoration treatments and a more successful agriculture. The thematic issues dealing with seed invigoration are broadly grouped as follows: (I) maintenance of seed viability by seed invigoration treatments; (II) seed invigoration for field performance and productivity; (III) modes of the protective action and possible mechanism of seed invigoration treatments involving (A) activation of a cellular repair system, (B) radioprotective action of seed invigoration treatments, (C) role of free radical controlling agents in seed deterioration, (D) control of lipid peroxidation by seed invigoration, (E) control of volatile aldehydes by seed invigoration; (IV) quantification of seed vigour by simple mathematical-statistical approaches, and (V) allied invigoration related miscellaneous issues. Our theme papers are selected publications, usually (but not always) the initial ones, representing a particular theme; generally 3 to 4 published papers, have been so considered.

A perusal of some of the selected abstracts of publications on seed invigoration by researchers in College of Agriculture, Calcutta University (as chronologically arranged in Chapter 4) would reveal that the number of abstracts of papers on certain themes are many, while on certain other themes there are fewer papers on some topics; presently greater emphasis is being given to field performance and productivity related papers.

Initially, the sole thrust was on the maintenance of seed viability of many of our crop cultivars under the prevailing hot and humid storage conditions. The very first study on stored rice and jute seeds, employing a short duration seed soaking followed by drying back (midstorage hydration-dehydration) proved highly effective for viability maintenance; very soon the same was confirmed in wheat and subsequently in a wide range of crops. We originally presumed that leaching out of undesirable metabolites was responsible for the beneficial effect but hydration-dehydration treatments that would not involve leaching such as spraying-drying, dipping-drying, moisture equilibration-drying (humidification-drying) also proved equally effective casting doubt on the leaching hypothesis. Obviously, some other mode(s) of action would be operative.

Besides viability, we studied the effect of seed treatments on seedling vigour (post-ageing root and shoot growth of seedlings, in particular) in a large number of crop seeds and ultimately took up field performance and productivity which are of great practical importance to our predominantly small and marginal farmers.

We focused on prevailing general theories on seed deterioration and studied their applicability towards interpretation of the effects of seed invigoration treatments. The discourse has covered the elucidation of possible roles of the cellular repair system lipid peroxidation and free radical reactions involving, in particular, cell membranes; radioprotective action of seed invigoration treatments with the use of diverse chemicals, radioprotectors and sensitizers, antioxidants and pro-oxidants (many of which act as free radical controlling agents) employing different treatment methods. To overcome the large effect of hydration and its masking effect on the relatively smaller effects of chemicals as in hydration-dehydration (H-D) treatments, besides aqueous solutions, the use of chemicals dissolved in non-aqueous organic solvents compatible with the seed being treated, and chemicals in vapour phase was started. Also, dry-dressing with finely powdered materials including pharmaceutical and botanical preparations was taken up simultaneously.

Our pursuits also entailed new methodological ideas on germination tests to suit our needs and circumstances more effectively leading to the development of the very useful inclined glass plate blotter method of seed germination (see the first item in the "Allied miscellaneous issues section V") for all small and medium sized seeds. The development of the first ever seed vigour bioassay method employing high vigour jute seed needs to be mentioned in this connection that would enable the quantification of vigour of a seed lot of unknown vigour (see Annexure No A-).

A mathematical-statistical approach was made to quantify the vigour status of different seed lots (including seeds from different seed invigoration treatments) using available data (usually published data from different sources) on various morphological, physiological and biochemical vigour parameters. The integration of germination factor (GF, germination per cent divided by 100) with different positive vigour parameters (which enabled us to separate the relative contributions of germinable and non-germinable seeds in respect of a particular vigour parameter) has proved to be very useful in understanding the influence of seed vigour on field performance and productivity. That also enabled us to mathematically evaluate the relative efficacy of different seed invigoration treatments in quantitative terms (in percentage values). More details on quantification of seed vigour are given in Chapter 6.

The miscellaneous section of theme papers also serves to cover several other relevant thematic issues such as germplasm conservation for a decade or so employing hydration-dehydration seed treatment followed by storage over calcium chloride for a short period and thereafter transferred to rubber-stoppered sealed small to medium sized glass bottles under ambient conditions. Also, issues like the relative roles of embryo and endosperm in deteriorative senescence and the effect of invigoration treatment on the same using rice as the test material have been taken up. The genesis and progressive development of seed invigoration studies in the University of Calcutta (as covered in a special lecture in Tamil Nadu Agricultural University at Coimbatore during the inauguration of the All India Seed Seminar in 1994 and published in Seed Tech News of the Indian Society of Seed Technology, New Delhi) the extract of which has been appended at the end of this chapter as a document of our sustained struggle for a foothold in the field of seed invigoration.

List of Thematic Issues and Theme Papers

The selected thematic issues covered in our studies on seed invigoration are arranged for the sake of convenience under the following heads along with respective theme papers. Some unavoidable overlapping of themes would be apparent but for an analytical discussion, the same has to be accepted. The themes are serially numbered and selected publications (theme papers) related with a theme are placed under that, occasionally a particular paper would come under two or more themes.

I. Maintenance of Seed Viability by Seed Invigoration Treatments

A. Hydration-Dehydration (H-D) Treatments of Stored Seed

Theme Papers

(i) Maintenance of seed viability in rice (*Oryza sativa* L.) and jute (*Corchorus capsularis* L. and *C. olitorius* L.). Basu, R.N., Chattopadhyay, K. and Pal, P. (1974). *Indian Agriculturist,* **18,** 75-79.

(ii) Control of seed deterioration in wheat (*Triticum aestivum* L.). Basu, R.N. and Dasgupta, M. (1974). *Indian Agriculturist,* **18,** 285-288.

(iii) Seed treatment for the maintenance of vigour and viability. Basu, R.N., Bose, T.K., Chattopadhyay, K., Dasgupta, M., Dhar, N., Kundu, C., Mitra, R., Pal, P., and Pathak, G. (1975). *Indian Agriculturist*, **19**, 91-96.

(iv) Physico-chemical control of seed deterioration. Basu, R.N. (1976). *Seed Research*, **4(1)**, 15-23.

B. Prestorage dry seed invigoration treatments

Theme Papers

(i) Iodine treatment of seed for the maintenance of vigour and viability. Basu, R.N. and Rudrapal, A.B. (1979). *Seed Research*,**7**, 80-82.

(ii) Iodination of mustard seed for the maintenance of vigour and viability. Basu, R.N. and Rudrapal, A.B. (1980). *Indian Journal of Experimental Biology*, **18**, 492-494.

(iii) Iodine treatment of mungbean seeds for the maintenance of vigour and viability. Rudrapal, A.B. and Basu, R.N. (1980). *Current Science*, **49**(8), 319-320.

(iv) Extension of wheat seed viability with plant leaf powders and an aspirin-containing pharmaceutical preparation. Pal, P. and Basu, R.N. (1994). *Indian Agriculturist*, **38**, 145-149.

II. Seed Invigoration for Field Performance and Productivity

A. Midstorage Hydration-Dehydration Treatment

Theme Papers

(i) Seed treatment for vigour, viability and productivity of wheat (*Triticum aestivum* L.). Dasgupta, M., Basu, P. and Basu, R.N. (1976), Indian Agriculturist, **20**, 265-273.

(ii) Seed treatment to maintain viability, vigour and yield potential of stored rice seed. Basu, R.N. and Pal, P. (1978). *International Rice Research Newsletter* (IRRN), **3**(2), 5.

(iii) Response of different wheat cultivars to hydration-dehydration treatment for the maintenance of vigour, viability and productivity. Mandal, A.K. and Basu, R.N. (1984). *Seeds and Farms*, **10,** 9-10, 39-42.

(Many such papers on majority of our cereal crops, pulses and vegetables have been published in national and international journals, see selected abstracts in Chapter 4).

B. Prestorage Dry Seed Invigoration Treatment for Field Performance and Productivity

Theme Papers

(i) Treatment of rice seed with iodine and chlorine for the maintenance of vigour, viability and productivity. Pal, P. and Basu, R.N. (1988). *Indian Agriculturist*, **32**(1), 71-75.

(ii) Dry physiological seed treatments for improved performance of wheat (*Triticum aestivum* L.) and sunflower (*Helianthus annuus* L.). Pal, P., Ray, I., Mitra, R. and Basu, R.N. (1996). *Indian Agriculturist*, **40**(4), 219-232.

(iii) Seed invigoration treatments for improved storability, field emergence and productivity of Soybean (*Glycine max* L.). Mandal, A.K., De, B.K., Saha, R. and Basu, R.N. (2000). *Seed Science and Technology*, **28**, 349-355.

III. Modes of Protective Action of Seed Invigoration Treatments

A. Involvement of Cellular Repair System in Seed Invigoration

Theme Papers

(i) Control of lettuce seed deterioration Basu, R.N., Pan, D. and Punjabi, B. (1979). *Indian Journal of Plant Physiology*, **22**, 247-253.

(ii) Control of age and irradiation induced seed deterioration in lettuce (*Lactuca sativa* L.) by hydration-dehydration treatments. Punjabi, B. and Basu, R.N. (1982). *Proceedings of the Indian National Science Academy*, B **48**(2), 242-250.

(iii) Absence of *de novo* protein synthesis in moisture equilibration – drying treatment for the maintenance of lettuce (*Lactuca sativa* L.) seed longevity. Pan, D. and Basu, R.N. (1985b). *Indian Journal of Experimental Biology*, **23**, 375-379.

B. Radioprotective Action of Seed Invigoration Treatments

Theme Papers

(i) Radioprotective action of seed invigoration treatments. Dasgupta, M., Chattopadhyay, K., Basak, S.L. and Basu, R.N. (1977). *Seed Research*, **5**, 105-118.

(ii) Control of age and irradiation induced seed deterioration in lettuce (*Lactuca sativa* L.) by hydration-dehydration treatments. Punjabi, B. and Basu, R.N.(1982). *Proceedings of the Indian National Science Academy*, **B 48**(2), 242-250.

(iii) Physico-chemical control of radiation and ageing damage in mustard seed. Dey (Pathak), G. and Basu, R.N. (1985). *Indian Journal of Experimental Biology*, **23**, 167-171.

C. Free Radical Controlling Agents in Seed Invigoration

Theme Papers

(i) Control of seed deterioration by free radical controlling agents. Basu, R.N. and Dasgupta, M. (1978). *Indian Journal of Experimental Biology*, **16**, 1070-1073.

(ii) Physico-chemical control of radiation and ageing damage in mustard seed. Dey (Pathak), G. and Basu, R.N. (1985). *Indian Journal of Experimental Biology*,**23**, 167-171.

(iii) Iodination of mustard seed for the maintenance of vigour and viability. Basu, R.N. and Rudrapal, A.B. (1980). *Indian Journal of Experimental Biology*,**18**, 492-494.

D. Control of Lipid Peroxidation by Seed Invigoration

Theme Papers

(i) Lipid peroxidation and membrane damage in deteriorating wheat and mustard seeds. Rudrapal, A.B. and Basu, R.N. (1982). *Indian Journal of Experimental Biology*,**20**, 465-470.

(ii) Iodination of mustard seed for the maintenance of vigour and viability. Basu, R.N. and Rudrapal A.B. (1980). *Indian Journal of Experimental Biology*, **18**, 492-494.

E. Control of Volatile Aldehyde Production by Seed Invigoration

Theme Papers

(i) Volatile aldehyde production in relation to seed vigour of rice (*Oryza sativa* L.). Pal, P. and Basu, R.N. (1989). *Indian Agriculturist*, **33**, 255-258.

(ii) Reduced volatile aldehyde production in wheat by seed invigoration treatments. Sur, K. and Basu, R.N. (1990a). *Current Science*, **59**, 799-800.

(iii) Seed invigoration for extended storability. Basu, R.N. (1993). *Seed Research* (**Special volume 1**), pp. 217-230.

(iv) Seed invigoration treatments for French bean (*Phaseolus vulgaris* L.) and soybean (*Glycine max* L.) for improved vigour and viability after storage. Mukhopadhyay, A., Pal, P. and Basu, R.N. (1997). *Indian Agriculturist*, **41**, 79-85.

(Results of bioassay of seed vigour have always been highly correlated with reduced volatiles emanation and would serve as a quantitative estimate of the vigour status; see Chapter 6 on bioassay of seed vigour, including autoassay of vigour for further details).

IV. Quantification of Seed Vigour by Simple Mathematical Approaches

Theme Papers

(i) Approaches to quantification of seed vigour. Basu, R.N. (2004). *Indian Biologist*, **36**(2), 1-23.

(ii) A holistic approach to quantification of seed vigour through integration of morphological and physiological–biochemical parameters. Basu, R.N. (2007). *Indian Biologist*, **39**(1), 1-44.

(Also see the rejoinder on the issue in Chapter 6).

V. Allied Relevant Issues (Selected Publications on Germination, Seedling Growth *in vitro*, and Invigoration Related Topics of Particular Significance and Relevance)

Selected Papers

(i) Testing germination and seedling growth by an inclined glass plate blotter method. Punjabi, B. and Basu, R.N. (1982). *Indian Journal of Plant Physiology*, **25**, 289-298.

(ii) Influence of hydration-dehydration treatments of rice seed on the growth of embryo-endosperm transplants. Sur, K. and Basu, R.N. (1985). *Current Science*, **54**(17), 862-864.

(iii) Storage of rice seed for more than a decade under ambient temperature and low humidity regimes. Pal, P. and Basu, R.N. (1989). *Indian Agriculturist*, **33**, 181-183.

(iv) Seed invigoration studies in the University of Calcutta – Glimpses of the past and present. Basu, R.N. 1994. Special Lecture, organized by Indian Society of Seed Technology (New Delhi) at TNAU, Coimbatore, pp. 16-22. (Subsequently also published in full in the *Seed Tech News* of ISST, New Delhi, 1994)

(v) Seed invigoration treatment in wheat (*Triticum aestivum* L.) as influenced by vigour status of seed at the time of treatment. Pal, P., Ray, I. and Basu, R.N. (2001). *Indian Agriculturist*, **45**, 1-9.

(vi) The role of initial seed vigour status in the expression of invigoration effect of wet and dry treatments in French bean (*Phaseolus vulgaris* L.). Rudrapal, D. and Basu, R.N. (2004). *Seed Science and Technology*, **32**, 205-212.

(vii) Seed invigoration treatments on different seed sizes of wheat (*Triticum aestivum* L.) for improved storability and field performance. De, B.K., Mandal, A.K. and Basu, R.N. (2003). *Seed Science and Technology*, **31**, 379-388.

(viii) Seed invigoration for seed saving by small and marginal farmers and ensuring food security. Basu, R.N., Mandal, A.K. and Pal, P. (2010). *Indian Biologist*, **Special Volume**, pp. 153-167.

General Discussion on Thematic Issues and Theme Papers on Seed Invigoration

This discussion primarily involves selected theme papers and other closely related papers on seed invigoration only.

I. Maintenance of Viability by Seed Invigoration Treatments

A. Hydration-Dehydration (H-D) Treatment of Stored Seed

The four theme papers are grouped together for a composite approach to the maintenance of seed viability and vigour. The first two papers on rice and jute, and wheat primarily deal with the maintenance of seed viability while the third paper on 15 crops put more emphasis on vigour. The overall effects of hydration with aqueous solution of a range of chemicals was compared with the effect of hydration alone for an approximate effect of chemicals on viability and vigour (post-ageing) germination per cent and root and shoot growth of seedlings *vis-à-vis* non-aged control. Further storage at a high humidity (98 per cent RH) for 3 months at room temperature and two jute cultivars (JRC 212 and JRO 878) after two months storage at 98.5 per cent RH at room temperature (at around 25°C) showed significant beneficial effect on viability maintenance (Tables 5.1a and b).

Table 5.1a: Effect of Water-Soaking Drying (S-D) Treatment of Stored Rice, Jute and Wheat Seeds on the Maintenance of Viability (Germination per cent at room temperature, around 25°C) following Re-storage under High Humidity (98 per cent RH) Storage Condition for Rice and Jute and Natural (Ambient) Ageing Conditions for 5 Months in Wheat (*vide* Basu, Chattopadhyay and Pal, 1974; Basu and Dasgupta, 1974)

Treatment	*Rice*	*Jute*		*Wheat*
		C. capsularis L.	*C. olitorius L*	
Control	30	7	0	62
Soaking-drying (S-D)	70	76	18	84

Note: Soaking duration for rice 6 h, jute 3 h, wheat 3 h; (treatments were given to stored seeds 6-12 months old) kept in gunny or cloth bags, under ordinary indoor conditions; hydrated seeds were dried back in a drying cabinet in a current of hot air at 35°C; the hydrated-dehydrated wheat seeds were re-stored in cloth bags under ambient (warm-humid) laboratory conditions for 5 months for natural ageing; rice and jute seeds were stored in perforated paper packets in glass desiccators at 98 per cent RH at room temperature (average 25°C – 26°C) for 3 months and 2 months respectively; the wheat cultivar was Sonalika and the rice cultivar was IR8; jute cultivars were JRC 212 (*Corchorus capsularis* L.) and JRO 878 (*Corchorus olitorius* L.).

* In numbering of tables in this chapter, '5' stands for the chapter number (chapter 5).

Tentatively it was suggested that the beneficial effect of soaking-drying treatment was attributable to leaching out of auto-toxic metabolic byproducts accumulated in the seed during the previous storage period.The dilute solution of chemicals at the effective concentrations might slow down catabolic reactions in the senescing seed or act as antimicrobial agents (Basu *et al.*, 1974; Basu and Dasgupta, 1974). However, as has already been mentioned earlier, the leaching hypothesis now stands cancelled.

In the theme paper covering seed treatment for vigour and viability of 15 species of crops, the effective chemical treatments have been summarized (Basu *et al.*, 1975); the same is reproduced here (Table 5.2). The effectiveness of the chemical treatments was rather arbitrarily estimated by adding 50 per cent of the value of final (post-ageing) germination percentage at 25 per cent of root length and 25 per cent of shoot length (respective non-aged control values taken as 100). See Table 5.2 with footnotes for details. The small seeded pulses (medium vigour lots) were less susceptible to soaking injury; subsequent studies showed that moisture equilibration-drying or moist sand conditioning-drying would be better seed invigoration treatments for legumes susceptible to soaking injury.

Response to chemicals differed in the different crops. In rice, for example, the effects of chemicals were less pronounced and hydration alone was very effective. In wheat, sodium phosphate (dibasic) at 10^{-4}M was more effective. A significant improvement of germinability was noted with the radioprotective chemicals, cysteine, *p*-aminobenzoic acid and ethylenediaminetetracetic acid (EDTA – Na salt).

In the soaking-drying seed treatment involving full immersion in water or aqueous solutions of chemicals for several hours, the possibility of leaching of toxic metabolites was indicated in our initial studies. It was subsequently noted that seed hydration, unaccompanied by leaching as with moisture equilibration (humidification)-drying, or light aqueous spray would also prolong viability besides very good retention of seed vigour (*vide* theme paper (iv) on physico-chemical control of seed deterioration, Basu, 1976). Thus, moisture equilibration-drying (ME-D) as well as spraying-drying proved to be effective seed invigoration treatments. In the latter case, stored 8-month-old wheat seeds, spread in a single layer were sprayed with water or dilute chemical solutions (0.1 ml solution per gram of seed, kept covered for one hour after spraying) followed by drying back to the original weight. The hydrated-dehydrated seeds were subjected to accelerated ageing at 100 per cent RH and 45°C for 8 days, prior to post-ageing germination test. The seed invigoration effect was very clearly reflected in all the treated materials after ageing but tests done before ageing did not show any noticeable positive effect suggesting a prophylactic (preventive) effect of the hydration-dehydration (H-D) treatments. Such effects would be difficult to explain on the basis of activation of cellular enzymatic repair system for which the level of hydration or time necessary for initiation of protein synthesis during the hydration phase would be inadequate (Bhatt and Padayatty, 1975, Dobrzanska *et al.*, 1975). Evidences put forward would

Table 5.1b: Effect of Soaking-Drying (S-D) Treatments of Stored Rice, Jute and Wheat Seeds on further Post-Ageing Seedling Growth in the Laboratory (*vide* Basu, Chattopadhyay and Pal, 1974 and Basu and Dasgupta, 1974)

Rice IR8				*Jute (JRC 212)*				*Wheat (Sonalika)*			
Treatment (T_1-T_{11})		*Total Seedling Length (TSL)*		*Treatment (T_1-T_8)*		*Total Seedling Length (TSL)*		*Treatment (T_1-T_{10})*		*Total Seedling Length (TSL)*	
		I (mm)	*II (mm)*			*I (mm)*	*II (mm)*			*I (mm)*	*II (mm)*
1	Control	20.4	6.1	1	Control	10.9	0.8	1	Control	106	66
2	Water	43.1	30.2	2	Water	33.1	25.2	2	Water	120	101
3	Sodium thiosulphate	43.9	34.2	3	Sodium thiosulphate	35.9	32.3	3	Sodium chloride	195	168
4	Potassium nitrate	57.1	45.7	4	*p*-Hydroxybenzoic acid	39.4	25.6	4	Sodium phosphate (dibasic)	201	173
5	Hydrogen peroxide	64.2	54.6	5	Cinnamic acid	48.9	46.5	5	Sodium phosphate (monobasic)	156	128
6	Indole acetic acid (IAA)	53.7	49.4	6	Pyrogallol	37.2	32.0	6	Sodium thiosulphate	154	140
7	Ethrel (10 ppm)	53.6	46.4	7	Gallic acid	31.3	23.8	7	Urea	152	111
8	*p*-Hydroxybenzoic acid	50.5	36.4	8	Tannic acid	41.2	35.0	8	Gallic acid	182	160
9	Cinnamic acid	45.7	38.9		LSD (*0.05P*)	8.4	10.0	9	Tannic acid	160	147

Rice IR8				*Jute (JRC 212)*				*Wheat (Sonalika)*			
Treatment (T_1-T_{11})		*Total Seedling Length (TSL)*		*Treatment (T_1-T_8)*		*Total Seedling Length (TSL)*		*Treatment (T_1-T_{10})*		*Total Seedling Length (TSL)*	
		I (mm)	*II (mm)*			*I (mm)*	*II (mm)*			*I (mm)*	*II (mm)*
10	Gallic acid	39.6	33.7					10	Penicillin + streptomycin (200 ppm each)	164	130
11	Tannic acid	47.4	41.2						LSD (*0.05P*)	20	22
	LSD (*0.05P*)	7.2	8.1								

Note: Concentration of chemicals (molar, M): Rice T_3, $T_{9\text{-}11}$ 10^{-4}M, T_4 10^{-2}M, T_5 3×10^{-2}M, T_6 10^{-3}M, T_8 10^{-5}M; for Jute T_3, T_4, T_6, T_8 10^{-5}M, T_5, T_7 10^{-4}M; for Wheat T_3, T_7 10^{-3}M, others 10^{-4}M. Total seedling length in mm is the sum of mean root length and shoot length (all seminal roots in case of rice and wheat), TSL I is total length calculated on the basis of germinated seedlings only, TSL II is based on all seeds taken for the test (germinated as well as non-germinated seeds) and is therefore equivalent to TSL × GF (where GF is the germination factor, germination per cent ÷ 100); other details same as in Table 5.1a.

Table 5.2: Some Effective Chemical Treatments for Maintaining Seed Viability (See footnotes for the explanation of abbreviations) (*vide* Basu, Chattopadhyay, Dasgupta, Dhar, Kundu, Mitra, Pal and Pathak, 1975)

*Material**	*Soaking Duration (at 25°C) Before Drying** (h)*	*Ageing Conditions****	*Effective Treatments other than Water*****	*Remarks******
Abelmoschus esculentus cv. Pusa Sawani	3	A_4	DSP^b, TU^b, U^b, MSP^b	Treatments also effective under A_1 and A_2
Allium cepa cv. Red Globe	6	A_1	$MSP^{a,b}$, $DSP^{a,b}$, FS^b, ST^b, $OA^{a,b}$, MC^c, $BA^{a,b}$	Also effective under A_2
Cajanus cajan cv. B4	3	A_1	DSP^b, U^b, SC^b, $GA^{a,b}$, MNC^a, CS^c	Also effective under A_2, A_4
Capsicum annum cv. Suryamukhi	3	A_4	DSP^b, MSP^b, $TA^{b,c}$	-
Cicer arietinum cv. S4	3	A_4	TU^c, TA^c, DSP^a	-
Corchorus sp. C. capsularis cv. JRC 212, 321 and *C. olitorius* cv. JRO 632.878	3	A_2	SC^{ab}, $DSP^{b,c}$, MSP^b, ST^b, TA^c, OA^b	Effect of chemicals less pronounced
Daucus carota cv. Nantes	2.5	A_1	ST^b, $MSP^{a,b}$, DSP^b, SC^a, FS^a	Also effective under A_2
Helianthus annus cv. EC 68414	3.5	A_1	SC^a, MSP^a, DSP^a, $ST^{a,b}$, FA^b, TA^b	Also effective under A_2 and A_4
Lathyrus sativus cv. Local	3	A_2	U^b, TU^b, ST^b	-
Lens culinaris cv. W.B.81	3	A_2	AA^a, OA^c, BA^a, SM^c, GAL^c	Also effective under A_4
Lycopersicon esculentum cv. Pusa Early Dwarf	3	A_4	DSP^b, MSP^b, TA^c, SC^b	-
Oryza sativa cv. Ratna	6	A_1	$MSP^{a,b}$, $DSP^{a,b}$, $SC^{a,b}$, ST^b, OA^a, TA^b, BA^a, MC^c	Hydration followed by drying very effective under A_1, A_2, A_3, A_4 and A_5; effect of chemicals less pronounced

*Material**	*Soaking Duration (at 25°C) Before Drying** (h)*	*Ageing Conditions****	*Effective Treatments other than Water*****	*Remarks******
Phaseolus aureus cv. B1	3	A_2	BA^a, MNC^b, ZS^a, SM^c	-
Solanum melongena cv. Mukhtakeshi	3	A_4	TA^c, TU^b, ST^c, DSP^b	-
Triticum aestivum cv. Sonalika	3	A_1	DSP^b, SC^a, ST^b, MSP^b, TA^b, BA^a, OA^b, CYS^a, $PABA^a$, $EDTA^a$	Chemical effects very pronounced; also effective under A_2, A_3 and A_5

* Seeds were stored for 3-10 months before treatment.

** Duration of soaking would depend on the ambient temperature, it would be more at lower and less at higher temperatures.

*** Ageing conditions: A_1 – ageing under ambient conditions for 6-10 months, A_2 – 45°C and 100 per cent RH for 7-14 days, A_3 – 45°C and 36 per cent RH for 1-3 months, A_4 – room temp. and 100 per cent RH for 1-5 months, A_5 – room temp. and 76 per cent RH for 3-10 months.

**** Simple hydration followed by drying is effective in all the materials; the chemicals further increase the effectiveness. Abbreviations for the chemicals: AA-ascorbic acid, BA-boric acid, CS-copper sulphate, CYS-cysteine, DSP-disodiumhydrogen phosphate, EDTA-ethylenediaminetetraacetic acid (Na-salt), FA-ferulic acid, FS-ferrous sulphate, GAL-gallic acid, MC-mercuric chloride, MNC-manganese chloride, MSP-monosodium dihydrogen phosphate, OA-oxalic acid, PABA-*p*-aminobenzoic acid, SC-sodium chloride, SM-sodium molybdate, ST-sodium thiosulphate, TA-tannic acid, TU-thiourea, U-urea, ZS-zinc sulphate. Superscripts a, b, c stand for molar concentration of chemicals: a-10^{-3}, b-10^{-4}, c-10^{-5}M.

***** The different chemicals and their concentrations have not yet been tested for all the materials under the different ageing conditions; the list of effective chemicals and their concentrations is, therefore, incomplete. The effectiveness of the chemicals was estimated by adding 50 per cent of the value of final germination percentage and 25 per cent of the mean root length and 25 per cent of the mean shoot length in millimetres.

indicate the involvement of free radical controlling agents in the maintenance of seed vigour and viability.

B. Prestorage Dry Seed Invigoration Treatments

Of the three selected theme papers, the first two are on iodination and the third is on treatments with several kinds of leaf powders and an aspirin containing pharmaceutical formulation 'aspro'. A number of related papers were published on halogenation that include besides iodination of mungbean seeds (Rudrapal and Basu, 1980), the use of two more halogens, chlorine and bromine for the maintenance of vigour and viability of mustard seeds (Rudrapal and Basu, 1981).

Data presented in the first theme paper (Basu and Rudrapal, 1979) showed that treatment of wheat, jute and mustard seeds with iodine in the vapour phase for 5, 15 and 24 h would very effectively maintain post-ageing vigour and viability of seeds (Table 5.3) with the optimum duration being 15 h for wheat and jute and 24 h for mustard. The response of mustard seed was the highest. Although the seeds used in this study were not harvest-fresh, they were well dried and stored in rubber stoppered wax sealed glass bottles that maintained a relatively high vigour status and responded very well to iodination.

Table 5.3: Effects of Iodine (Vapour) Treatment on the Germinability of Wheat, Jute and Mustard Seeds after Accelerated Ageing at 100 per cent RH and 42°C (*vide* Basu and Rudrapal, 1979)

Materials	*Hours of Iodination*	*Germination (per cent)*	*Mean Root Length (mm)*	*Mean Shoot Length (mm)*
Wheat	0 (Control)	65	194 ± 4.3	22 ± 1.7
	5	80	218 ± 6.6	23 ± 2.0
	15	85	230 ± 7.3	25 ± 2.9
	24	60	199 ± 4.1	24 ± 2.4
Jute	0 (Control)	58	26 ± 2.2	16 ± 2.0
	5	67	25 ± 2.4	16 ± 1.6
	15	76	28 ± 3.4	18 ± 1.2
	24	65	26 ± 3.2	16 ± 1.4
Mustard	0 (Control)	55	49 ± 2.7	26 ± 2.1
	5	75	57 ± 2.8	33 ± 1.3
	15	85	57 ± 2.7	33 ± 3.0
	24	91	71 ± 4.2	34 ± 3.6

Treatments were given to 6-month-old wheat, 10-month-old jute and 12-month-old mustard seeds. Accelerated ageing was done for 7, 8 and 14 days for wheat, jute and mustard seeds respectively. Data on germination and seedling growth were recorded after 120 hr in wheat and mustard and 96 hr for jute, root length in wheat included all the seminal roots.

Subsequent studies of prestorage 'dry' seed invigoration treatments showed that much of the observed differences in response could be attributed to seed vigour at the time of treatment with the harvest-fresh high vigour seeds giving the optimum effect; very often there was no effect at all with low vigour seeds (in which H-D treatment could be better). This issue has been discussed elsewhere by Pal *et al.* (2001) and Rudrapal and Basu (2004).

Table 5.4a: Effect of Iodination on Germination Percentage of Mustard Seeds under different Ageing Conditions (Within a column values scored by the same letter are not significantly different at *0.05 P*) (*vide* Basu and Rudrapal, 1980)

Hours of Idodine Treatment	*Ageing Conditions*				
	Before Ageing (A)	*Ageing at 100 per cent RH and 40°C for 13 Days (B)*	*Ageing at 30 per cent RH and 45°C for 3 Months (C)*	*Storage Under Ambient Conditions for 3 Months (D)*	*Heated at 100°C for 30 min (E)*
0 (Control)	97^{a}	51^{c}	64^{c}	75^{c}	52^{c}
8	98^{a}	75^{b}	88^{b}	85^{b}	80^{b}
16	96^{a}	85^{a}	90^{b}	96^{a}	85^{ab}
24	98^{a}	91^{a}	96^{a}	98^{a}	90^{a}
48	92^{a}	76^{b}	85^{b}	83^{b}	81^{b}

Note: Data were recorded after 6 days of placement for germination at 24 ± 1°C.

Table 5.4b: Total Seedling Length (Root length + shoot length in mm) as Influenced by Iodine Treatment and Ageing Conditions (ageing conditions, *viz.*, A, B, C, D and E and other details same as in Table 5.4a) (*vide* Basu and Rudrapal, 1980)

Hours of Idodine Treatment	*Ageing Conditions*				
	Before Ageing (A)	*Ageing at 100 per cent RH and 40°C for 13 Days (B)*	*Ageing at 30 per cent RH and 45°C for 3 Months (C)*	*Storage Under Ambient Conditions for 3 Months (D)*	*Heated at 100°C for 30 min (E)*
0 (Control)	101^{a}	84^{c}	74^{c}	76^{c}	65^{c}
8	103^{a}	93^{b}	84^{b}	85^{ab}	78^{b}
16	100^{a}	96^{b}	92^{ab}	89^{a}	85^{a}
24	104^{a}	103^{a}	96^{a}	89^{a}	88^{a}
48	99^{a}	92^{b}	86^{b}	82^{b}	78^{b}

Note: Data were recorded as in Table 5.4a.

The physiology of iodine treatment of mustard seed has been analysed by Basu and Rudrapal (1980) (see Tables 5.4a and 5.4b, and Figures 5.1 and 5.2). Iodination of mustard seeds did not improve their immediate germinability. However, germination percentage and seedling growth were significantly higher in treated seed after accelerated, natural ageing or a short duration heat stress (see Tables 5.4a and 5.4b).

Iodination imparted great resistance to heat injury and the iodine treated mustard seeds remarkably withstood the damaging effect of very high temperature (100°C for 30 min.) on germination and seedling growth. The beneficial effect of iodine treatment was reflected upon the activities of amylolytic and dehydrogenase enzymes in the seed. The electrical conductance of the seed-steep water and leaching of sugars and amino acids were lower in the iodine treatments. Lipid peroxidation in the iodinated seeds was also considerably lower than in the control which would partly account for the better membrane functions of the treated seeds. These findings would strongly suggest the role of iodine acting as a controlling agent against lipid peroxidation and free radical reactions in the seed, besides, may be other possible roles. The highly beneficial effects of iodination in mungbean seeds (Rudrapal and Basu, 1980) and of the two other halogens, chlorine and bromine in mustard (Rudrapal and Basu, 1981) would suggest a broadly similar operative mechanism for halogenation of seed for the maintenance of vigour and viability, most likely through a counteraction of lipid peroxidation and free radical reactions. Rudrapal and Basu(1980) opined that the stabilization of olefinic bonds of unsaturated fatty acid moieties of lipoprotein component of cell membranes might perhaps render them less susceptible to lipid peroxidation and free radical reactions.

The significant retention of seed viability of wheat with finely powdered 'aspro' (a common aspirin-containing pharmaceutical formulation, an off the counter widely used anti-inflamatory medicine in tablet form) and dried and finely powdered leaves of some common plants would add another dimension to prestorage 'dry' seed invigoration treatments (Pal and Basu, 1994). Just a year before that, treatment of harvest-fresh wheat seed with red chilli fruit powder, turmeric rhizome powder and neem leaf powder was found to slow down the loss of vigour and viability after accelerated and natural ageing (Pal and Basu, 1993). In that study volatile aldehyde emanation was significantly reduced in the beneficial treatments accompanied by increased seed vigour bioassay values. Besides certain important active ingredients such as curcumin, capsicin, azadirachtin, *etc.* in the aforesaid plants, salicylic acid and its derivatives would be common in most such plant materials. Aspirin (*ortho*-acetylsalicylic acid) perhaps the most widely used medicine in the world itself is a plant derivative whose role in plants and medicinal mechanism of action are still not fully elucidated. As such, the role of such phytoproducts on deteriorative senescence is as yet speculative. Nevertheless, the seed germination and seedling growth data show that wheat seed treated with dried leaf powders of bael (*Aegle marmelos* Corr.), mango (*Mangifera indica* L.), tamarind

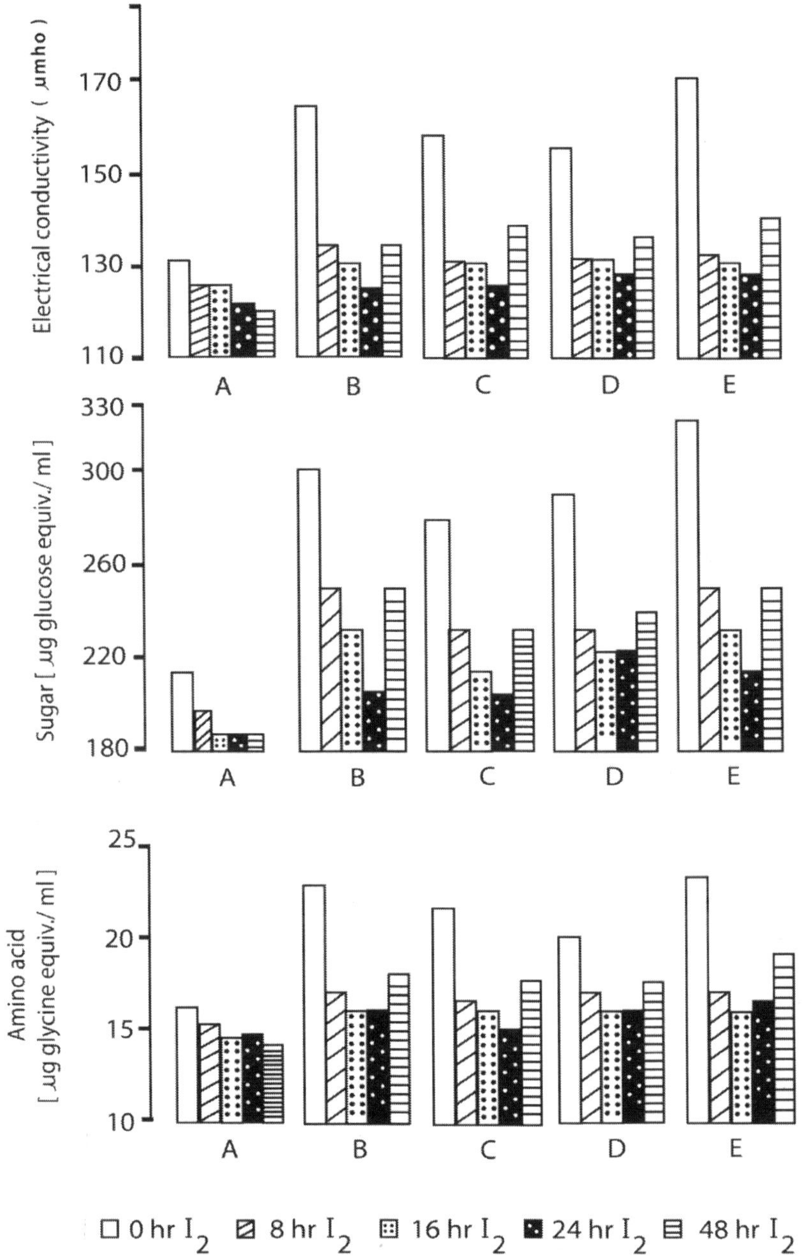

Figure 5.1: Effect of iodination of mustard seed for different durations on electrical conductivity of seed-steep water (Upper row), leaching of sugars (Middle row) and amino acids (Lower row) under different ageing conditions: A, before ageing (immediately after treatment); B, accelerated ageing at 100 per cent RH and 40°C for 13 days; C, 30 per cent RH and 45°C for 3 months; D, natural ageing under ambient conditions (av. RH 60.3 ± 3.2 per cent, temp. 29.9° ± 1°C) for 3 months; and E, heated at 100°C for 30 mins. (*Vide* Basu and Rudrapal, 1980)

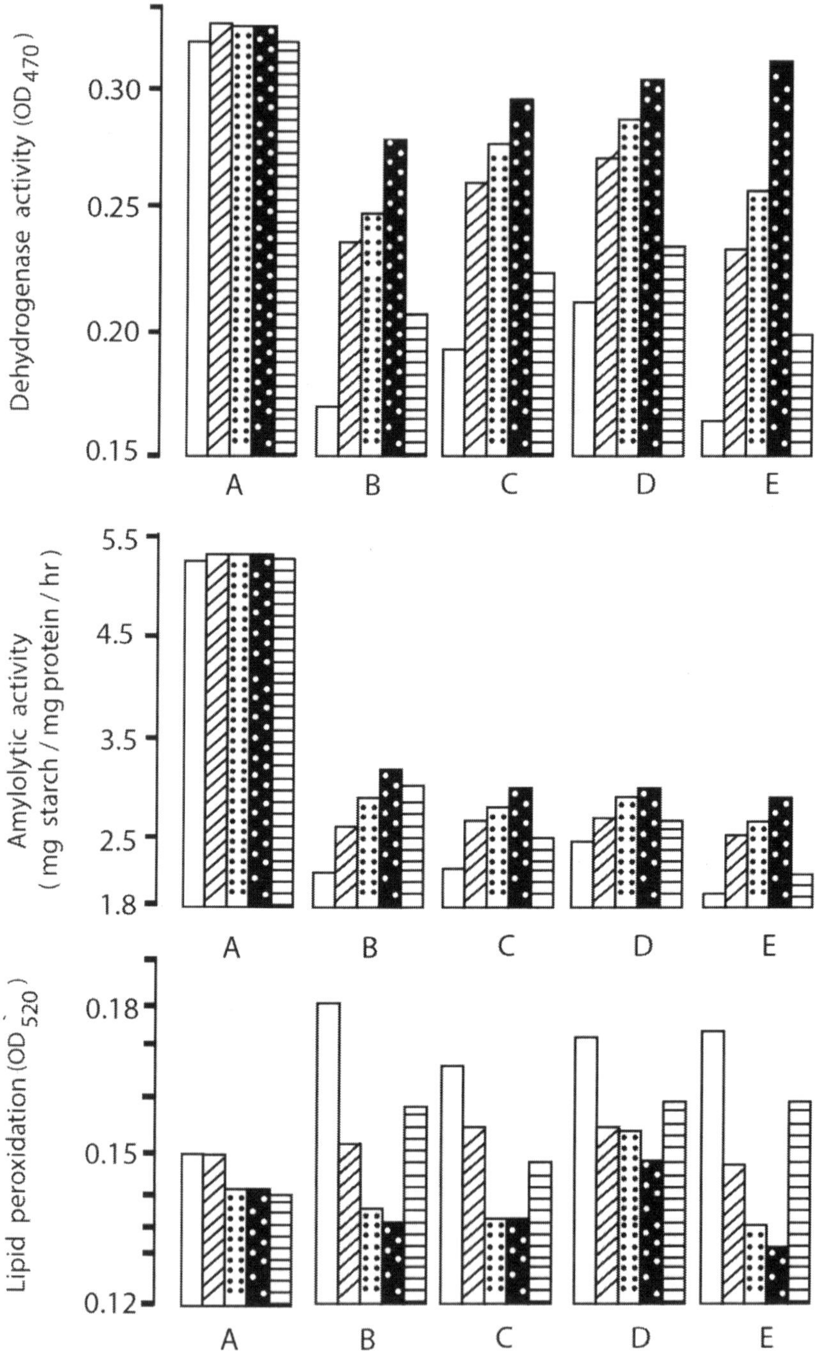

Figure 5.2: Effect of iodination of mustard seeds on dehydrogenase activity, amylolytic activity and lipid peroxidation under different ageing conditions; other details same as in figure 5.1. (*Vide* Basu and Rudrapal, 1980)

(*Tamarandus indica* L), nisinda (*Vitex negundo* L.), kalke (*Thevetia peruviana* L.) at 1-2 g leaf powder per kg of seed and 'aspro' at 20-100 mg effectively maintained seed viability following accelerated ageing at 98 per cent RH and 40°C for 9 days. Pal and Basu, 1994 confirmed similar beneficial effects with dried chilli fruit powder, turmeric rhizome powder and neem leaf powder (Table 5.5).

Table 5.5: Germinability of Wheat Seeds cv. Sonalika after Treatment[(a)] with Powdered Leaf Preparations and Accelerated Ageing at 98 per cent RH and 40°C for 9 Days (*vide* Pal and Basu, 1994)

Treatment[(a)] (Dry-dressing with leaf powder g/kg of seeds)	*Germination[(b)]*		*Mean Root Length[(b)] (mm)*	*Mean Shoot Length[(b)] (mm)*
	Per cent	*Arc-sin value*		
Control	55	47.9	261	40
Bael (2 g)	78	62.0	283	44
Mango (2 g)	86	68.0	256	47
Tamarind (2 g)	77	61.3	267	42
Nisinda (1 g)	78	62.0	293	47
Thevetia (2 g)	71	57.4	292	46
Neem (2 g)	83	65.6	300	50
LSD at *0.05 P*	-	6.4	12.8	3.0

(a) Treatments (dry-dressing) were given to 4-month-old wheat seeds stored in rubber-stoppered glass bottle after harvest, nisinda was used at 1g/kg, others at 2g/kg of seed. (b) Germination tests were done by the glass plate blotter method and data on germination percentage and seedling length were taken after 96 h at 30 ± 1°C.

Table 5.6: Effects of Dry-Dressing with Powdered Aspro[(a)] on the Germinability of Wheat Seeds after Accelerated Ageing at 98 per cent RH and 40°C for 8 Days (*vide* Pal and Basu, 1994)

Treatment		*Germination*		*Mean Root Length (mm)*	*Mean Shoot Length (mm)*
		Per cent	*Arc-sin value*		
Control		62	51.9	220	36
Aspro	20 mg/kg	87	68.9	267	42
"	50 "	88	69.7	274	42
"	100 "	92	73.6	275	43
"	200 "	89	70.6	268	41
"	500 "	74	59.3	234	35
LSD at *0.05 P*		-	11.2	28.1	4.0

(a) Aspro: a pharmaceutical formulation containing orthoacetyl salicylic acid (aspirin) of Nicholas Pharma India Ltd., dry-dressing with powdered Aspro was given to 5-month-old wheat seeds stored in rubber-stoppered glass bottles. Germination conditions same as in Table 5.5.

'Aspro' powder at 100 mg/kg of wheat seed showed 92 per cent viability against 62 per cent in control after 8 days ageing at 98 per cent RH and 40°C (Table 5.6; Pal and Basu, 1994). Soil emergence, after natural ageing for three months was greatly improved by neem leaf powder and 'aspro' (aspirin) treatment; while the soil emergence in control was 39 per cent, the respective values for neem leaf powder and 'aspro' were 65 per cent and 75 per cent respectively.

II. Seed Invigoration for Vigour, Viability, Field Performance and Productivity

A. Midstorage Hydration-Dehydration Treatments

Studies on the successful maintenance of seed viability started with rice and jute and it was considered appropriate to verify how the field performance and productivity of rice, the most important crop of eastern India would be influenced by midstorage soaking-drying treatments. In fact, with seeds of most of our major crops we have gone upto the productivity level, but only one theme paper separately covered here would be on rice. All others have been covered in the chronologically arranged abstracts of papers on seed invigoration. The results of the first theme paper on rice is reproduced hereunder (Table 5.7).

Table 5.7: Effect of different Physico-Chemical Seed Treatments on Germination and Yield of Rice[a] (*vide* Basu and Pal, 1978)

Treatment[b]	*Germination[c] Per cent*	*Mean Root Length after 120 h (mm)*	*Mean Shoot Length after 120 h (mm)*	*Grain Yield[d] (t/ha)*
Control	68b	52b	15b	4.6c
Water	88a	62a	23a	6.1b
Sodium phosphate (dibasic, 10^{-4}M)	91a	66a	25a	6.7a

(a) In any column, means that are followed by a common letter are not significantly different at the 5 per cent level. (b) Nine-month-old Ratna seed, which had been stored under ambient conditions (RH 67 ± 11 per cent, temp. 27.1 ± 4°C), were treated on 22 Aug., 1976 and were subsequently sown in Dec., 1976. (c) Germination tests were conducted in the laboratory at 28°C before sowing in the seedbed on 21 Dec., 1976. (d) Yield of crop at 135 days' duration (sowing to transplanting, 40 days; transplanting to harvest, 95 days); differences in field (nursery) emergence of seedlings were eliminated during transplanting.

B. Prestorage Dry Seed Invigoration Treatments for Field Performance and Productivity

In the theme paper dealing with productivity of rice with halogenation using iodine and chlorine, the study was carried out with iodinated French chalk (prepared by thoroughly mixing 1 g of solid iodine in 20 g French chalk and 1 g and 2 g of the same were used per kg of rice seed). Rice seeds so treated were stored along with untreated control seeds in separate cloth bags prior to transplantation.

For chlorination, rice seeds were treated with 2 g of bleaching powder per kg of seed and stored in the same manner as for iodination.

Germination test in the laboratory was done following the inclined glass plate blotter technique before sowing in the nursery prior to transplantation in the field. The higher germination percentage and seedling vigour (growth of seedlings) were partly reflected in the yield of the transplanted *aman* (kharif) crop. However, the increase in yield, with iodinated French chalk over the control was only 6.4 per cent, with bleaching powder the same was 7 per cent and though the values were statistically significant, they were lower than in other crops, or with direct seeded rice crop (not transplanted as above). A consistent increase (around 16 per cent) in rice production with halogen treatment of rice in farmers' fields in Tamil Nadu has been reported (Dharmalingam *et al.* (1997)).

In the second theme paper, the 'dry' invigoration treatment in wheat and sunflower, the same could not be given to harvest-fresh seeds, instead the seeds were 3 months old (high medium vigour, HMV) at the time of treatment. As such, besides finely powdered plant powders, aspirin, "Aspro","Celin" and bleaching powder, a water soaking-drying treatment was also included.

Table 5.8: Effects of Prestorage Treatments of Wheat (cv. Sonalika) Seed on the Field Performance and Productivity (*vide* Pal, Mitra and Basu, 1996)

Treatments (Dry-dressing with fine powder) (g/kg seed)[(a)]	*Plant Population/m²*	*Number of Shoots/ m²*	*Grain Yield (g/m²)*	*Length of Panicle (cm)*	*Grain Number/ Ear*	*1000 - Grain Weight (g)*
Control (untreated)	98	171	284	23	31	29.4
Turmeric (1 g)	115	189	320	26	34	33.6
Neem (2 g)	121	213	340	27	38	33.6
Nisinda (1 g)	119	195	330	26	33	32.8
Aspirin (0.1 g)	122	204	358	27	38	33.6
Aspro (0.2 g)	124	216	360	28	39	35.0
Celin (0.5 g)	123	211	344	29	36	34.4
Bleaching powder (3 g)	123	209	348	28	37	34.8
Water-soaking (3h) drying	120	231	370	30	40	36.0
LSD at *0.05 P*	6.4	14.9	20.6	1.5	2.5	2.3

(a) Treatments with the powdered materials, bleaching powder and water-soaking-drying were given to 3-month-old wheat seed in the month of May 1995, treated seeds and untreated control were kept for 7 months in glass bottles sealed with rubber cork till sowing in the field.Aspro is an aspirin (orthoacetyl salicylic acid) - containing pharmaceutical acid, Celin is vitamin C (ascorbic acid) a pharmaceutical product.

Data on wheat (Table 5.8) showed significantly higher plant population/ m² in all the treatments over the control and interestingly number of shoots/m² was highest in watersoaking-drying along with grain yield and most yield related characters. Of course yield and yield attributes of all the dry treatments were significantly higher than in the control. In harvest-fresh HV seed, soaking-drying would not normally give such effects as has been shown in our studies on vigour status versus 'wet' and 'dry' seed invigoration treatments (Pal, Ray and Basu, 2001; Rudrapal and Basu, 2002, 2004).

In high vigour seeds of sunflower all the 'dry' treatments (nisinda leaf powder, bleaching powder, aspirin, 'aspro' and 'celin') significantly surpassed the control values; water soaking-drying however, though a bit higher than control, failed to reach the level of significance at 0.05 P.(Table 5.9)

Table 5.9: Effects of Pre-storage Treatments of Sunflower (cv. Morden) on the Field Performance and Productivity (*vide* Pal, Mitra and Basu, 1996)

Treatments (Dry-dressing with fine powder) (g/kg seed) (a)	*Field Emergence (Per cent)* (b)	*Height of Plants (cm)* (c)	*Number of Leaves/ Plant* (d)	*Seed Yield (g/m²)* (e)	*1000-seed Weight (g)* (f)
Control (untreated)	40[c]	52	27	21.3	34
Nisinda (1 g)	46[b]	63	25	26.0	38
Aspirin (0.1 g)	49[a]	68	29	23.7	40
Celin (0.5 g)	50[a]	66	27	25.3	39
Bleaching powder (2.5 g)	49[a]	63	25	24.3	38
Water-soaking (2h) - drying	42[c]	72	25	22.8	39
LSD at *0.05 P*		14	NS	2.4	3.5

(a) Treatments with the powdered materials, bleaching powder and water-soaking-drying were given to 3-month-old sunflower seed in the month of May 1995, after treatment all the seeds were kept for 7 months in glass bottles sealed with rubber cork till sowing in the field. (b) Data were recorded for each replication two weeks after sowing. (c) and (d) data were recorded two weeks before harvesting. (e) and (f) data were recorded after harvest.

In the third theme paper on soybean, prestorage 'dry' treatments were given to harvest-fresh seeds employing bleaching powder (2 g/kg) as a source of chlorine, two pharmaceutical formulations, aspirin-containing 'aspro' (100 mg/kg), 'ibucon' (100 mg/kg) and red chilli fruit powder (1 g/kg), *Catharanthus* leaf powder (2 g/kg) besides a mid-storage (using 5 month-old seed) moist sand (6 per cent m.c)-conditioning soaking-drying (MSC-SD) as a reference treatment (usually very effective in soaking injury-prone leguminous seed). All those treatments showed significantly higher germination percentage and mean root and shoot lengths than the control (untreated). Also, natural ageing for 90 days at average 66.7 per cent RH and 27°C confirmed significant retention of vigour and viability (Table 5.10). The same was associated with reduced leaching of electrolytes, sugars, amino acids and by an increase in dehydrogenase enzyme activity in the invigorated seed. The aforesaid treatments also showed greater field emergence and productivity. Aspro proved to be a very good treatment closely followed by chilli powder (Table 5.11). Seed yield per plant, however, just reached the level of significance (at 0.05 P) in the reference treatment MSC–SD.

Table 5.10: Effects of Dry Treatments and Moist Sand Conditioning Soaking-Drying (MSC-S-D) on the Germinability and Seedling Lengths (5 days after germination) of Soybean Seeds Subjected to Natural Ageing under Ambient (average 66.7 ± 3.4 per cent RH and 26.9 ± 1.3°C) Conditions for 90 days (*vide* Mandal, De, Saha and Basu, 2000)

Treatment (mg or g/kg seed)	*Germination*		*Mean Root Length (mm)*	*Mean Shoot Length (mm)*
	Per cent	*Arc-sin value*		
Control (untreated)	53	46.7	37	32
Bleaching powder (2 g)	75	60.0	48	41
Aspro (100 mg)	79	62.7	62	49
Ibucon (100 mg)	70	56.8	54	43
Red chilli powder (1 g)	83	65.6	58	46
Catharanthus leaf powder (2 g)	85	67.2	56	45
MSC-S-D (W)	73	58.7	54	38
LSD at *0.05 P*	-	4.5	5.0	4.0

Note: Dry-dressing was done in harvest-fresh soybean (cv. Soyamax), the midstorage moist sand conditioning water soaking-drying was given to seeds stored for 5 months under ambient conditions and used as a reference treatment.

Table 5.11: Effect of Wet and Dry Seed Treatments on Soybean Field Emergence and Productivity (*vide* Mandal, De, Saha and Basu, 2000)

Treatments	*Seedling Emer-gence/m²*	*Number of Pod/Plant*	*Length of Pod (mm)*	*Seed Yield (g/plant)*	*1000-Seed Weight (g)*
Control	21	84	29	18.3	70.1
Bleaching powder	36	99	40	22.2	71.9
Aspro	36	150	36	24.6	72.0
Ibucon	34	147	32	23.7	72.1
Red chilli powder	40	156	31	24.0	71.2
Catharanthus leaf powder	33	138	33	22.3	71.0
MSC-S-D (W)	38	120	38	21.1	71.9
LSD at *0.05 P*	5	13	7	2.7	0.8

Note: Other details same as in Table 5.10.

A review of the field performance and productivity data in all our seed invigoration studies, irrespective of 'wet' or 'dry' treatment, as covered in the chronologically arranged selected papers (*vide* Chapter 4), would clearly show a vigour and viability linked increase in all such invigoration treatments, may be somewhat differing in the extent of yield increase. Even a relatively stable character like 1000 seed weight showed a small but consistent increase in all cases. We may, therefore, very confidently conclude that if properly carried out there should not be any doubt about the overall positive response on productivity. Vigour, viability and field performance and productivity are all positively correlated characters – that is the basic axiom, things cannot be otherwise under normal circumstances.

III. Modes of Protective Action of Seed Invigoration Treatments

A. Involvement of Cellular Repair System in Seed Invigoration

The three selected theme papers deal with dormant lettuce seeds as in this material, rather unique in orthodox seeds, the cellular repair hypothesis for prolonging seed viability was developed by Villiers and coresearchers (Villiers 1974, 1975; Villiers and Edgcumbe, 1975). According to the Villiers' hypothesis, the capacity of the fully imbibed seed to retain viability is due to the effective operation of a biochemical repair mechanism as against an accumulation of damage to cellular organelles in the air-dry state (5-20 per cent seed moisture). Villiers and Edgcumbe further showed that post-irradiation damage to chromosomes was much less in seeds kept fully imbibed after γ-irradiation.

In the first theme paper on cellular repair in lettuce seed, it has been shown that short-term water soaking-drying treatment of stored high medium vigour dormant lettuce seed greatly maintained seed vigour and viability. It is of particular interest to note that even dipping seeds in water for a minute with vigorous stirring at 27°C followed by drying back reduced loss of post-ageing (accelerated ageing at 100 per cent RH and 40°C for 6 days) viability and vigour. There was no difference in germinability before the seeds were subjected to accelerated ageing. Hydration by moisture equilibration-drying as well as by antiageing chemicals such as *p*-aminobenzoic acid, *p*-hydroxybenzoic acid, potassium iodide, *etc.*, by dry permeation employing acetone as the solvent reduced loss of lettuce seed viability to a great extent indicating the physico-chemical rather than biochemical nature of viability maintenance (see Table 5.15 in III,B).

The aforesaid observations in thermodormant lettuce seed have been confirmed in a wide range of orthodox nondormant crop seeds investigated by the researchers in the agriculture faculty of the University of Calcutta. We have, however, confined our discussion on lettuce only because of the emphasis on Villiers' hypothesis.

In the second theme paper on the control of age- and irradiation-induced seed deterioration by different hydration-dehydration methods such as soaking-drying (S-D), dipping-drying (D-D) and moisture equilibration-drying (ME-D) in stored one-year old medium vigour lettuce (cv, Suttons A1 thermodormant prechilling requiring variety) the great efficacy of all the treatments on post-ageing germinability has been clearly demonstrated. The treatments did not show any significant effect on immediate germinanability (before ageing), the anti-ageing effects were recorded only after ageing, initially the ageing effect in lettuce was apparent only on root length; germination percentage and shoot length after longer storage period. It was of interest to note that dipping for two minutes was almost as effective as S-D and ME-D for alleviating pre- and post-irradiation damage to seed germinability. Broadly similar results were obtained with X- and γ-irradiation as well as with pre-irradiation and post-irradiation hydration-dehydration treatments (*i.e.*, Non-irrad = 0 kR; Pre-irrad = H-D + 5kR; Post-irrad = 5kR + H-D, where R is X- or γ-irradiation and 0 kR is no X- or γ-irradiation).

The development of biological after-effects (as measured by ratios of before and after storage values for different parameters for germination) was considerably minimized by all the three hydration-dehydration treatments. The after-effect was most pronounced on root growth of seedlings of X- and γ-irradiated seeds that did not receive any hydration-dehydration treatment. The results clearly reveal a broad parallelism between the effects of H-D treatments in the counteraction of age- and irradiation-induced seed deterioration in lettuce confirming similar observation in several other crop seeds studied in the University of Calcutta. The relatively low moisture level and a very brief period available for correction of cellular lessions may not be possible with very short-term dipping-drying treatments. Kesavan *et al.* (1978) favoured a physico-chemical mechanism. Afzal and Kesavan (1979) in interpreting the differential effects of seed moisture content on the development of oxygen-dependent damages, also pointed out that the oxic effect was certainly of a physico-chemical nature.

In the third theme paper on the question of protein synthesis in the hydration phase of ME-D treatment of lettuce seed in relation to viability maintenance it has been demonstrated that *de novo* protein synthesis may not be a major reason of extension of lettuce seed viability by moisture equilibration-drying treatment.

In one experiment the protein synthesis inhibitor cycloheximide was introduced into the seed by the dry permeation technique employing acetone as the solvent using a range of concentrations of the inhibitor. Thereafter moisture equilibration treatment was given to protein synthesis inhibitor pretreated seeds keeping appropriate controls and before ageing as well as post-ageing germination of seeds taken for evaluating the role of protein synthesis in viability maintenance by ME-D. In another experimental set up the lettuce seeds were pretreated with aqueous solutions of the same cytoplasmic protein synthesis inhibitor cycloheximide (different concentrations because of much greater inhibitory effect in the aqueous solutions). After drying back the pretreated seeds were subjected to basic ageing to make them receptive to further moisture equilibration-drying treatment.

Germination test immediately after dry permeation of cycloheximide showed reduction of root length of seedlings but shoot length and germination percentage were unaffected. However, after accelerated and natural ageing the moisture equilibrated and dried seeds showed highly significant beneficial effects under all the different concentrations of the inhibitor. The same was especially magnified when expressed as percentages of the non-moisture equilibrated counterparts. The same was also noted in case of the aqueous permeation-cum-basic ageing set. The results would clearly indicate that the cytoplasmic protein synthesis inhibitor cycloheximide did not reduce the beneficial effect of ME-D in the maintenance of viability and vigour of the thermodormant prechilling requiring variety (Suttons A1) of lettuce. For further elucidation of the role of protein synthesis in viability maintenance see extracts of this paper at the end of Chapter 5.

It is, however, necessary to mention in this connection that cellular repair is certainly important in presowing priming of seeds for improved field performance and productivity as has been discussed in the relevant section of this treatise.

B. Radioprotective Action of Seed Invigoration Treatments

The three theme papers show that chemicals which are acknowledged radioprotectors or sensitizers are analogous in respect of their behaviour as seed invigoration promoters or inhibitors. The same analogy would be seen between radioprotectors and free radical controlling agents; radiosensitizers and free radical stimulators and with enhanced lipid peroxidation as a consequence – no wonder that at the basic molecular level there is considerable overlapping of thematic issues.

The first theme paper on wheat (and also jute) principally focussed on the use of radioprotective chemicals such as cysteine, *p*-aminobenzoic acid, potassium iodide, *etc.*, which would render effective protection against X- or γ- radiation damage when used before and after the physico-chemical seed treatments (pre- and post-irradiation treatments); on the other hand, chemicals such as sodium azide, 2,4-dinitrophenol, caffeine, *etc.*, would accentuate the damage. However, when used in aqueous solutions, the masking effect of hydration could be a problem in the actual expression of radiation protection or sensitization.

The results of experiments on seeds with different chemicals in aqueous solution indicated that hydration alone very effectively slowed down the loss of vigour and viability. Spraying stored seeds with water or dipping in water for a few minutes, followed by usual drying back would serve the purpose. In wheat, chemical effects were a bit more pronounced than in jute; the list of chemicals would include some well known radiation protecting chemicals (many of which are acknowledged free radical controlling agents) such as cysteine, *p*-aminobenzoic acid, potassium iodide, *etc.*, as mentioned before. Commonly known radiosensitizing agents did not show any additional advantage over hydration, in fact the hydration effect on vigour and viability was considerably reduced.

The activities of the enzymes amylase and dehydrogenase which are positively correlated with seed vigour status were significantly increased in wheat; dibasic sodium phosphate, cysteine and *p*-aminobenzoic acid showed higher enzyme activities but the radiosensitizer 2,4 - DNP showed a decline in enzyme activity.

Pre-irradiation physico-chemical treatments of 9-month-old wheat seed (invigoration + 30kR X-ray) showed significantly greater root and shoot length of seedlings compared to the untreated (control) but X-irradiated seeds. However, apart from conventional radioprotectors, the positive effect of sodium chloride (common salt) and disodium phosphate would deserve especial mentioning. Pre-irradiation seed treatment with other chemicals such as sodium thiosulphate and oxalic acid also significantly counteracted the X-irradiation effects on the growth of seedlings.

Post-irradiation seed treatments (30kR X-ray + invigoration treatment) employing 3-month-old (ambient storage) wheat seed also very effectively counteracted the radiation effect but chemical effects were less pronounced and only minor additional protection was noted with sodium chloride, cysteine, and *p*-aminobenzoic acid.

With both pre-and post-irradiation physico-chemical treatments, the radio-sensitizers either accentuated the damage or failed to show any improvement over water. It needs to be re-emphasized that the effect of chemicals, have to be judged after subtracting the water effect since hydration alone (followed by drying) itself has the major radioprotective action.

Pre- and post-irradiation physico-chemical treatments in wheat also showed a vigour oriented response as activities of two enzyme systems, amylase and dehydrogenase were better under conventional radioprotective chemicals as well as with other chemicals such as disodium phosphate. The radiosensitizers, sodium azide and caffeine reduced the activities of the aforesaid enzymes.

A very noticeable parallelism between the effects of accelerated ageing (100 per cent RH and 44°C) of physico-chemically treated wheat seed and post-irradiation response (30kR X-ray + seed treatment) with the same treatment with respect to electrical conductance of seed leachate in the two sets (Table 5.12) would clearly reveal the analogy between ageing response and response to radiation. This is of great interest because a basic function, in deteriorative senescence, the loss of semi-permeability of the cell membrane is involved.

The cytologial studies on chromosomal aberrations as observed during wheat root-tip mitosis following post irradiation physico-chemical treatment (30kR X-ray + seed treatment) clearly demonstrated the significant protection over water alone afforded by chemicals like sodium phosphate (dibasic) and EDTA (Na salt). Radiosensitizers like caffeine curtailed the positive action of water, in fact it virtually erased the radioprotective effect of water soaking-drying (Table 5.13). Regarding radioprotection by water soaking-drying, as early as 1967 Haber and Randolph showed that in irradiated lettuce seeds 92 per cent of the radiation induced Electron Spin Resonance (ESR) signals were eliminated when such seeds were wetted and redried. However, no clear mechanism of that quenching and radioprotection could be given. Regarding chemical effects, cysteine has long been shown to protect cells against damage by $H^{\bullet}$ free radicals. While $OH^{\bullet}$ free radicals

Table 5.12: Effect of Physico-Chemical Treatments on the Membrane Permeability of Wheat Seed Subjected to Accelerated Ageing and X-irradiation (*vide* Dasgupta, Chattopadhyay, Basak and Basu, 1977)

Treatment	*Electrical Conductance of Seed Leachate* (μmho)*	
	Accelerated Ageing (100 per cent RH and 44°C)	*Post-irradiation (30 kR) Treatments*
Control	214	285
Water	154	152
Sodium chloride	154	160
Sodium phosphate (dibasic)	148	152
Sodium thiosulphate	153	157
Oxalic acid	136	153
Cysteine	151	156
p-Aminobenzoic acid	149	154
Potassium iodide	154	153
Sodium azide	166	-
Caffeine	-	167

*Leachate of 5 g seed in 25 ml water for 0.5 h at 25°C.

Table 5.13: Effects of Post-irradiation Physico-Chemical Treatments on the Chromosomal Aberrations in Wheat Seed as Observed during Root-Tip Mitosis (*vide* Dasgupta, Chattopadhyay, Basak and Basu, 1977)

Treatments	*Total No. of Cells Studied*	*Total No. of Breaks Observed*	*No. of Breaks/Cell*	*Modulation Index*
Control	53	618	11.6	1.00
Water	91	576	6.3	0.54
Sodium phosphate (dibasic)	100	414	4.1	0.36
EDTA (Na-salt)	112	621	5.5	0.48
Caffeine	92	1102	11.9	1.03
Non-irradiated control	127	1	0.008	0.0007

Note: Modulation index is the number of breaks per cell in a particular treatment divided by the number of breaks in the control.

are effectively scavenged by chemicals like *p*-aminobenzoic acid and potassium iodide; a number of antioxidants and metal chelating agents have been shown to play a significant role in controlling free radical chain reactions in various systems (Demopoulos, 1973). Depending on the cellular environment and its concentration, sodium chloride has been suggested to be an antioxidant (Nickerson, 1967) and a free radical controlling agent (Heckly and Dimmick, 1967). Regarding the role of phosphates, they could act as antioxidant synergists (Joslyn and Timmons, 1967) and therefore would exert an indirect controlling effect on free radical chain reactions. Such effects as well as the accentuation of radiation effects and loss of viability by radiosensitizers such as caffeine, 2, 4-dinitrophenol and sodium azide would further suggest a parallelism of free radical pathology, radiosensitivity and deteriorative seed senescence and the counteraction of the same by appropriate radioprotective agents; the concept of radioprotective action of seed invigoration treatments, as such is based on a very strong foundation.

As regards the beneficial effects of pre-irradiation soaking-drying treatments, that would very likely quench the naturally produced free radicals thereby rendering the seed less radiosusceptible; the chemicals through as yet unknown mechanisms counter the free radical generation and chain propagation and reduce lipid peroxidation and as a consequence, would help to retain seed vigour and viability. Our observations would further indicate that a time gap between physico-chemical treatment and irradiation in pre-irradiation (seed treatment + irradiation) as well as post-irradiation (irradiation + seed treatment) treatments would increase the damage as the ageing damage and irradiation damage would be additive (compounding of biological after-effects of irradiation) (Table 5.14).

Table 5.14: Development of Biological after-effects[(a)] of γ-irradiation on Germinability of Rice Seed (*vide* Sur and Basu, 1986)

Vigour Status		*No Irradiation (NR)*	*Pre-irradiation (H-D + R)*	*Post-irradiation (R + H-D)*
HV	Control	10	12	13
	SD	19	24	04
	ME-D	01	02	06
MV	Control	14	17	19
	SD	06	06	07
	ME-D	08	11	13
LV	Control	18	25	24
	SD	12	15	18
	ME-D	13	18	20

(a) Biological after-effect was estimated by taking the difference between germination percentage before ageing (x) and germination percentage after ageing (y) and expressing the difference as the percentage of germinability before ageing *i.e.* $(x - y) \div x \times 100$ = biological after effect. The extent of biological after-effect would depend on the vigour status of the seed at the time of treatment, the efficacy of the particular seed treatment on the vigour status as well as whether irradiation is before or after the physico-chemical seed treatment.

In the second theme paper on lettuce, soaking-drying treatment was given to stored (one-year-old) lettuce seed cv. Sutton's A1 (a wet pre-germination chilling requiring cultivar which otherwise would not germinate at ordinary room temperature of 25-30°C) and can be kept fully viable but dormant in an imbibed state at around 25°C for a very long period. The soaking-drying treatment (with or without a range of chemicals in aqueous solutions, 10^{-5},10^{-4}M) very effectively maintained seed viability after accelerated ageing at 100 per cent RH and 40°C for 7 days after treatment. However, as with many other crop seeds, there was no additional advantage with any of the chemicals that included sodium phosphate, sodium thiosulphate, oxalic acid, *p*-hydroxybenzoic acid, *etc.* In fact, the very large effect of hydration (water-soaking) alone masked the chemical effects. In order to confirm the promoting effects of chemicals by separating them from the hydration effect, some of the chemicals were incorporated into the seed by the dry permeation technique using acetone as the solvent. The results so obtained very clearly brought out the beneficial effects of acetone soluble chemicals like *p*-aminobenzoic acid, tannic acid, gallic acid and potassium iodide. Caffeine, a radiosensitizer did not check the deterioration of lettuce seeds (Table 5.15). It has been of great interest to record in this connection a vast improvement in correlations between germination and total seedling length when the latter was based on all seeds placed for germination test instead of that based on the number of germinated seeds only (TSL I). TSL II is, in fact, equal to TSL I × GF where GF is the germination factor equal to germination per cent divided by 100 (see Chapter 6 on quantification of seed vigour for further details).

The results with acetone soluble radioprotecting and sensitizing chemicals on seed deterioration in lettuce would confirm the findings of Dasgupta *et al.* (1977) in wheat and strongly suggest the involvement and overlapping of functioning of radioprotective free radical controlling and lipid peroxidation preventive agents in seed invigoration.

In the third theme paper on physico-chemical control of radiation and ageing damage in mustard seed (Dey and Basu, 1985), pre- and post-irradiation physico-chemical treatments, besides water soaking-drying, were done on high vigour (HV), high medium vigour (HMV) and low medium vigour (LMV) mustard seeds, through dry permeation of a number of acetone soluble chemicals such as potassium iodide, *p*-aminobenzoic acid, hydroquinone and caffeine. The first three, which are acknowledged radioprotectors, exhibited better germinability on accelerated ageing than the solvent control only. But caffeine, the radiosensitizer, increased radiation damage. Pre-and post-irradiation seed treatment showed differential behaviour in HV, HMV and LMV seeds. It would be worthwhile to mention here that the high vigour mustard seed was less susceptible to irradiation and less responsive to pre-irradiation treatment. Except in HV non-irradiated seeds water soaking-drying showed better germinability in all cases; dry permeation via acetone of potassium iodide, *p*-aminobenzoic acid and hydroquinone very effectively counteracted the adverse effects of ageing and irradiation. The beneficial effects of these chemicals were particularly prominent in the post-irradiation treatments of HMV and pre-irradiation set of LMV seeds. The seedling length also showed similar effects of the physico-chemical treatments; soaking-drying as well as dry permeation of the

Table 5.15: Effect of Chemicals, Introduced into the Seed by the Dry Permeation Technique using Acetone, on the Germinability of Lettuce Seeds Subjected to Accelerated Ageing at 100 per cent RH and 40°C for 7 Days after Treatment (*vide* Basu, Pan and Punjabi, 1979)

Treatment[(a)]	*Germination*		*Mean Root Length (mm)*	*Mean Shoot Length (mm)*	*Total Seedling Length (mm)*	
	Per cent	*Arc-sin Value*			*TSL I*	*TSL II*
Control 1[(b)] (untreated)	42	40.4	32	32	64	27
Control 2[(c)]	37	37.5	34	27	61	23
p-Aminobenzoic acid	60	50.8	33	25	58	35
Tannic acid	63	52.5	36	27	63	40
p-Hydroxybenxoic acid	64	53.1	33	28	61	39
Gallic acid	56	48.4	32	27	59	33
Potassium iodide	70	56.8	34	29	63	44
Caffeine	37	37.5	32	30	62	23
LSD at *0.05 P*	-	6.8	NS	NS	NS	6.1

(a) Treatments were given to 18-month-old seeds; chemicals were introduced through acetone; concentration of chemicals, *p*-aminobenzoic acid, potassium iodide and caffeine, 10^{-4}M; others, 10^{-5}M. (b) and (c) Control 1 was the untreated control and Control 2 was soaked in acetone solvent only (without chemical); both the controls were aged at 100 per cent RH and 40°C along with the other treatments. NS - not significant; *New correlations involving germination and total seedling length (TSL I and TSL II) are included in the table for further clarification. Correlations involving germination and total seedling length (TSL): TSL I is based on germinated seeds only, TSL II is based on all seeds placed for the laboratory germination test.

Correlations Involving the Variates	*Correlation Coefficient (r)*	*Coefficient of Determination (r^2)*
Germination (per cent), TSL I	-0.1262	0.0159
Germination (per cent), TSL II	0.9892*	0.9784
Germination (arc-sin), TSL I	-0.1209	0.0146
Germination (arc-sin), TSL II	0.9899*	0.9799

Note: Little difference in correlations due to germination per cent values and their arc-sin conversions; correlations vastly improved in TSL II *vis-à-vis* TSL I.

radio-protective chemicals exhibited better seedling growth than control. Besides water soaking-drying, the dry permeated beneficial physico-chemical treatments potassium iodide, *p*-aminobenzoic acid and hydroquinone effectively minimized the leakage of electrolytes from seeds but caffeine showed an increase in leaching with adverse effect on germinability (Figure 5.3).

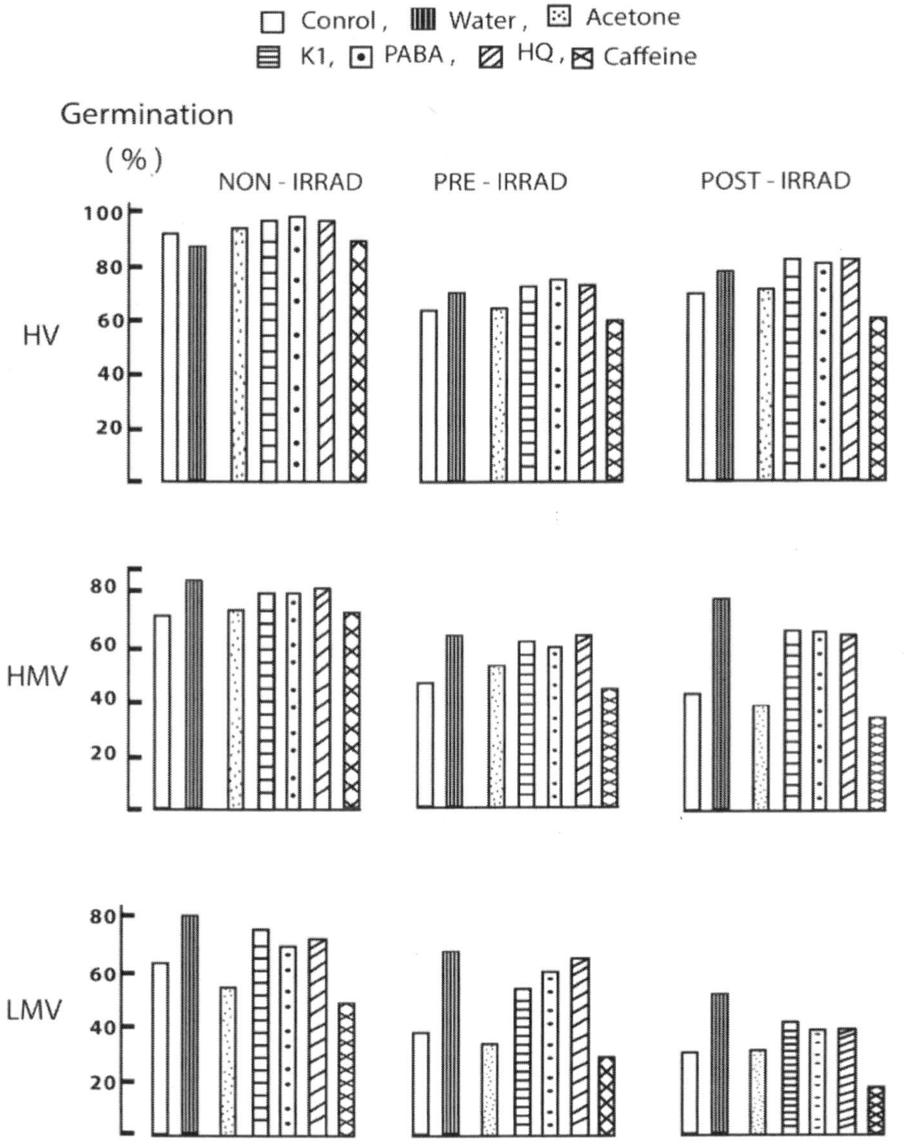

Figure 5.3: Germinability of High Vigour (HV, 2- month-old) High-Medium Vigour (HMV, 14-month-old) and Low Medium Vigour (LMV, 26-month-old) Mustard Seeds as Influenced by Seed Treatment and γ-irradiation and 6 Months of Natural Ageing after Treatment. (*Vide* Dey (Pathak) and Basu, 1985)

Lipid peroxide formation was closely related to the loss of germinability and all physico-chemical treatments, including water soaking-drying and the three radioprotective chemicals were characterized by substantially reducing lipid peroxidation. In keeping with the low germinability of caffeine treated seeds, lipid peroxidation was higher in seeds dry permeated with caffeine (Nadkarni and Kesavan, 1975). The extent of radio-sensitization by caffeine would depend on its concentration; accentuation of radiation damage may often be modified.

The results of this study on mustard would also confirm the magnification of irradiation effect upon storage (ageing). Immediately after irradiation, germination percentage was not much reduced and it was seedling growth, that was primarily affected. The development of biological after-effects (attributable to a substantial extent to the indirect and latent effect of radiation) would continue when the seed is stored in the dry state. In germination test before ageing, hydration would check much of the latent effect of radiation but storage and ageing (accelerated as well as natural for different periods, the latter depending on ambient conditions) of irradiated dry seeds would allow the latent effects to show up in post-ageing germination tests. A considerable part of the irradiation damage has long been proposed to be due to generation of free radicals and their chain reactions. Seed hydration would cut off free radical chain propagation reactions (Grenberg, 1961, Haber and Randolph, 1967) and would reduce cellular injury. Thus apart from enzymatic repair, before ageing germination test (requiring hydration) as well as post-irradiation soaking-drying would reduce the harmful effects of irradiation by quenching of free radicals. Such a quenching would explain the post-irradiation recovery of seeds with even small rises in seed moisture content (Wolff and Sicard, 1961, Joshi, Gaur and Notani, 1969).

The significant effect of the free radical controlling agents cum radioprotectors, supplied through a non-aqueous medium like acetone, would lend strong support to a physico-chemical mechanism of ageing and irradiation damage and its control by radioprotective agents (Kesavan, 1975, Kesavan *et al.*, 1978). It was not possible for us to determine the free radical levels in the seed but we could study lipid peroxidation which in the dry seed was most likely coupled with free radical reactions (see Chapter 3, mode of action of mid- and prestorage treatments). Lipid peroxidation was considerably lower in the dry-permeated beneficial physico-chemical treatments; dry permeation with caffeine on the other hand showed a substantial increase in lipid peroxidation (Figure 5.4). Buchvarov and Gantcheff (1984) clearly showed the involvement of free radicals in seed ageing through direct experimental verification of higher free radical levels in embryonic axes of soybean. Though Priestley and Leopold (1983) did not assay free radicals, they also showed greater lipid peroxidation in naturally aged soybean.

Our results with different crop seeds lend strong support to the concept that the damaging effects of ageing and irradiation include, besides other effects, free radical induced lipid peroxidation that can be effectively reduced by appropriate seed invigoration treatments which operate mainly through a physico-chemical

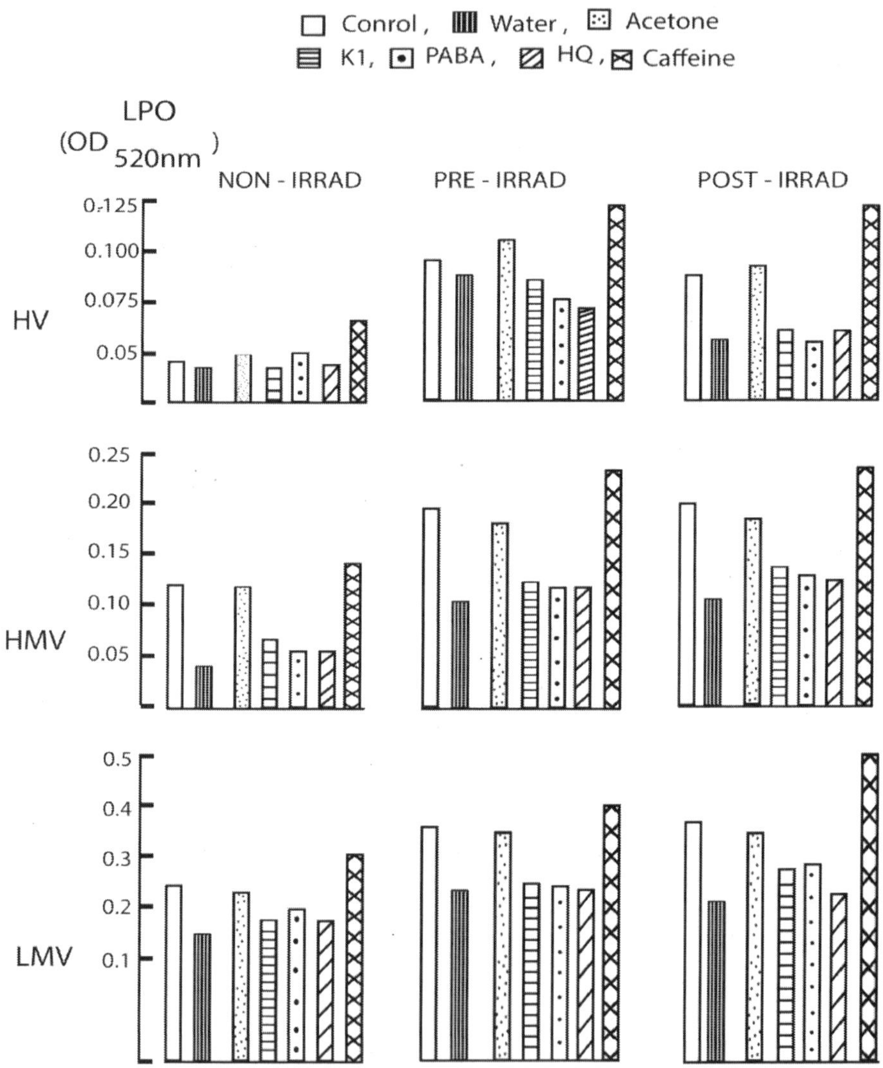

Figure 5.4: Lipid Peroxidation in Mustard Seeds of different Treatments as Influenced by Vigour Status, γ-irradiation and Seed Treatments; estimations were made after 6 Months of Natural Ageing following Treatment and Irradiation. (*Vide* Dey (Pathak) and Basu, 1985).

mechanism. If lipid peroxidation is allowed to go unchecked the loss of selective permeability of the embryonic cell membranes would at the end result in loss of seed vigour and viability.

C. Free Radical Controlling Agents in Seed Invigoration

As already mentioned in the thematic issues discussed above, many of the radioprotective agents are free radical controlling agents but radiation damage is not solely attributable to free radical generation; free radical induced damage would be part of the total irradiation effect of which a considerable share would be more direct. In any case as mentioned earlier, an overlapping of themes involving irradiation, free radical generation and propagation, their manifestation in lipid peroxidation, volatile aldehyde generation in deteriorative senescence of seed cannot simply be avoided as they are interconnected and often sequential. Only for the sake of convenience they are taken up separately for discussion but with repetition of similar ideas as a consequence.

The findings of the first theme paper on free radical controlling agents in relation to seed invigoration of wheat would be very close to that discussed earlier for radio-protective agents (Dasgupta *et al.*, 1977) as the chemicals were almost the same, only the age of the seed and ageing conditions were different with a few other differences in procedural aspects. Data presented (Tables 5.16 and 5.17) for germinability of 6-month-old wheat seed treated with a range of free radical controlling agents showed little effect soon after treatment but after accelerated ageing for 8 days at 100 per cent RH and 44°C germination fell to 67 per cent in control against 87 per cent in water-soaking-drying (water S-D) with highly significant (P=0.01) increments in root and shoot length in the latter (water S-D) confirming the very well acknowledged role of hydration in free radical quenching. But with the other free radical controlling agents (syn. radioprotective chemicals) the effect of hydration with water alone was significantly exceeded not only by the conventional free radical controlling agents such as *p*-aminobenzoic acid, cysteine, potassium iodide but also by disodium phosphate, sodium thiosulphate, sodium chloride, *etc.* and seedling length (root length in particular) as a measure of vigour was always higher in the beneficial treatments. Sodium azide and caffeine in their aqueous solutions exceeded control values, because of hydration effect, but always reduced the effect of hydration by water alone (difference with water S-D in case of germination was over 10 per cent, same with seedling vigour). Pre- and post-irradiation physico-chemical treatment effects were also broadly similar to those obtained by Dasgupta *et al.* (1977) confirming the synonymous nature of free radical controlling/counteracting agents and radioprotective substances (Table 5.17).

All radioprotective agents in the second theme paper by Dey (Pathak) and Basu (1985) as discussed in item C above are acknowledged free radical controlling agents. While in aqueous solution the free radical controlling agents often failed to exceed the effects of hydration alone (water alone being a very effective quencher of free radicals), in nonaqueous medium (organic solvent infusion *via* acetone)

Table 5.16: Effect of different Physico-chemical Treatments on the Germinability of 6-Month-Old Wheat Seeds before and after Accelerated Ageing (*vide* Basu and Dasgupta, 1978)

Treatment	*Before Ageing*			*After Accelerated Ageing*		
	Germination (Per cent)	*Mean Root Length (mm)*	*Mean Shoot Length (mm)*	*Germination (Per cent)*	*Mean Root Length (mm)*	*Mean Shoot Length (mm)*
Control	94	218	31	67	131	27
Water	94	233	32	87	178	32
Sodium chloride	98	232	33	89	197	35
Disodium phosphate	95	237	32	92	201	39
Sodium thiosulphate	95	234	31	89	194	36
Oxalic acid	93	227	31	90	196	34
Cysteine	95	243	32	92	199	36
p-Aminobenzoic acid	95	229	32	93	203	38
Potassium iodide	96	232	33	91	197	37
Sodium azide	94	227	31	74	162	29
Caffeine	93	216	29	75	159	30
LSD at *0.05 P*	-	11.0	3.1	-	8.2	2.2
LSD at *0.01 P*	-	15.7	NS	-	11.6	3.2

Accelerated ageing was done for 8 days at 100 per cent RH and 44°C. Concentrations of chemicals, disodium phosphate and sodium thiosulphate, 10^{-4}M; others 10^{-3}M. Data on per cent germination and root and shoot length were recorded after 96 h of germination.

NS – Not significant.

Table 5.17: Effect of Pre-irradiation and Post-irradiation Physico-chemical Treatments on the Germinability of 7-month-old Wheat Seed (*vide* Basu and Dasgupta, 1978)

Treatments	*Pre-Irradiation Set H-D Treatment + 30 kR X-ray*			*Post-Irradiation Set (30 kR X-ray + H-D treatment)*		
	Germination (Per cent)	*Mean Root Length (mm)*	*Mean Shoot Length (mm)*	*Germination (Per cent)*	*Mean Root Length (mm)*	*Mean Shoot Length (mm)*
Control	78	61	21	79	60	20
Water	93	80	25	93	98	29
Sodium chloride	95	89	28	94	120	32
Disodium phosphate	94	94	28	95	123	31

Treatments	*Pre-Irradiation Set H-D Treatment + 30 kR X-ray*			*Post-Irradiation Set (30 kR X-ray + H-D treatment)*		
	Germination (Per cent)	*Mean Root Length (mm)*	*Mean Shoot Length (mm)*	*Germination (Per cent)*	*Mean Root Length (mm)*	*Mean Shoot Length (mm)*
Sodium thiosulphate	94	90	28	93	108	30
Cysteine	94	94	29	94	120	32
p-Aminobenzoic acid	94	95	28	96	111	32
Potassium iodide	94	90	27	95	109	29
Sodium azide	89	68	23	95	89	29
Caffeine	91	67	21	91	78	25
LSD at *0.05 P*	-	7.9	3.5	-	8.7	4.5
LSD at *0.01 P*	-	10.9	4.8	-	12.0	6.2

Physico-chemical treatments were given in pre-irradiation set before subjecting seeds to X-irradiation of 30 kR, and in post-irradiation set immediately after X-irradiation of 30 kR. Germination test was done 10 days after irradiation treatment. Other details same as in Table 5.16.

chemicals like cysteine, *p*-aminobenzoic acid, potassium iodide, *etc.* acted as very effective free radical controlling agents for radioprotection and maintenance of vigour and viability of mustard seed most likely through a significant reduction in lipid peroxidation reactions (free radical mediated) and reducing damage to vital cell membranes as discussed below.

D. Control of Lipid Peroxidation by Seed Invigoration

In the first theme paper on lipid peroxidation and membrane damage in wheat and mustard seeds (Rudrapal and Basu, 1982), a significant negative correlation between germinability and electrical conductance of seed steep water was noted in wheat and mustard under all ageing conditions (Table 5.18). However, despite the broad similarity in the correlations all negative correlations are significant at 0.05 P level. The regression lines and the slopes of the lines were influenced to a certain extent by the kind of seed (wheat or mustard), and ageing conditions determining the rate of loss of germinability.

The significant positive correlation between lipid peroxidation and electrical conductivity of seed leachate would indicate the dependence of membrane permeability on lipid peroxidation. In mustard, besides other ageing conditions, that under a very high heat stress at 100°C for 15 min to 1 h was of particular interest (Figures 5.5 and 5.6).

Table 5.18: Relationship between Lipid Peroxidation, Electrical Conductance and Germination of Wheat and Mustard Seeds under different Ageing Conditions (*vide* Rudrapal and Basu, 1982)

Variables	*Materials*	*Correlation Coefficients Ageing Conditions*			
		A	*B*	*C*	*D*
Lipid peroxidation and germination	Wheat	-0.955*	-0.940*	-0.940*	-
	Mustard	-0.855*	-0.912*	-0.936*	-0.995*
Electrical conductance and germination	Wheat	-0.864*	-0.788*	-0.914*	-
	Mustard	-0.787*	-0.728*	-0.706*	-0.582*
Lipid peroxidation and electrical conductance	Wheat	+0.678*	+0.842*	+0.848*	-
	Mustard	+0.901*	+0.812*	+0.735*	+0.839*

*$P < 0.05$.

Ageing conditions: A, 100 per cent RH and 42°C for 3-12 days for wheat and 7-30 days for mustard; B, 33 per cent RH and 44°C for 3-9 months for wheat and 4-12 months for mustard; C, Natural ageing under ambient conditions (2-10 months for wheat and 3-15 months in case of mustard); D, Heat stress (100°C for 15 min to 1 h).

Hydration-dehydration treatment of stored wheat seed by soaking drying (S-D) dipping-drying (D-D) and moisture equilibration-drying (ME-D) very effectively controlled the loss of vigour and viability of seeds. The greater post-ageing germinability of the treated seeds was associated with better membrane integrity as is evident from reduced electrical conductance of seed-steep water and lesser leakage of sugars and amino-acids from the seed and considerably lower lipid peroxidation in wheat embryos as well as in whole seeds of wheat (Table 5.19). As regards leaching or leakage of electrolytes and metabolites from the seed it needs to be pointed out here that during soaking and dipping (*i.e.*, soaking-drying and dipping-drying treatments) there could be some initial leakage of electrolytes and metabolites from the seed. However, a comparison of the pre- and post-ageing levels of leaching in such treatments would reveal a considerably better membrane integrity in those treatments. The ME-D treatments of wheat and mustard seeds (Tables 5.19 and 5.20) in which there was no initial leakage of electrolytes and metabolites, would further bear out such a contention (Figure 5.7).

The close relationship between loss of germinability and lipid peroxidation in wheat could be noticed in the moisture equilibration-drying treatment employing different durations of moisture equilibration prior to drying back. The progressive improvements in storability was accompanied by a near equivalent reduction in lipid peroxidation (Figure 5.8).

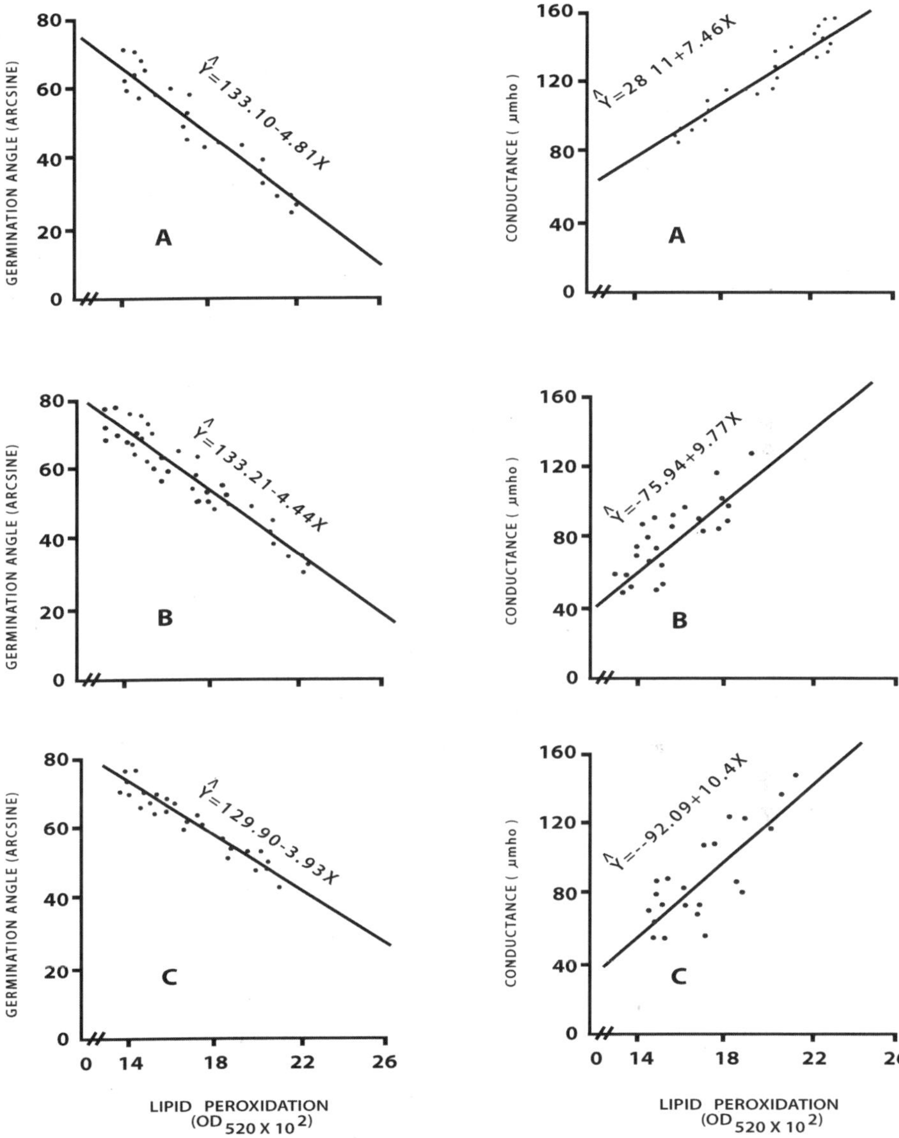

Figure 5.5: Regression lines for lipid peroxidation and germination (Left column) and, lipid peroxidation and electrical conductance (Right column) of wheat seed after ageing under different conditions: A, accelerated ageing at 100 per cent RH and 42°C for 3-12 days; B, ageing at 33 per cent RH and 44°C for 3-9 months; and C, natural ageing under ambient conditions for 2-10 months. (*Vide* Rudrapal and Basu, 1982)

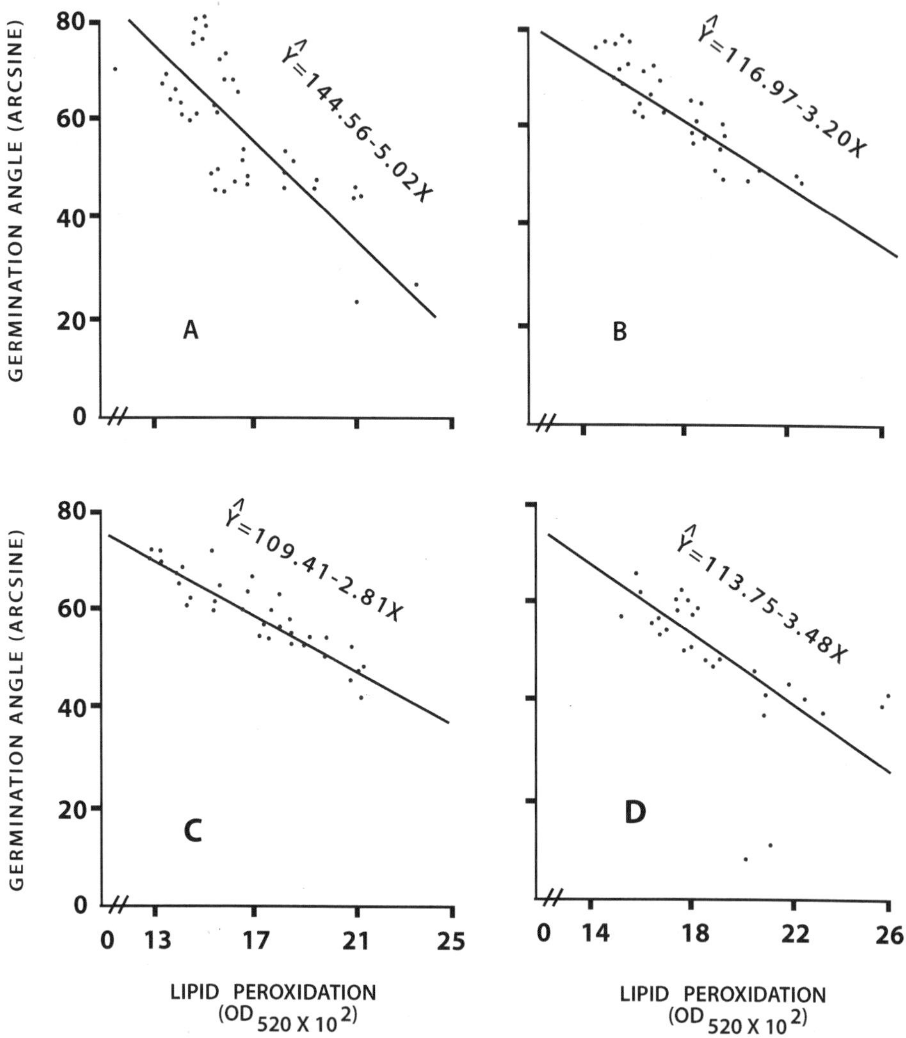

Figure 5.6: Regression lines for lipid peroxidation and germination of mustard seed under different ageing conditions: A, accelerated ageing at 100 per cent RH and 42°C for 7-30 days; B, ageing at 33 per cent RH and 44°C for 4-12 months; C, natural ageing under ambient conditions for 3-15 months; and D, heated at 100°C for 15 min. to 1 hr. (*Vide* Rudrapal and Basu, 1982)

Table 5.19: Effect of Hydration-Dehydration Treatments on Germinability and Membrane Functions of Wheat Seeds and Lipid Peroxidation in Embryo and Whole Seed Before (A) and After Accelerated Ageing (B) (*vide* Rudrapal and Basu, 1982)

Treatments	*Germination per cent*		*Electrical Conductance of Leachate after 1 h (µmho)*		*Leakage of Sugar (µg glucose equiv./ml)*		*Leakage of Amino Acid (µg glycine equiv./ml)*		*Lipid Peroxidation (OD at 520 nm)*			
									In Embryo		*In Whole Seed*	
	A	*B*	*A*	*B*	*A*	*B*	*A*	*B*	*A*	*B*	*A*	*B*
Control	93.1	42.3 (-50.8)	72	119 (47)	120	300 (180)	13.5	33.5 (20.0)	0.050	0.127 (.077)	0.145	0.223 (.078)
Dipping-drying	94.0	73.1 (-20.9)	56	81 (25)	86	147 (61)	10.1	17.3 (7.2)	0.050	0.075 (.025)	0.140	0.172 (.032)
Soaking-drying	95.4	81.4 (-14.0)	47	67 (20)	70	107 (37)	6.8	11.2 (4.4)	0.050	0.062 (.012)	0.138	0.155 (.017)
Moisture equilibration drying	95.1	76.0 (-19.1)	70	89 (19)	99	168 (69)	12	20.0 (8.0)	0.048	0.072 (.024)	0.142	0.160 (.018)

(A) Measurements were made immediately after treatment, before subjecting to ageing. (B) Estimations were done after ageing at 100 per cent RH and 42°C for 7 days. Electrical conductance readings and leaching of metabolites (sugars and amino acids) were made on 25 ml (5 g seed) after soaking for 1 h and 4 h respectively. For lipid peroxidation measurements, 100 embryos and 500 mg dry whole seed powder were taken, absorbance was recorded at 520 nm. Values in parenthesis represent the differences in various parameters between A and B for the respective treatments.

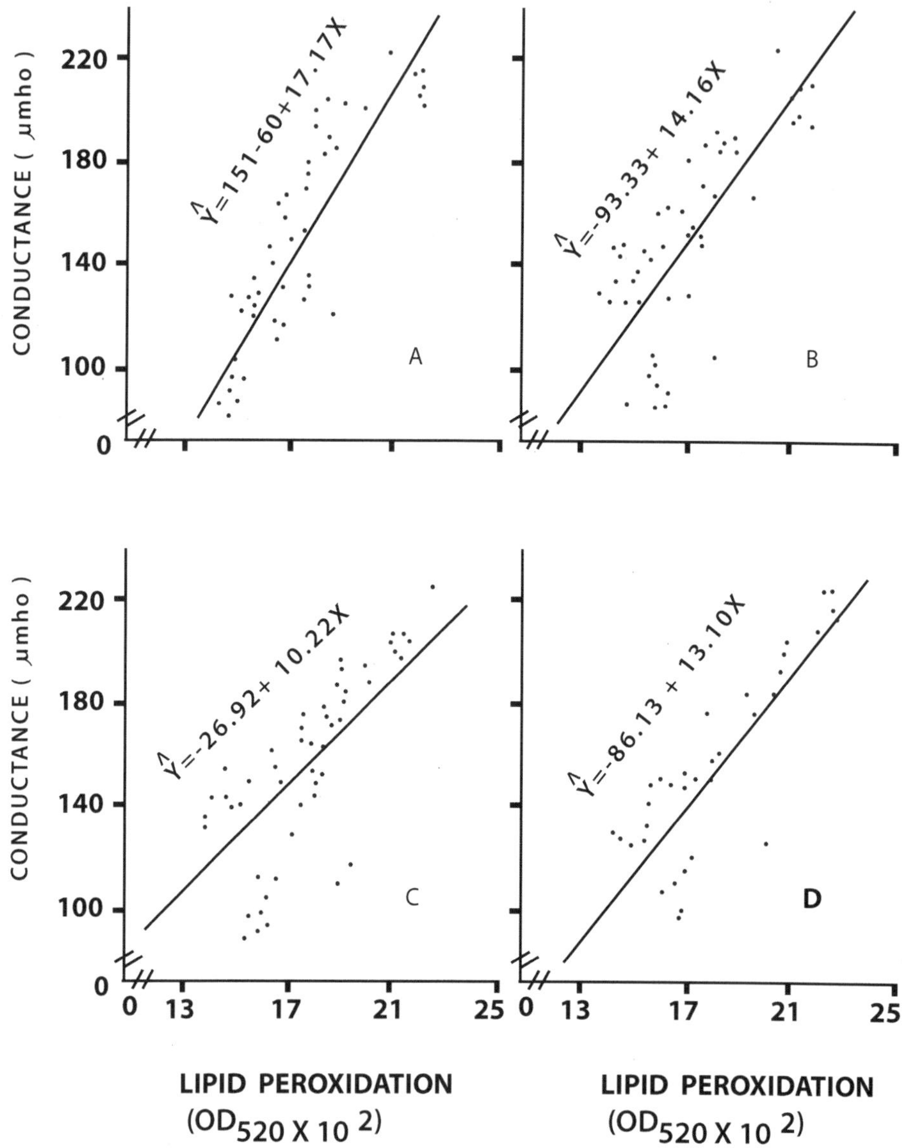

Figure 5.7: Regression lines for lipid peroxidation of and electrical conductance of seed-steep water of mustard seeds under different ageing conditions; other details same as in figure 5.2. (*Vide* Rudrapal and Basu, 1982)

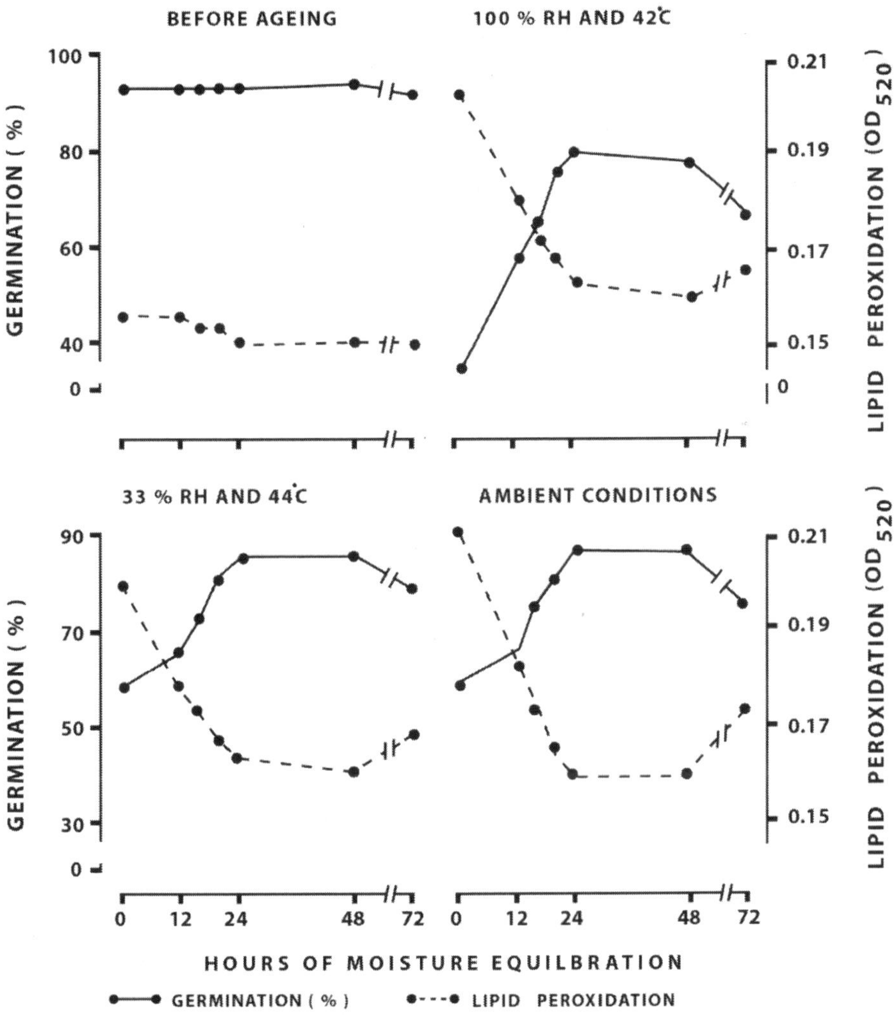

Figure 5.8: Effect of moisture equilibration for different durations followed by drying on germinability and lipid peroxidation of wheat seeds under various ageing regimes. (*Vide* Rudrapal and Basu, 1982).

Table 5.20: Effect of Durations of Moisture Equilibration followed by Drying on Germinability, Membrane Permeability and Lipid Peroxidation of Mustard Seeds Subjected to Accelerated Ageing (100 per cent RH and 42°C) for 19 days* (*vide* Rudrapal and Basu, 1982)

Hours of Moisture Equilibration (followed by drying)	*Germination (Per cent)*	*Electrical Conductance of Leachate (μmho)*	*Leakage of Sugars (μg glucose equiv./ml)*	*Leakage of Amino Acids (μg glycine equiv./ml)*	*Lipid Peroxidation (OD at 520 nm)*
0	46.7	203	335	29.0	0.220
8	52.0	190	321	27.3	0.208
12	59.7	179	293	25.0	0.200
24	68.5	150	250	22.0	0.180
48	75.3	138	239	18.7	0.170
96	65.0	135	245	20.3	0.176

*Electrical conductance and leaching of metabolites were determined after soaking seeds in distilled water (500 mg seed/25 ml) at 27°C for 1 h and 4 h respectively. For lipid peroxidation studies, 250 mg dry whole seed powder were taken; other details same as in Table 5.19.

Moisture equilibration for different durations (0-96h) followed by drying of mustard seeds also very clearly implicated the post-ageing loss of vigour and viability with corresponding loss of membrane integrity (Table 5.20). The highly significant positive correlation between electrical conductance of seed leachates and lipid peroxidation (the same between leaching of sugars or amino acids and lipid peroxidation) would suggest that the damage to the semipermeable nature of the cell membranes was caused by oxidation of lipid constituents. It would be most obvious that the unsaturated fatty acid moieties of lipoprotein membranes of the cells of embryonic tissues were the primary sites of peroxidative degradation leading to the loss of vigour and viability.

There was a minor difference in the relative response of wheat and mustard seeds in respect of loss of germinability and lipid peroxidation. In wheat, significant increase in lipid peroxidation was noted only when there was an actual fall of germination percentage, but in mustard, even differences in vigour (germination per cent unaffected but rate of germination and growth of seedlings affected) were clearly reflected in lipid peroxidation.

The viability maintaining property of the seed invigoration treatments was always associated with a significantly consistent lowering of lipid peroxidation. The free radical chain reactions in the stored seed were possibly terminated through quenching or scavenging mechanisms as shown in a good number of radiobiological experiments.

As mentioned earlier under radioprotective agents, Haber and Randolph (1967) demonstrated that 92 per cent of radiation induced electron spin resonance

(ESR) signals of γ-irradiated lettuce seeds were eliminated by wetting and drying of the seeds (8 per cent of the signals were resistant to such hydration-dehydration treatment). Thus, if free radical reactions were involved in seed ageing, other such protective agents, antioxidants and membrane stabilizing drugs, *etc.*, would also be effective in controlling the loss of seed vigour and viability and that possibility has been successfully confirmed in our studies. The concentration dependent nature of such antioxidant and related agents and their entry to the site of action are important issues as depending on concentration an antioxidant may act as anti- or pro-oxidant; some are toxic even at low concentrations and often interact with the endogenous controlling agents. H-D, however, is very effective and safe but except ME-D, and moist sand conditioning-drying, MSC-D, others particularly soaking-drying would not be effective in high vigour seed. It is also suggestive of the fact that harvest-fresh seed may have lost some vigour since attaining physiological maturity (PM) while still on the mother plant. However, threshing, cleaning, drying, *etc.*, may be responsible for some loss of vigour in harvest-fresh seeds also.

The TBA reaction (thiobarbituric acid as the reagent) for assaying lipid peroxidation, is not very specific as the same is influenced by several compounds. But in our studies, involving large number of samples it worked effectively and gave a broad and consistent information on the maintenance of membrane integrity through reduced lipid peroxidation by seed invigoration treatments.

The second theme paper on iodination of mustard seed has already been covered under free radical controlling agents and here we would just highlight the lipid peroxidation controlling effect of the halogen.

The effect of iodine treatment (vapour phase) for 0 (control) 8, 16, 24 and 48 h on germination and seedling growth of mustard was studied before ageing and after accelerated ageing at 100 per cent RH and 40°C for 13 days, ageing at 30 per cent RH and 45°C for 3 months, natural ageing under ambient conditions (average RH 60.3 per cent, temp. 29.9°C) for 3 months, and also after subjecting to a heat stress of 100°C for 30 minutes. Following seed ageing and germination test, seed samples were analysed for lipid peroxidation, membrane functions (electrical conductivity of seed-steep water and leaching of sugars and amino acids) and dehydrogenase and amylolytic enzyme activity, *etc.* We are giving below in tabular form the correlations involving lipid peroxidation, seed germination, seedling growth, and vigour parameters such as electrical conductivity of seed steep water. Also correlations between lipid peroxidation (x) and electrical conductivity (y) are given in Table 5.21.

Further, a new approach especially those between LPRS (lipid peroxidation resistance score) and ECRS (electrical conductivity resistance score) and LPRS × GF and ECRS × GF (calculation based on a concept developed in 2004) is provided in Chapter 6 on quantification of seed vigour.

The correlation between lipid peroxidation and germination was statistically not significant before ageing but after ageing under different conditions all the correlations (negative correlations) were significant at 0.05 P; that under ageing under high humidity and high temperature was also significant at 0.01P.

Table 5.21: Iodine Treatment of Mustard Seed for Invigoration Correlations between germination and lipid peroxidation and electrical conductance, and between lipid peroxidation and electrical conductivity (*vide* Basu and Rudrapal, 1980)

Hours of Iodine Treatment	*Germination*		*Lipid Peroxidation (OD at 620 nm)*	*Electrical Conductivity (μmho)*
	Per cent	*Arc-sin*		
0 (Control)	64	53.13	0.167	162
8	88	69.73	0.157	132
16	90	71.56	0.137	130
24	96	78.46	0.137	126
48	85	67.21	0.155	138
Mean	84.6	68.018	0.1506	137.6

Correlations between the Variates	*Correlation Coefficient (r)*	*Coefficient of Determination (r^2)*
Germination, Lipid peroxidation	-0.8670	0.7517
(arc-sin) (OD)	(-0.8414)	
Germination, Electrical conductivity	-0.9779	0.9562
(arc-sin) (μmho)	(-0.9950)	
Lipid peroxidation, Electrical conductivity	0.8287	0.6868
(OD) (μmho)		

Note: The correlation coefficient values within brackets denote the 'r' values obtained with germination percentages before their arc-sin transformation; also see the revised correlations obtained with the integration of germination factor (GF) in Table 5.22.

The extent of correlation between lipid peroxidation (*x*) and seedling growth (y) after ageing was still higher and in all cases the negative correlations were significant at 0.01 P with coefficient of determination turned in percentage values ranging from 85.9 per cent to 92.1 per cent.

All the correlations between lipid peroxidation (x) and electrical conductance were positive and significant with that under natural ageing reaching an 'r' value of 0.999 (with 99.8 per cent coverage of $r^2 \times 100$).

Very interestingly, based on a different approach (see Chapter 6, Quantification of seed vigour) we converted only one set, that between lipid peroxidation versus electrical conductance under 30 per cent RH and 45°C ageing for 3 months to lipid peroxidation resistance score (LPRS) versus electrical conductivity resistance score

(ECRS), and calculated their correlations without or with germination factor (GF) incorporation (LPRS × GF versus ECRS × GF).

Table 5.22: Iodine Treatment of Mustard Seed for Invigoration Revised correlations involving germination, lipid peroxidation and electrical conductivity (*vide* Basu and Rudrapal, 1980)

Hours of Iodine Treatment	*Germination Per cent*	*GF*	*LPRS Per cent*	*ECRS Per cent*	*LPRS Per cent × GF*	*ECRS Per cent × GF*
0	64	0.64	83.3	83.8	53.3	53.6
8	88	0.88	84.3	86.8	74.2	74.4
16	90	0.90	86.3	87.0	77.7	76.3
24	96	0.96	86.3	87.4	82.8	83.9
48	85	0.85	84.5	86.2	71.8	73.3
Mean	84.6		84.94	86.24	71.96	72.30

Revised Correlations between Variates	*Correlation Coefficient (r)*	*Coefficient of Determination (r^2)*
Germination per cent, LPRS per cent	0.8414	0.7080
Germination per cent, ECRS per cent	0.9950	0.9901
LPRS per cent, ECRS per cent	0.8287	0.6868
Germination per cent, LPRS per cent × GF	0.9981	0.9963
Germination per cent, ECRS per cent × GF	0.9963	0.9927
LPRS per cent × GF, ECRS per cent × GF	0.9983	0.9966

Note: All correlations have been transformed to positive values, GF incorporation has greatly improved the correlations to very high 'r' values (almost reaching the highest values in some instances), see Chapter 6 for quantification and more details.

The data presented in the Table 5.22 would categorically support the theory that lipid peroxidation is a basic reason for cellular membrane damage and loss of semi-permeability affecting vital cellular functions of the seed during germination and seedling growth and the seed invigoration treatments like halogenation, would resist that enabling improved performance in the field.

E. Control of Volatile Aldehyde Production by Seed Invigoration

The three theme papers on the issue have been taken up together, two of them deal with rice and wheat and the third one on two legumes, soybean and French bean. As volatile aldehyde production based on optical density (OD) measurements during the deteriorative senescence of the seed is an extension of lipid peroxidation process, there would obviously be a broad similarity in respective seed vigour related responses.

Table 5.23: Germinability of Rice Seeds (cv. Ratna) of different Vigour Levels and Volatile Aldehyde Production during the First (0-24h), Second (24-48h) and Third (48-72h) Days of Germination (*vide* Pal and Basu, 1989)

Vigour Levels of Rice Seed (Obtained by accelerated ageing of harvest-fresh seed)	*Germination after 72h*		*Mean Seedling Length (mm) after 72h*	*Volatile Aldehyde Production during 0-72h (OD x 10^3) 635 nm*		
	Per cent	*Arc-sin Value*		*0-24h*	*24-48h*	*48-72h*
High vigour (HV)	100	90.0	70 (26.4)	0	39	15
High -medium vigour (HMV)	92	73.6	48 (40.1)	0	50	20
Medium vigour (MV)	84	66.4	39 (53.8)	0	63	31
Low-medium vigour (LMV)	82	64.9	29 (67.5)	0	68	38
LSD at *0.05 P*	-	5.1	8.6	-	5.4	4.8
LSD at *0.01 P*	-	7.5	13.0	-	8.1	7.2

Correlations involving germination and seedling length with volatile aldehyde production between 24-48 h and 48-72 h.

Pairs of Variates (n=4)	*Correlation Coefficient (r)*	*Coefficient of Determination (r^2)*
Germination, Volatile aldehyde OD		
G per cent, 24-48 h	-0.9968	(0.9936)
G per cent, 48-72 h	-0.9640	(0.9293)
Total seedling length (mm), Volatile aldehyde OD		
TSL, 24-48 h	-0.9774	(0.9553)
TSL, 48-72 h	-0.9419	(0.8872)

Note: The different vigour levels were obtained after subjecting the high vigour (HV) seed to accelerated ageing at 100 per cent RH and 40°C for 5, 7 and 10 days for HMV, MV and LMV respectively; coefficient of variation (*cv*) of seedling length given within brackets.

A negative correlation between seed vigour and volatile aldehyde was demonstrated in 1982 by Harman *et al.* (1982) in soybean and pea seed of relatively poor quality. Fielding and Goldsworthy (1982) reported the evolution of volatiles during the ageing of dry wheat seeds. A suitable method of chemically assaying such aldehydes was reported for soybean seed vigour by Wilson and McDonald in 1986. Pal and Basu (1989) further modified the chemical assay method and using four vigour categories, high vigour (HV), high medium vigour (HMV), medium vigour (MV) and low medium vigour (LMV) seeds confirmed the production of higher quantities of volatiles in low vigour rice seeds than in high vigour seeds. (Table 5.23)

Table 5.24: Effect of Soaking Drying and Dipping-Drying Treatments of Rice Seeds (cv Ratna) on Post-ageing Germinability and Volatile Aldehyde Production (*vide* Pal and Basu, 1989)

Hydration-Dehydration Treatments	*Germination*		*Total Seedling Length (mm) After 72 h at 27 ± 1°C*	*Volatile Aldehyde Production of Rice during: 0-48 h (O.D. x 10^3) 635 nm*
	Per cent	*Arc-sin Value*		
Control (untreated)	62	52.0	17	111
Soaking-drying (S-D)	91	73.1	38	40
Dipping-drying (D-D)	77	61.4	31	76
LSD at *0.05 P*	-	4.7	3.2	22
LSD at *0.01 P*	-	7.1	4.8	33

Note: Five-month–old rice seeds were soaked in water for 6 h followed by drying back (S-D), and dipped in water for 5 minutes taken out and wet seed kept covered for 6 hours before drying back (D-D), natural ageing under ambient conditions was done for another 5 months by storing in cloth bags.

During the first 24 hour (0-24 hours) no evolution of aldehyde production could be detected with the method used employing 3-methyl-2-benzothiazolinone hydrazone(MBTH) reagent. (Table 5.24)

Seed invigoration by soaking-drying (which is very effective in medium vigour seeds) gave very high post-ageing germination and seedling vigour (total seedling length, root + shoot) clearly showing the negative relation between evolution of volatile aldehydes and maintenance of vigour and viability.

In the second theme paper in *Current Science* on 'Reduced volatile aldehyde production in wheat by seed invigoration treatments', Sur and Basu (1990) showed that immediately after hydration-dehydration treatment (before ageing) volatile aldehyde production from germinating wheat seeds was only slightly lower than control but after ageing a considerable increase in volatile aldehyde production took place. The invigoration treatments effectively reduced the production of volatile aldehyde and soaking-drying was very successful. Highly significant negative correlations between volatile aldehyde production and germination and seedling growth were recorded.

In the third theme paper on volatile aldehyde production in relation to seed invigoration in prestorage seed (high vigour) treatment of French bean and soybean, the beneficial effects of invigoration were closely associated with significant reduction of volatile aldehyde emanation in both the materials in almost the same manner. (Table 5.25a and 5.25b).

All the dry prestorage treatments showed significantly higher post-ageing (2 years natural ageing) vigour and viability than the untreated control but MSC-D though showed some improvement but the same did not reach statistical significance (P = 0.05), except for vigour bioassay (and that too marginally). Most interestingly the consistently reduced volatile aldehyde emanation in the effective seed invigoration treatment resulted in highly significant correlations (P = 0.01) involving the vigour parameters with coefficient of determination (r^2) multiplied by 100 (*i.e.*, $r^2 \times 100$) reaching well over 90 per cent and in case of GF (germination factor; GF = G per cent ÷ 100) incorporation going over 99 per cent in many cases (see Table 5.25a, b, for greater details on the issue see Chapter 6). We have all the evidences in support of the contention that reduction of volatile aldehyde emanation (which is an extension of lipid peroxidation process) is certainly involved in seed invigoration.

Table 5.25a: Effects of Seed Invigoration Treatment of Soybean on the Germinability along with Seedling Growth of Jute (Bioassay material) and Volatile Aldehyde Production from the Stock Seeds after Storage in Glass Bottles under Ambient Temperature for 24 Months (*vide* Mukhopadhyay, Pal and Basu, 1997)

Invigoration Treatments (Dry-dressing of seeds with fine powder, g/kg of seed) (a)	*Germination (b)*		*Total Seedling Length (mm) (b)*	*Seed Vigour Bioassay (c)*	*Volatile Aldehyde Production After 48 h (OD at 635 nm)*
	Per cent	*Arc-sin Value*			
Control (untreated)	40	39.2	78	74	0.30
Nisinda (2g)	69	56.2	106	83	0.18
Aspro (0.1g)	82	64.9	144	90	0.12
Celin (0.2g)	77	61.3	140	88	0.14
p-Aminobenzoic acid (0.1g)	68	55.5	115	86	0.17
Bleaching powder (3 g)	73	58.5	124	88	0.16
Moist sand-conditioning (10 per cent m.c., 24 h) drying	47	43.3	91	81	0.25
Non-aged control	90	71.6	167	93	0.09
LSD at *0.05 P*	-	6.4	16.8	5.3	0.05

(a) Invigoration treatments were given to 1-month-old soybean (cv. Soyamax) seed stored in sealed glass bottles under laboratory conditions. (b) Germination percentage and seedling length were taken after 5 days at 26 ± 1°C. (c) Volatile gases produced by the twenty soybean seeds (stock material) influenced the growth of jute seedlings (bioassay material); total seedling length of jute (root + shoot) after 48 h was expressed as percentage of blank.

Correlations involving germinabilty and other vigour parameters

Pairs of Variates (n=8)		*Correlation Coefficient (r)*	*Coefficient of Determination*	
(x)	*(y)*		*(r^2)*	*($r^2 \times 100$)*
G per cent,	Volatile OD	-0.9947	0.9894	98.9 per cent
TSL I,	Volatile OD	-0.9613	0.9241	92.4 per cent
TSL II,	Volatile OD	-0.9653	0.9318	93.2 per cent
G per cent,	JSBS per cent	0.9591	0.9199	92.0 per cent
TSL I,	JSBS per cent	0.9551	0.9123	91.2 per cent
TSL II,	JSBS per cent	0.9468	0.8964	89.6 per cent
Volatile OD,	JSBS per cent	0.9788	0.9580	98.5 per cent
G per cent,	VERS per cent	0.9947	0.9894	98.9 per cent

Note: The transformation of negative correlation in the first row to exactly the same positive correlation values in the last row, employing simple mathematical approaches (see Chapter 6 for details).

Abbreviations: G per cent, germination per cent of soybean; volatile OD, volatile aldehydes trapped by MBTH solution, optical density at 635 nm; TSL I, total seedling length of soybean (root + shoot in mm) based on germinated seeds only, TSL II, total seedling length based on all seeds placed for germination tests, JSBS, jute seedling bioassay score in per cent value; VERS, volatile (aldehyde) emanation resistance score in per cent value being the transformation of volatile aldehyde OD values. (negatively correlated with germination per cent).

Table 5.25b: Effect of Seed Invigoration Treatments on the Germinability of French Bean Seed along with Seedling Growth of Jute (Bioassay material) and Volatile Aldehyde Production from the Stock Seed after Storage in a Glass Bottle under Ambient Laboratory Conditions for 24 Months (*vide* Mukhopadhyay, Pal and Basu, 1997)

Invigoration Treatments (Dry-dressing of seeds with fine powder, g/kg of seed) (a)	*Germination (b)*		*Total Seedling Length (mm) (b)*	*Seed Vigour Bioassay (c)*	*Volatile Aldehyde Production After 48 h (OD at 635 nm)*
	Per cent	*Arc-sin Value*			
Control (untreated)	68	55.5	121	67	0.30
Turmeric (2g)	80	63.4	143	73	0.18
Nisinda (2g)	88	69.7	145	75	0.13
Aspro (0.2g)	89	70.6	156	79	0.09
Celin (0.5 g)	87	68.9	138	76	0.16
p-Aminobenzoic acid (0.1 g)	88	69.7	154	77	0.09
Bleaching powder (3 g)	84	66.4	148	74	0.13

Invigoration Treatments (Dry-dressing of seeds with fine powder, g/kg of seed) (a)	*Germination (b)*		*Total Seedling Length (mm) (b)*	*Seed Vigour Bioassay (c)*	*Volatile Aldehyde Production After 48 h (OD at 635 nm)*
	Per cent	*Arc-sin Value*			
Moist sand-conditioning (15 per cent m.c., 24 h)-drying	75	60.0	144	75	0.20
Non-aged control	100	90.0	197	91	0.04
LSD at *0.05 P*	-	4.0	14.5	4.1	0.06

(a) Invigoration treatments with the finely powdered materials, bleaching powder and moist sand-conditioning-drying were given to 4-month-old French bean (cv. Contender) seeds stored in a sealed glass bottle under laboratory conditions. (b) Germination percentage and seedling growth were taken after 7 days at 26 ± 1°C. (c) Volatile gases produced by the fifteen French bean seeds (stock materials) influenced the growth of jute seedlings (bioassay material); total seedling length of jute (root + shoot) taken after 48 h was expressed as percentage of blank (taking length of jute without French bean = 100).

Correlations involving germinabilty and other vigour parameters

Pairs of Variates (n=9)		*Correlation Coefficient (r)*	*Coefficient of Determination*	
(x)	*(y)*		*(r^2)*	*($r^2 \times 100$)*
G per cent,	Volatile OD	-0.9536	0.9094	90.9 per cent
TSL I,	Volatile OD	-0.8547	0.7306	73.1 per cent
TSL II,	Volatile OD	-0.9023	0.8141	81.4 per cent
G per cent,	JSBS per cent	0.8884	0.7892	78.9 per cent
TSL I,	JSBS per cent	0.9740	0.9487	94.9 per cent
TSL II,	JSBS per cent	0.9774	0.9552	95.5 per cent
G per cent,	TSL I	0.8570	0.7344	73.4 per cent
G per cent,	TSL II	0.9359	0.8760	87.6 per cent

Note: Besides confirming the observations on soybean, the correlations were better involving TSL II in which length of seedling in the germination test was based on all seeds, both germinable and non-germinable rather than that based on germinated seedlings only as in virtually all materials so far studied.

IV. Quantification of Seed Vigour by Simple Mathematical-Statistical Approaches

(See Chapter 6 and also Annexure No A-2)

V. Allied and Relevant Issues

Summaries of the seven selected papers (out of a total of 8 papers) covered in this section have been given in Chapter 4 (except for the last one, a special lecture at TNAU, Coimbatore) and we shall just discuss salient points of each hereunder. Only for the last one the full text has been provided.

The paper "Testing germination and seedling growth by an inclined glass plate blotter method" was first presented in the 19th ISTA (International Seed Testing Association) Congress in Vienna in 1980 by one of us (R.N. Basu). The paper was circulated as a preprint (preprint No.27) to the ISTA members and delegates, but along with discussion points and important editorial notes, the manuscripts were lost in transit and could not be finally printed in the official publication of ISTA (Seed Science and Technology). It was ultimately published in the Indian Journal of Plant Physiology in 1982. Nevertheless, the method has been of great help to us in developing the base of seed invigoration research in the University of Calcutta because of various technical advantages over existing methods.

It has been put on record how a team of Soviet Scientists led by academician Professor N.V. Turbin who came to New Delhi to participate in an Indo-Soviet Cereal Improvement Seminar in 1978 made an unscheduled 1500 km trip just for a day to our laboratory in Calcutta to know more about our work on seed invigoration and was so fascinated with the inclined glass plate blotter method that they took some plates with seeds placed on them for germination along with the plastic racks.

In the inclined glass plate blotter method of germination, seeds are placed on moist blotters on glass plates which are kept in polythene packets containing 20-30 ml of water and provided with two rows of holes punched at the back side (see extracts of the paper with illustrations at the end of this chapter). The plates are arranged at 66° angles on plastic dish-stands which can be placed inside temperature controlled incubators or in the room for the different periods depending on the kind of seed. Daily observations on the progress of germination and seedling growth can be very conveniently done without any disturbance to the germinating seeds (Figure 5.9).

For small seeded materials like mustard, jute, lettuce, carrot, *etc.* longer plates can also be used but for space saving inside incubators or on laboratory shelves 15 cm × 15 cm plates would be better. (Table 5.26)

As all the crop seeds were HV there was little difference in germination percentage that ranged from 96 to 100. A comparison of seedling growth (a very good indicator of seed vigour) of 6 crop seeds under the inclined glass plate and two other common practices namely, petri dish method and rolled towel method (widely practiced in public seed testing laboratories) would show the relative advantage of the inclined glass plate blotter method. Standard deviation of root and shoot length and coefficient of variation expressed as a percentage of mean, in particular, were consistently lower in the inclined glass plate method than in the other two methods. That would be of considerable advantage in vigour rating of seed lots of the same crop of different provenances or invigoration treatments of various kinds. (See Table 5.26 with notes)

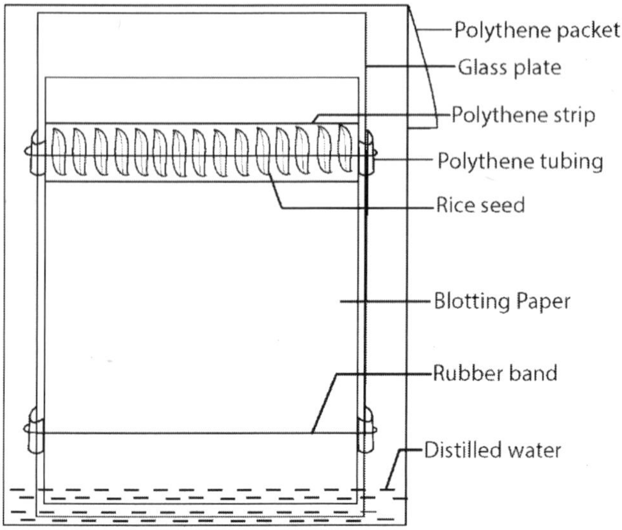

Figure 5.9 a: Diagram showing the placement of rice seeds on moist blotter spread on 25 × 15 cm glass plates kept in polythene packets (35 × 18 cm, made of 0.15 mm thick polythene sheets) containing 30 ml of Water (vide Punjabi and Basu, 1982).

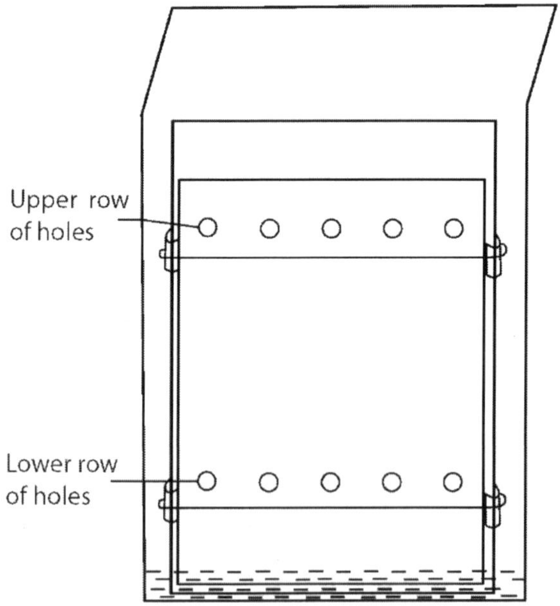

Figure 5.9b: The back side of the polythene packet showing two rows of holes (each 3 mm diam.). (vide Punjabi and Basu, 1982)

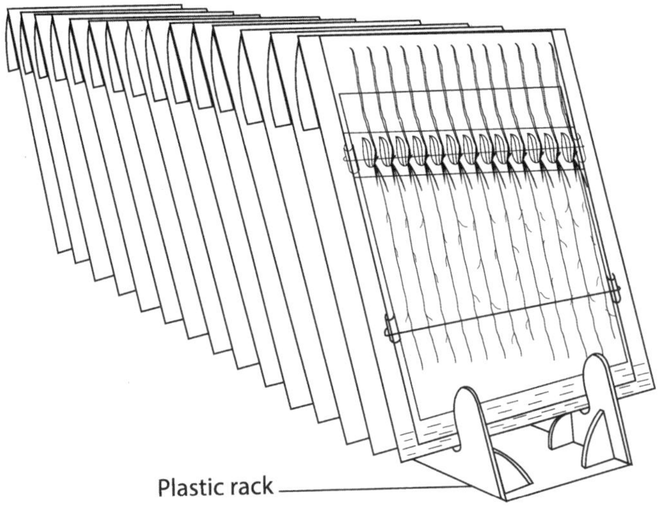

Figure 5.9c: The arrangement of plates on a plastic dish-stand showing the root and shoot growth of rice seedling after five days. Further details are given in the text. (vide Punjabi and Basu, 1982)

Table 5.26: Seedling Growth (Root and shoot length in mm) in different Methods of Germination (*vide* Punjabi and Basu, 1982)

Materials		*Germination Methods*		
		Blotter on Glass Plate	*Petri Dish*	*Rolled Towel*
Mustard	Root	114 ± 1.2 (1.05)	77 ± 5.7 (7.40)	107 ± 4.1 (3.83)
	Shoot	55 ± 1.5 (2.72)	40 ± 0.6 (1.50)	66 ± 3.4 (5.15)
Jute	Root	62 ± 1.7 (2.74)	42 ± 0.6 (1.43)	77 ± 2.1 (2.72)
	Shoot	56 ± 0.3 (0.53)	51 ± 0.3 (0.58)	59 ± 1.2 (2.03)
Lettuce	Root	47 ± 0.9 (1.91)	34 ± 1.2 (3.53)	48 ± 2.1 (4.37)
	Shoot	40 ± 0.3 (0.75)	32 ± (3.12)	42 ± 0.6 (1.43)
Rice	Root	120 ± 1.8 (1.50)	63 ± 1.2 (1.90)	129 ± 3.2 (2.48)
	Shoot	59 ± 1.2 (2.03)	47 ± 0.6 (1.28)	64 ± 2.2 (3.44)
Soybean	Root	104 ± 8.5 (8.17)	42 ± 5.2 (12.38)	76 ± 9.6 (12.63)
	Shoot	97 ± 3.3 (3.40)	63 ± 6.0 (9.52)	97 ± 9.5 (9.79)
Sunflower	Root	219 ± 4.0 (1.83)	92 ± 12.8 (13.91)	187 ± 15.2 (8.13)
	Shoot	104 ± 3.4 (3.27)	99 ± 4.4 (4.44)	126 ± 6.2 (4.92)

Germination tests for jute, mustard and lettuce were done on 15 × 15 cm plates and for rice, soybean and sunflower on 25 × 15 cm plates; tests were made with high-vigour (HV) seeds of 96-100 per cent germination; data on seedling growth were recorded after germination at 24 ± 1°C for ten days in soybean and after seven days for the other materials. Values in parenthesis denote the respective coefficient of variation (*cv*, standard deviation expressed as percentage of respective mean values).

In the second theme paper of the miscellaneous section Sur (Mandal) and Basu (1985) made embryo-endosperm reciprocal transplants of rice seed by transplanting treated embryo (hydrated-dehydrated embryo of 6-month-old seed) on endosperm of untreated (control) seed and the control embryo on endosperm of treated seed and *vice-versa* (see table below). When transplants were made from non-aged materials no significant effects on germinability could be seen but transplants made after accelerated ageing of the seed at 100 per cent RH and 40°C statistically significant differences on germinability and vigour (root and shoot length) were recorded (Table 5.27).

Table 5.27: Effect of Hydration-Dehydration on Reciprocally Transplanted Embryo and Endosperm of Rice Seed after Accelerated Ageing at 100 per cent RH and 40°C for 10 Days (*vide* Sur and Basu, 1985)

Transplants (Embryo + Endosperm)	*Germination*		*Mean Shoot Length (mm)*	*Mean Root Length (mm)*
	Per cent	*Arc-sin Value*		
Cont + Cont	63	52.5	6	10
Cont + SD	66	54.3	9	14
SD + Cont	84	66.4	13	18
SD + SD	86	68.0	16	26
Cont + MED	64	53.1	8	14
MED + Cont	80	63.4	12	17
MED + MED	83	65.6	15	24
CD at *0.01 P*	-	3.71	3.00	3.77
CD at *0.05 P*	-	2.73	2.22	2.27

Abbreviations: Cont - Control; SD - Soaking-drying; MED - Moisture equilibration-drying.

Reciprocal embryo-endosperm transplantations employing treated embryos on untreated endosperms and untreated embryos on treated endosperms clearly showed that hydration-dehydration treatment (S-D and ME-D) not only retained the vigour and viability of the embryo but also significantly improved the capacity of the endosperm to nourish the growing embryo transplanted on it. In an earlier study, Mandal (1981) showed ageing adversely affected amylolytic enzyme synthesis by the aleurone cells of the endosperm. The reciprocal embryo-endosperm transplantation study clearly showed that the hydration-dehydration invigoration treatment besides benefitting the embryo, effectively prevented a decline in functional activity of the aleurone cells of the endosperm during ageing. Increased amylase production and solubilisation of starch reserves would provide more nourishment to the embryo transplanted on it (Table 5.28).

Table 5.28: Effect of Hydration-Dehydration on α-amylase Activity[(a)] of Rice Seed after Accelerated Ageing at 100 per cent RH and 40°C for 10 days (*vide* Sur and Basu, 1985)

Effect of Hydration-Dehydration	*Starch Hydrolyzed/h in μg*			
	Full Seed[(b)]		*Half Seed[(c)]*	
	Without GA	*With GA*	*Without GA*	*With GA*
Control	80	161	5.9	10.2
Soaking-drying	125	320	9.3	20.2
Moisture equilibration-drying	122	302	9.0	19.0

(a) Values expressed as μg of starch hydrolyzed by enzyme equivalent one half-seed. (b) Full seed was allowed to sprout without GA (in water) or with GA (5 x 10^{-6}M) for 48 h at 25 ± 1°C and then one-half of the seed (embryo-less half) was taken for enzyme extraction. (c) One-half of the dry seed (embryo-less) was incubated for 48 h at 25 ± 1°C with or without GA (5 x 10^{-6}M) followed by extraction and assay of enzyme.

Table 5.29: Germinability of Control and Soaked-Dried Rice Seeds (cv. Ratna) Stored in a Desiccator over Fused Calcium Chloride (*vide* Pal and Basu, 1989)

Storage Period (Years in desiccator)	*Control*			*Soaking-drying*		
	Germination Per cent	*Mean Root Length (mm)*	*Mean Shoot Length (mm)*	*Germination Per cent*	*Mean Root Length (mm)*	*Mean Shoot Length (mm)*
0	100	82 ± 3.0	52 ± 1.6	100	86 ± 1.2	53 ± 2.2
3	91	70 ± 6.6	46 ± 2.3	99	85 ± 1.5	51 ± 2.0
6	74	64 ± 4.9	41 ± 2.2	98	79 ± 5.4	44 ± 2.5
9	54	45 ± 7.3	26 ± 3.0	-	-	-
12	12	40 ± 5.1	24 ± 2.7	85	66 ± 6.3	34 ±2.8

Note: Average stabilized seed moisture content was 7.01 ± 0.16 per cent, all germination tests were done following the inclined glass plate blotter method of Punjabi and Basu (1982) at a temperature of 28 ± 1°C in an incubator; data were recorded after 7 days.

The third theme paper on over a decade long storage of rice seed given midstorage soaking-drying treatment by Pal and Basu (1989) would open up the possibility of storage of genetic stocks under ambient temperature regimes for a prolonged period. (Table 5.29)

In the first week of July 1976, 7-month-old Ratna rice seed (stored in a gunny bag under ambient temperature and humidity conditions since harvest (average RH 65 per cent ± 9.6 per cent and temperature 27.5°C ± 3.8°C) was given a soaking-drying (S-D) treatment (1 kg of seed soaked in 2 litres of water for 6 h, then dried back to the original weight in a drying cabinet in a current of hot air at 36° ± 1°C); one kg of untreated seed was taken as the control. Control and S-D seeds were taken in 20 perforated paper packets 10 for control and 10 for S-D and placed in a 10-litre capacity glass desiccator containing 1.5kg of fused calcium chloride at the bottom. The desiccator was made air-tight and kept in the laboratory under ordinary (no temperature control) indoor conditions and samples were drawn at requisite intervals for germination tests and moisture determination. Care was taken to avoid entry of humid air from outside during the entire period of storage, the calcium chloride in the desiccator remained in the fused state all throughout. Moisture stabilisation of the seeds was accomplished within a fortnight (7.0 ± 0.16 per cent).

The control seeds kept over fused calcium chloride maintained very good germinability upto 3 years although a small reduction of growth of seedlings was noticed. The soaked-dried seed at that stage showed little change from the initial value. After 6 years, there was a noticeable fall in the germination percentage of control seeds but the soaked-dried seeds retained very good germinability. Germination percentage and seedling growth in control showed further decline when tested after 9 years of storage in the desiccator and by the end of 12 years viability of control seeds dropped to 12 per cent with vigour of surviving seedlings further reduced. The S-D seed treatment at that stage retained 85 per cent viability with above average vigour of seedlings.

Incidentally, Pal and Basu (1978) in their report in the International Rice Research Newsletter (IRRN) confidently stated that the method of seed treatment could be used to maintain germplasm collection under ordinary storage conditions for a considerable period and would eliminate the need to regrow the genetic stocks frequently.

The theme papers on wheat and French bean (as also another on mustard on the same aspect) deal with differential response of seeds of high, medium and low vigour status to wet and dry seed invigoration treatments.

Dry seed invigoration treatment of high vigour wheat seed with 'aspro' (an aspirin containing pharmaceutical) at 0.1g per kg of seed and bleaching powder (chlorine releasing, calcium oxychloride as the active ingredient) at 3g per kg of seed showed significantly higher post-storage viability and vigour than untreated and aged control, the same treatment gave smaller positive effects in medium vigour wheat seed but did not show any improvement over control in case of low vigour seeds. Wet seed treatment by soaking-drying (S-D) showed significantly higher post storage vigour and viability in medium and low vigour seeds but failed to do so in case of high vigour seeds (Table 5.30).

The interactions between seed treatment and viability and between treatment and vigour (as manifested by seedling growth) were statistically significant at 0.05 P as well as 0.01 P. (Table 5.31, field performance and productivity)

Table 5.30: Influence of Seed Vigour on the Efficacy of Seed Invigoration Treatments on Post-Storage Germinability of Wheat (*vide* Pal, Ray and Basu, 2001)

Treatments	*Seed Vigour Status*						*Mean (Invigoration treatment)*	*LSD Values for Main Effects and Interaction*		
	High Vigour (HV)		*Medium Vigour (MV)*		*Low Vigour (LV)*					
(a) Germination angles (arc-sin transformations for germination percentage values)										
Control (untreated)	62.0	(78)	56.8	(70)	52.1	(64)	56.9	Source	*P=0.05*	*P=0.01*
Aspro (0.1 g/kg seed)	72.2	(91)	62.0	(78)	54.3	(66)	62.8	Treatment (T)	2.9	4.0
Bleaching powder (3 g/kg seed)	71.6	(90)	63.4	(80)	54.9	(67)	63.3	Vigour (V)	3.2	4.6
Water soaking (2h)-drying	62.7	(79)	70.6	(89)	62.0	(78)	65.1	Interaction (T×V)	5.4	7.6
Mean (vigour levels)	67.1		63.2		55.8					
LSD at *0.05 P**	4.1		5.5		6.0					
LSD at *0.01 P**	6.2		8.3		9.1					

Treatments	*Seed Vigour Status*			*Mean (Invigoration treatment)*	*LSD Values for Main Effects and Interaction*		
	High Vigour (HV)	*Medium Vigour (MV)*	*Low Vigour (LV)*				
(b) Total seedling length (mm)							
Control (untreated)	224	182	155	187	Source	*P=0.05*	*P=0.01*
Aspro (0.1 g/kg seed)	285	191	160	212	Treatment (T)	3.1	4.4
Bleaching powder (3 g/kg seed)	276	187	165	209	Vigour (V)	3.6	5.0
Water soaking (2h)-drying	218	221	181	207	Interaction (T×V)	6.1	8.7
Mean(vigour levels)	250	195	165				
LSD at *0.05 P**	14.1	12.8	14.6				
LSD at *0.01 P**	21.3	19.3	22.1				

HV– High vigour; MV – Medium vigour; LV – Low vigour; LSD – Least significant difference; P – Probability

After treatment wheat seeds were stored in cloth bags under ambient conditions (average RH 62.3 ± 3.1 per cent and temperature 27.3 ± 1.7°C for 115 days).

Germination percentage and measurement of seedling growth were recorded after 120 h at 27.3 ± 1.7°C, 10 days before sowing in the field.

Values in parentheses have been denoted as original germination percentage values.

* For comparison of treatment means within a vigour lot.

Table 5.31: Influence of Seed Vigour on the Efficacy of Seed Invigoration Treatments for Improvement of Field Performance (Plant population, grain yield and 1000-grain weight) (*vide* Pal, Ray and Basu, 2001)

Treatments	*Seed vigour status*			*Mean (Invigoration treatment)*	*LSD Values for Main Effects and Interaction*		
	High Vigour (HV)	*Medium Vigour (MV)*	*Low Vigour (LV)*				
(a) Plant population m^{-2}							
Control (untreated)	178	141	114	144	Source	P=0.05	*P=0.01*
Aspro (0.1 g/kg seed)	208	174	125	169	Treatment (T)	3.5	4.0
Bleaching powder (3 g/kg seed)	207	172	128	169	Vigour (V)	3.9	5.4
Water soaking (2h)-drying	186	178	137	167	Interaction (T×V)	6.9	9.6
Mean(vigour levels)	195	166	126				
LSD at *0.05 P**	14.8	12.4	14.5				
LSD at *0.01 P**	22.2	18.7	21.9				
(b) Grain yield (m^{-2})							
Control (untreated)	260	253	210	241	Source	P=0.05	*P=0.01*
Aspro (0.1 g/kg seed)	301	279	213	264	Treatment (T)	6.6	9.4
Bleaching powder (3 g/kg seed)	294	287	216	266	Vigour (V)	7.6	10.8
Water soaking (2h)-drying	264	294	226	261	Interaction (T×V)	13.3	18.7
Mean(vigour levels)	280	278	216				
LSD at *0.05 P**	9.2	12.8	11.8				
LSD at *0.01 P**	13.9	19.3	17.8				

Treatments	*Seed vigour status*			*Mean (Invigoration treatment)*	*LSD Values for Main Effects and Interaction*		
	High Vigour (HV)	*Medium Vigour (MV)*	*Low Vigour (LV)*				
(c) 1000-grain weight (g)							
Control (untreated)	36,2	35.8	35.1	35.7	Source	P=0.05	*P=0.01*
Aspro (0.1 g/kg seed)	39.1	38.5	36.8	38.2	Treatment (T)	1.7	2.4
Bleaching powder (3 g/kg seed)	38.7	38.9	36.8	38.1	Vigour (V)	2.0	2.8
Water soaking (2h)-drying	37.1	39.2	38.4	38.2	Interaction (T×V)	3.5	NS
Mean (vigour levels)	37.7	38.1	36.8				
LSD at *0.05 P**	1.3	1.5	1.8				
LSD at *0.01 P**	1.9	2.3	2.7				

HV– High vigour; MV– Medium vigour; LV– Low vigour; LSD– Least significant difference; P– Probability.

* For comparison of treatment means within a vigour lot.

Plant population were counted three weeks after sowing in the field. Yield data were recorded 115 days after sowing. Other details same as in Table 5.30.

Field emergence of seeds followed a similar pattern as with germinability and vigour. Dry seed invigoration treatments showed significantly higher plant population per unit area in high vigour seed and to a moderate extent in medium vigour seed but in medium and low vigour seed water soaking-drying was much more effective. Grain yield/m^2 was greater in dry invigoration treatment of HV wheat with 'aspro' showing the maximum response of 16 per cent increase and bleaching powder 13 per cent over the control (HV control). In medium vigour seed also 'aspro' and bleaching powder gave 10 per cent higher yield with aspro and 13 per cent with bleaching powder over the control (MV control). It needs to be pointed out that irrespective of treatments the yield of HV was highest (HV control was 260 g/m^2 against 253 g/m^2 in MV control and 210 g/m^2 in LV control). Most importantly, the interaction between treatment and vigour level (T × V) was significant at 0.01 P. This point will be taken up again while discussing the effect of seed size and on the efficacy of different seed invigoration treatments in wheat.

As mentioned earlier the findings on wheat (Pal *et al.,* 2001) were confirmed by Rudrapal and Basu (2002) in mustard but more significantly the confirmation of the interaction effect by the same authors in the relatively bold seeded (soaking-susceptible leguminous crop French bean (*Phaseolus vulgaris* L.) is briefly discussed here with emphasis on the interaction of treatment and vigour status on seed vigour and viability. (Rudrapal and Basu, 2004)

French bean seeds of cultivar Pusa Pravati used in the study were collected 2 months after harvest and stored at –10°C in rubber stoppered wax-sealed glass bottles and served as the high vigour (HV) lot. A part of the refrigerator stored (–10°C) seed was taken out after 4 months and was stored under ambient laboratory conditions in an unsealed rubber stoppered bottle for 6 months and was used as the medium vigour (MV) lot and those stored for 10 months as the low vigour (LV) lot. Accepting (theoretically) that there was no noticeable ageing at –10°C, physiological ages of seeds were 2, 8 and 12 months respectively for HV, MV and LV seeds (chronologically all were 12 months old). The seeds were dry dressed with finely ground powders of dried leaves of bitter gourd (*Momordica charantia* L.) and amrul (*Oxalis corniculata* L.) at the rate of 2 g and 2.5 g per kg of seed respectively. Seed dressing was also done with finely powdered pharmaceuticals namely 'aspro' (0.2g/kg) an aspirin (*ortho*-acetyl salicylic acid) containing formulation, 'celin' (0.5g/kg) an ascorbic acid containing formulation (Vitamin C) and common bleaching powder (major active ingredient calcium hypochlorite that would slowly release chlorine) at 3g/kg of French bean seeds. Thorough mixing of the dry powders with the seed was done in rubber stoppered glass bottles that were kept at room temperature with occasional shaking for 7 days before the germination tests (pre-ageing tests). Besides the aforesaid dry treatments, a slow hydration treatment (followed by dehydration), was given to the soaking injury prone legume seed by moisture equilibration with a water saturated atmosphere for 24h and then slowly drying back in a drying cabinet in a current of hot air at 35°C (moisture equilibration-drying, ME-D). Pre-ageing germination tests were done following the inclined glass plate blotter method (Punjabi and Basu, 1982) as discussed in the first paper of this section. Requisite quantities of treated and untreated (control) seeds were taken in perforated paper envelopes for accelerated ageing or natural ageing under ambient laboratory (indoor) conditions; for the latter, paper packets were stored in cloth bags for different periods. In this brief discourse we would deal only with vigour and viability maintenance under natural ageing.

Germination tests of seeds before ageing showed no difference in germination percentage and only a marginal improvement in seedling growth in a few cases was seen.

The post-ageing germinability of French bean seeds (after natural ageing for 8 months, average RH 65.2 per cent and temperature 27.5°C) of the different treatments, clearly showed that the HV seed maintained higher germinability under the dry treatments which however, failed to invigorate the LV seed and showed marginal improvements in MV seeds with the same dry treatment. The MV seeds nevertheless benefitted considerably from ME-D treatment which was of very little advantage to HV or LV seeds (Figure 5.10).

The values shown are arc-sin transformations of germination percentage; in actual germination percentage scale the difference was more, it was from 43 per cent in control to 60 per cent in ME-D. Dry dressing of HV seed with powders of bitter gourd leaf and 'aspro' was found to be more effective than the others and in general over 25 per cent higher germinability over HV control seeds was recorded; in MV seed 'aspro' showed some increase in germinability over MV control. However, LV seeds in general showed no benefit from the seed treatments; even ME-D was not that effective as seen in some other crops, perhaps ME-SD or MSC-D (moist sand conditioning-drying) would have shown some improvement in a difficult leguminous material like French bean (see Mandal *et al.,* 2000 in soybean).

Post-ageing seedling length was similarly affected by the different seed treatments with the 'dry' treatments very consistently maintaining the vigour status of the HV seeds. Within the MV seeds maximum seedling length was noted under ME-D (40 per cent increase over MV control) however, unlike LV seeds which were almost unaffected by dry seed treatments the medium vigour (MV) seeds positively responded to dry treatments with 17-29 per cent more seedling growth over MV control. As such, the same sort of interaction (between seed treatment and vigour status of stock seed TxV) was recorded with statistical significance at both the levels of probability (0.05P and 0.01 P).

In this connection it would be pertinent to discuss the results of seed invigoration treatments on different sizes of wheat seed for improved storability, field emergence and productivity (De, Mandal and Basu, 2003) in which harvest-fresh (1-month-old high vigour, HV) wheat seed (cv. Sonalika) was graded into big, medium and small sizes by using appropriate sieves. The seeds of all three sizes were then dry-dressed with the pharmaceutical formulation 'aspro' (active ingredient ortho-acetylsalicylic acid) at 100 mg/kg of seed, salicyclic acid 50 mg/kg,

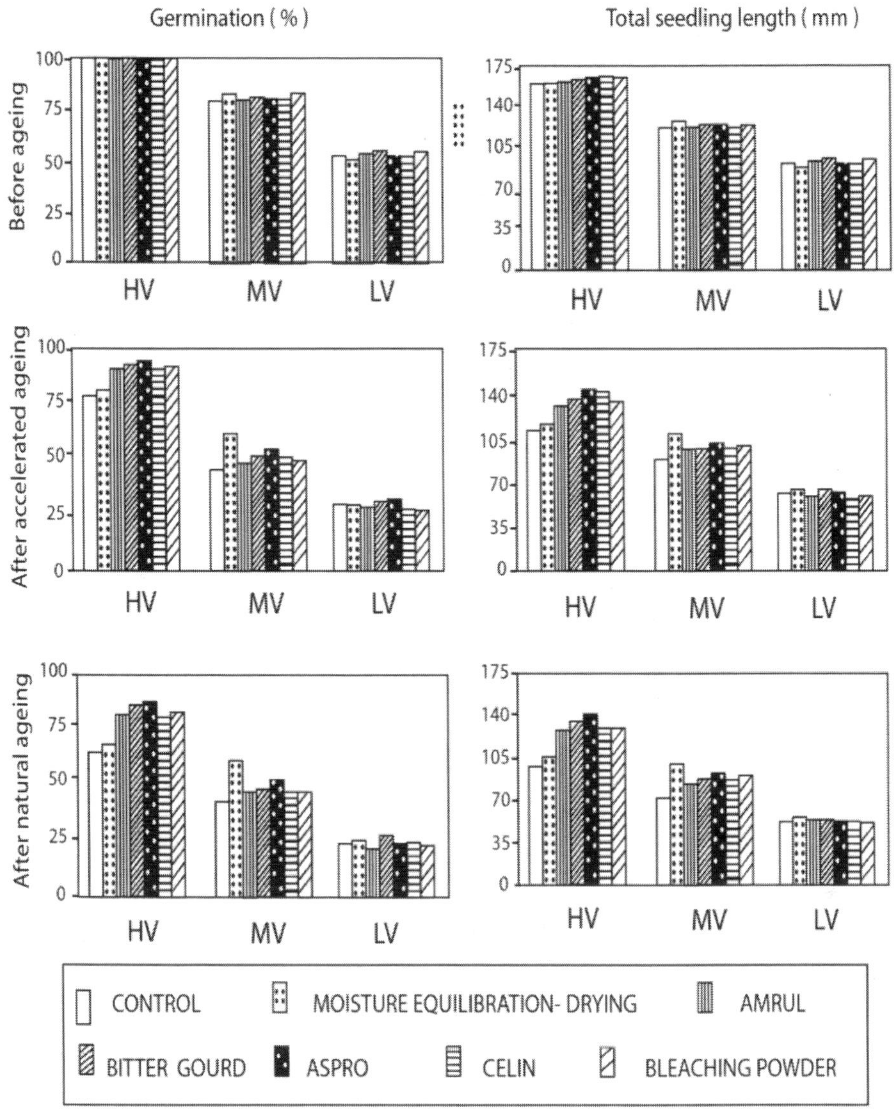

HV = HIGH VIGOUR, MV = MEDIUM VIGOUR, LV = LOW VIGOUR

Figure 5.10: Germinability of french bean seed as influenced by wet (moisture equilibration-drying, ME-D) and dry treatments of seeds of different vigour status (High, medium and low) before ageing and after accelerated ageing (93 per cent relative humidity and temperature 40°C for 2 days and natural ageing (average relative humidity 65.2 per cent and temperature 27.5 per cent °C) for 8 months after treatment. (*Vide* Rudrapal & Basu, 2004)

common bleaching powder (active ingredient calcium oxychloride 30 per cent) at 2g/kg, finely powdered red (hot) chilli powder and *Trigonella* seed powder each at 1g/kg of seed. The seeds were then stored in rubber stoppered glass bottles. A 'wet' midstorage soaking-drying treatment (2 h soaking in water followed by drying) was given as a reference treatment to all the three categories of seed (big, medium and small) which were stored under ambient conditions prior to midstorage treatment for comparison with the aforesaid prestorage dry treatments.

Natural ageing for another 90 days was done for all the treatments as shown in the table (Table 5.32 and 5.33) for which requisite quantities of treated and control seeds were placed in perforated paper packets (very small perforations) kept in cloth bags.

Table 5.32: Effect of Seed Invigoration Treatments on Germination Percentage of different Sizes of Wheat Seed after Natural Ageing under Ambient Conditions (Average RH 81 ± 1.88 per cent and temperature 29.58 ± 0.29°C) for 90 Days (*vide* De, Mandal and Basu, 2003)

Physiological Treatments (Dry-dressing at mg or g/kg of seed)	*Seed Size*			*Mean*
	Small	*Medium*	*Large*	
Control (untreated)	42	51	54	49
Aspro (100 mg/kg)	58	67	68	64
Salicylic acid (50 mg/kg)	50	61	71	61
Bleaching powder (2 g/kg)	50	60	69	60
Red chilli powder (1 g/kg)	51	61	65	59
Trigonella seed powder (1 g/kg)	58	65	76	66
Soaking-drying (midstorage)	53	61	68	61
Mean	52	61	67	-
	Probability level			
LSD values for mean effects	0.05 P		0.01 P	
Physiological treatments	4.3		5.8	
Seed size	2.8		3.8	
Interaction	NS		NS	

Note: Dry treatments were given to harvest-fresh one-month-old Sonalika wheat seeds; midstorage hydration-dehydration was given to 5-month-old seed (soaking in water for 2 h followed by drying back as a reference treatment).

Germination test before ageing (just after treatment) did not show any significant difference in germinability in the treated and untreated seeds; seed size also did not influence germination percentage but large differences in seedling growth were attributed to seed size irrespective of ageing or treatment. However, after natural ageing under ambient conditions, there were significant differences in germination percentage of treated and untreated seeds of different categories. In all seed sizes all the 'dry' invigoration treatments as well as the 'wet' midstorage soaking-drying treatment gave significantly better germinability than their respective untreated controls. The treatment effects were, however, independent of seed size and as such, the interaction between seed size and seed treatment was statistically not significant (Table 5.33).

Table 5.33: Effect of Seed Invigoration Treatments on Seedling Length (mm) of different Sizes of Wheat Seed after Natural Ageing under Ambient Conditions (Average RH 81 ± 1.88 per cent and temperature 29.58 ± 0.29°C) for 90 Days (*vide* De, Mandal and Basu, 2003)

Physiological Treatments	*Seed Size*			*Mean*
	Small	*Medium*	*Large*	
Control	145	215	236	199
Aspro	217	231	262	237
Salicylic acid	195	235	243	224
Bleaching powder	200	229	245	225
Red chilli powder	173	214	299	229
Trigonella seed powder	198	223	247	223
Soaking-drying (midstorage)	193	235	262	230
Mean	189	226	256	-
	Probability level			
LSD values for mean effects	0.05 P		0.01 P	
Physiological treatments	17		23	
Seed size	11		15	
Interaction	30		41	

Note: Significance of the interaction effect was possibly due to the unusually low seedling length in the control of small sized seed, the reasons for which are not obvious; in no other parameters the same has been recorded. An inadvertent error somewhere cannot be over ruled.

The field performance of treated and untreated seeds in terms of effective tillers per square meter of land and the grain yield (g/m^2) for the three seed size categories are shown in Table 5.34. Both field establishment and yield per unit area were significantly increased by 'dry' invigoration treatments as well as midstorage S-D treatment in all the three size categories of seed.

Among the 'dry' treatments 'aspro', red chilli powder and trigonella seed powder showed an average (mean of three seed sizes) of 18, 17 and 16 per cent increase in grain yield over the untreated control mean. The usually stable character, 1000-grain weight, was also significantly improved by the seed invigoration treatments (Table 5.35). The efficacy of the seed treatments was more or less similar in all the three seed sizes. Thus, interactions involving seed size and seed treatment were statistically not significant for the above productivity related characters (ear bearing effective tillers, grain yield per sq. m and 1000-grain weight).

Table 5.34: Effect of Seed Invigoration Treatments on Grain Yield (g)/m² of different Sizes of Wheat (cv. Sonalika) Seed (*vide* De, Mandal and Basu, 2003)

Physiological Treatments	*Seed Size*			*Mean*
	Small	*Medium*	*Large*	
Control	167	178	230	192
Aspro	199	215	263	226
Salicylic acid	189	219	246	218
Bleaching powder	198	211	244	218
Red chilli powder	196	228	250	225
Trigonella seed powder	189	214	263	222
Soaking-drying (midstorage)	192	208	247	216
Mean	190	210	249	-

	Probability level	
LSD values for mean effects	0.05 P	0.01 P
Physiological treatments	12	16
Seed size	8	10
Interaction	NS	NS

Data on grain yield were recorded from fixed plant population (100/m²), other details same as in Table 5.32.

Table 5.35: Effect of Seed Invigoration Treatments on 1000-Seed Weight (g) of different Sizes of Wheat (cv. Sonalika) Seed (*vide* De, Mandal and Basu, 2003)

Physiological Treatments	*Seed Size*			*Mean*
	Small	*Medium*	*Large*	
Control	33.48	34.13	34.92	34.18
Aspro	34.12	34.88	35.50	34.83
Salicylic acid	33.78	34.72	35.34	34.61
Bleaching powder	34.27	34.71	35.80	34.93
Red chilli powder	33.75	34.70	35.20	34.55
Trigonella seed powder	33.65	34.52	35.43	34.53
Soaking-drying (midstorage)	33.66	34.72	35.18	34.52
Mean	33.81	34.62	35.34	-

	Probability level	
LSD values for mean effects	0.05 P	0.01 P
Physiological treatments	0.25	0.33
Seed size	0.17	0.22
Interaction	NS	NS

Data on grain yield were recorded from fixed plant population (100/m²)

The literature pertaining to the influence of seed size on crop establishment in the field and productivity would with some exception show a positive effect of using large seed for sowing. Our data on wheat would also demonstrate the consistent statistically significant positive contributions of physiological seed invigoration treatments. It is considered necessary to understand whether the large, medium and small seeds would be physiologically synonymous with high vigour (HV), medium vigour (MV) and low vigour (LV) seeds. A critical examination of the data on germinability and field performance presented above would show that in all cases the mean values for small, medium and large seeds are expressed as percentage of respective control, the different seed sizes would show more or less similar sort of response to the treatments which would be contrary to response shown by seeds of different vigour status obtained by differential ageing of seed lot of the same provenance.

The overall performance of large sized seeds was better than medium and small seeds irrespective of treatments. The seed invigoration treatments gave added benefits (additive effect) which was manifested in statistical non-significance of the interaction effect (seed size × invigoration treatment). We may recall the scenario in our experiments on HV, MV and LV seeds as obtained by accelerated or natural ageing for basic seed lot. The differential response of the vigour categories was characterised by highly significant interaction between seed vigour status and seed invigoration treatment. LV seeds were not responsive to 'dry' seed invigoration treatments that were of great advantage to HV seeds for germinability as well as field performance and productivity. Non-leguminous LV seeds would greatly benefit by soaking-drying (S-D) but the same adversely affected the post-ageing performance of harvest-fresh HV seeds. Our main point of contention is that the seed size and seed vigour are not synonymous; a large sized seed may be of low vigour and a small seed a highly vigorous one. In this study the size grading by sieving was done soon after harvest therefore all the size categories were of high vigour (HV) status (the respective size categories may, however, be made MV or LV by accelerated or natural ageing). Conceptually seed vigour is a physiological character, and seed size a morphological character and seed vigour may or may not be related to weight or size of seed depending on storage conditions (physiological ageing).

In practical agriculture, in the farmers' field, however, selecting large seeds by seiving is strongly recommended for many crop seeds as the additional nutritional reserves in the large seed(most likely with a bigger embryo) would be of considerable advantage in the early seed reserve-dependent stage (pre-autotrophic phase) of seedling growth. Dry dressing of the high vigour (HV) seed or soaking-drying for the medium vigour (MV) seeds (moist sand conditioning–drying, MSC-D, for leguminous seeds) would further augment field establishment and crop productivity. In view of the high cost of labour and inputs the use of low vigour seeds for sowing should be avoided as far as possible.

The last selected paper on "Allied and Relevant Issues" namely "Seed invigoration for seed saving by small and marginal farmers and ensuring food security" by Basu, Mandal and Pal (2010) broadly outlines the findings on seed invigoration treatments with emphasis on prestorage dry seed treatments with the ultimate social objective of improving crop productivity specifically to safeguard food security of the predominantly small and marginal farmers with insignificant material resources. They feed the nation of over 1250 million people but majority of them are poor and malnourished and are dejectedly apprehending perhaps a still gloomier future. The green revolution brought a temporary relief, but it was short lived; its legacy has left an array of agroecological problems that would require a long time to eradicate. Seed is the backbone of crop-based agriculture. Green revolution itself has been a megascale seed based agricultural-cum-sociopolitical approach that in due course has started to usher in corporate sector-dominated seed-centered industrial agriculture in which the farm operations would be done by machines (even by remote controlled ones), in the long run replacing much of the human involvement, if necessary. These are not mere fantasies rather strong possibilities; but what may happen to the millions of poor small and marginal farmers of a thickly populated developing country is a matter of utmost concern and uncertainty.

In any case in the Indian perspective seed security of the farmer is closely linked with food security; farmers and their families comprise the largest sector of consumers of farm products and income from agriculture and allied branches still remains the main economic support system of rural India. These issues and their possible remedial measures have been broadly taken up in the supplement to thematic and allied issues (chapter 7).

RELEVANT EXTRACTS OF SOME SELECTED PAPERS

Punjabi, Bina and Basu (R.N. 1982). Testing germination and seedling growth by an inclined glass plate blotter method. *Indian Journal of Plant Physiology*, 25(3), 289-298.

A method of studying germination and early seedling growth should give highly reproducible results besides being easily practicable, inexpensive and suitable for a range of laboratory conditions. For uniformity in germination, controlling the moisture is very important as even minor differences may give large variations in the germination of certain seeds.

In many Indian laboratories, germination tests for common agricultural and horticultural seeds are done in petri dishes on moist blotters. Besides the problem of moisture control the recording of seedling growth, especially root growth, is difficult in the petri dish method because roots of most materials get entangled. Disentangling of roots results in breakage and erratic recording of data. Further, in old partially deteriorated seed, germination percentage and seedling growth is often badly effected due to leaching out of inhibitory substances into the germination medium. The rolled towel method of seed germination successfully eliminates many of these disadvantages but the results would depend to a large extent on the skill of the individual worker and the availability of paper towel of desired quality. Observation of the progress of germination is also difficult in this technique.

The "slanted substrate" method was developed by Jones and Cobb (1963) to overcome many of the aforesaid difficulties. Since then a number of workers have adopted a similar technique with modification to suit their individual requirements (Oelke *et al.*, 1969; Pihakaski, 1977; Peterson *et al.* (1978).

Although the Jones and Cobb (1963) method as well as that of Peterson *et al.* (1978) seem to overcome many of the disadvantages of the petri dish or rolled towel methods, they have their specific limitations. In the present paper, a simple and inexpensive method to test the germination and seedling growth is described which eliminates many of the disadvantages of the previously reported methods. The methodology is based on that briefly outlined by Dasgupta *et al.* (1976) and Dasgupta (1978) for wheat seeds and incorporates some useful modifications. A comparative assessment of the technique has been attempted to show its superiority over other conventional methods of germination.

Two mm thick glass plates were cut to a size of 15 x 15 cm. The blotting paper sheets (thick 57 × 45 cm 'Absorbo' white blotting sheet of Titaghur Paper Mills Co. Ltd., Calcutta) were previously washed overnight in running tap water followed by final washing in two changes of distilled water. When Whatman No. 1 sheets were used, a single washing with distilled water served the purpose. The sheets were dried in air and then cut to size of 15 ×12 cm and 2 ×15 cm. The bigger piece of blotting paper was wetted with distilled water and placed on the glass plate leaving 3 cm from the top. The seeds were placed embryo side down in a straight line by means of a pair of forceps 2 cm down the upper end of the paper (Figure 1a). After placing the seeds the 2 × 15 cm blotting paper strip was wetted with distilled water and placed on the row of seeds. Care was taken to provide a gap in between the two blotters on the upper side. This ensured a free access of air to the seeds and also the seeds could be viewed from the top. In case of rice, wheat and barley, a thin polythene strip, 2 × 15 cm, was used. However, blotter strips of the same size may also be used. In case of soybean, a slight modification of the placement technique was made. The smaller blotting strip (wet) was first placed on the glass plate and seeds were arranged in a linear fashion on that strip. The larger piece of blotter was then placed over the seeds. This was done to prevent the tendency of the roots to rise over the moist blotter and dry up. Two, 2 cm long with 1 cm diameter polythene tubings slit longitudinally on one side, were fixed on two sides of the glass plate pressing the ends of the blotters. A rubber band was then fixed across the plate, over the blotter strip and the polythene tubes, keeping them firmly in position. The rubber band should be placed in such a way that it does not obstruct in any way the upward growth of the shoot. For that it should be placed just below the line of the seed rather than above or over the seed. Another rubber band was fixed similarly over two side tubes about 4 cm from the lower end of the plate (Figure 1b). The plate was then placed in a polythene packet (0.15 mm thickness of sheet) measuring 23 ×18 cm and containing 20 ml of distilled water. When glass plate of 25 ×15 cm was used, the size of the bigger blotter was 18 × 15 cm, and the polythene packet was 33 ×18 cm. It needs to be pointed out that the thickness of the polythene sheet should not be thin as otherwise the inner side of the packet may stick to the blotter especially in case of longer plates. In fact, the lower rubber band was fixed on the plate to avoid adhering of the polythene to the wet blotting paper which would impede the direct downward growth of the roots. The plate was then placed on a plastic dish-holding rack ('Prince' or 'Mutual' dish-stand) at an angle of 66° for germination (Figure 1c). The whole set-up could be kept at the desired temperature in germinators, temperature-controlled rooms or even in ordinary ambient conditions in the laboratory. During the first one or two days, rough handling of the plates may disturb the alignment of the seeds but later on, when the roots emerge, the seeds become securely attached to the blotter and do not fall.

When germination test is done in light (artificial illumination or diffused day-light in the room) the racks should be placed in such a manner that the plates are inclined towards the source of light. This would enable the root and shoot to grow along the surface of the plate.

During the initial stages of germination, the polythene packet was kept folded and only when the shoots pushed against the polythene fold in certain cases the ends of the packets were straightened. To eliminate the possibility of oxygen tension and carbon dioxide accumulation within the packets, a row of small holes

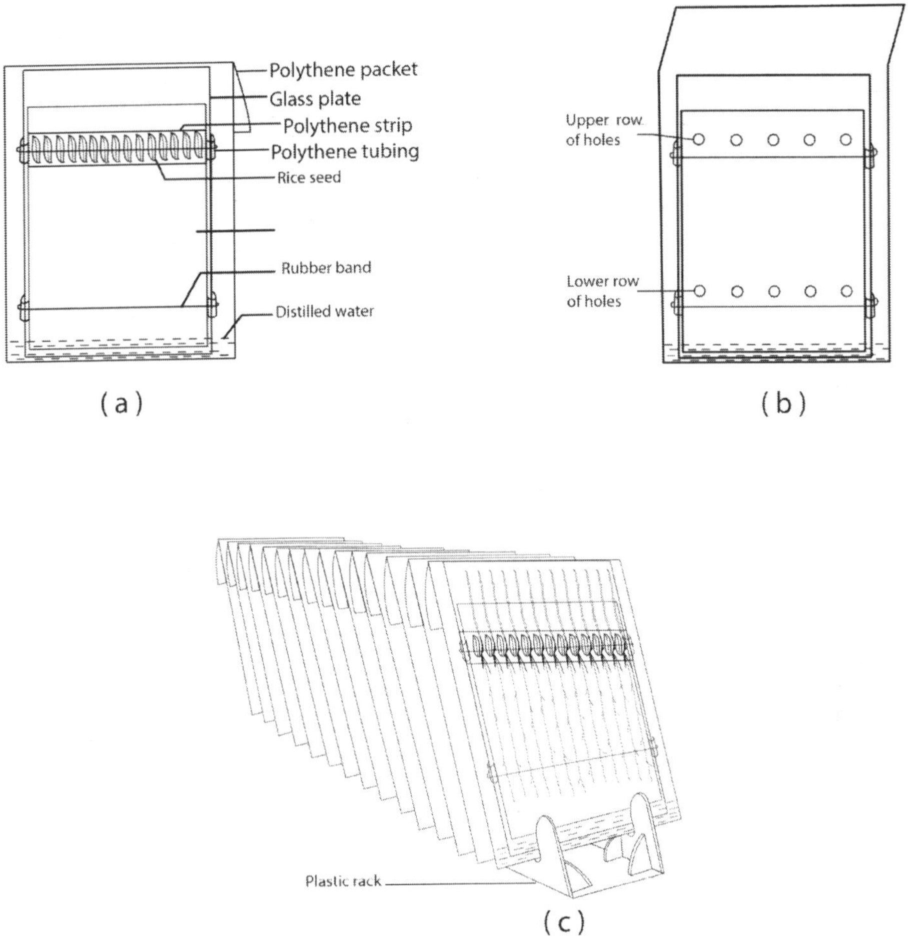

***Figure 1: (a) Diagram showing the placement of rice seeds on moist blotters spread on 25 x 15 cm glass plates kept in polythene packets (35 x 18 cm; made of 0.15 mm thick polythene sheets) containing 30 ml of water. (b) the back side of the polythene packets showing two rows of holes (each 3 mm diam.) and (c) the arrangement of plates on a plastic dish-stand showing the root and shoot growth of rice seedlings after five days. Further details are given in the text. (*Vide* Punjabi & Basu 1982)**

*Figure 1 in original publication.

(5 holes of 3mm diameter in each row) was made at the two ends of the packet as shown in Figure 5.1b. The holes were always positioned behind the germination plate to avoid the possibility of any direct contact of fungal spores with the seeds. It was noted that in closed packets without holes growth of root and shoot was considerably reduced.

The results of germination test with high vigour seeds of few representative materials such as rice (*Oryza sativa*), sunflower (*Helianthus annuus*), mustard (*Brassica juncea*), lettuce (*Lactuca sativa*), jute (*Corchorus capsularis*) and soybean (*Glycine max*) in petri dishes, rolled towels and glass plates in polythene packets with two rows of holes are presented in Table 5.36. The plates were kept in a temperature-controlled room at 24± 1°C. Data on germination and growth of seedlings were recorded after seven days for all the materials except soybean which was taken after ten days.

***Table 5.36: Seedling Growth (Root and Shoot Length in mm) in Different Methods of Germination**

Materials		*Germination Methods*					
		Blotter on Glass Plate	*CV*	*Petri Dish*	*CV*	*Rolled Towel*	*CV*
Mustard	Root	114 ± 1.2	1.05	114 ± 1.2	7.40	114 ± 1.2	3.83
	Shoot	55 ± 1,5	2.72	40 ± 0.6	1.50	114 ± 1.2	5.15
Jute	Root	62 ± 1.7	2.74	42 ± 0.6	1.43	77 ± 2.1	2.72
	Shoot	56 ± 0.3	0.53	51 ± 0.3	0.58	59 ± 1.2	2.03
Lettuce	Root	47 ± 0.9	1.91	34 ± 1.2	3.58	48 ± 2.1	4.37
	Shoot	40 ± 0.3	0.75	32 ± 1.0	3.12	42 ± 0.6	1.43
Rice	Root	120 ± 1.8	1.50	63 ± 1.2	1.90	129 ± 3.2	2.48
	Shoot	59 ± 1.2	2.03	47 ± 0.6	1.28	64 ± 2.2	3.44
Soybean	Root	104 ± 8.5	8.17	42 ± 5.2	12.38	76 ± 9.6	12.63
	Shoot	97 ± 3.3	4.43	63 ± 6.0	9.52	97 ± 9.5	9.79
Sun-flower	Root	219 ± 4.0	1.83	92 ± 12.8	3.91	187 ± 15.2	8.13
	Shoot	104 ± 3.4	3.27	99 ± 4.4	6.22	126 ± 6.2	4.92

Germination tests for jute, mustard and lettuce were done on 15 × 15 cm plates and for rice, soybean and sunflower on 25 × 15 cm plates; tests were made with high-vigour (HV) seeds for 96-100 per cent germination; data on seeding growth were recorded after germination at 24 ± 1°C for ten days in soybean and after seven days for the other materials.

*Table 1 in original publication.

The germination percentage of seeds did not differ significantly and ranged from 96 to 100 but seedling growth as measured by root and shoot length was poorest in the petri dishes irrespective of the material used. Root growth was most affected. In the rolled towel method, the shoots were, in general, longer but a bit thinner than in the glass plate. A high degree of variation in root and shoot growth existed between the replications compared to the plate method. We further noted that in sunflower and soybean, the germination percentage of relatively low-vigour seeds in roll towels was lower and there was higher fungal infection compared to that in the glass plates. In general, all the materials germinated better and showed uniform and good seedling growth on the glass plates. In most cases, less variation was noted between the seedlings on a plate or between plate replicates as evident from the lower standard deviation values (Table 5.36).

Of the different methods tested, the aforesaid technique can be used for various kinds of seeds by varying the length of the plate from 15 to 25 cm. For soybean, sunflower and rice longer plates (25 × 15 cm) and polythene packets (35 × 18 cm) were used. The shorter plates have been successfully employed for most vegetable and flower seeds.

The present method has basic similarities with the Jones and Cobb (1963) method and that of Peterson *et al.* (1978) but has a number of additional advantages. In the Jones and Cobb (1963) method and that of Peterson *et al.* (1978), the plates were placed in a tray containing a common water pool. Since the plates are open, the fungal infection, if and when it occurs, will spread rapidly from one plate to another. In one tray, only one type of material can be placed because the leaching out of the substances into the water may influence seedling growth in the adjacent plates. Greater attention would be necessary to check the water level and a day-to-day visual observation of the process of germination and seedling growth, which is very important from a physiological point of view, would be difficult in such an arrangement. The plastic wrap may also result in oxygen tension and accumulation of carbon dioxide.

In our laboratory, this method is being successfully used over the last about eight years for seeds of cereals, oil and fibre crops, pulses, vegetables and flower crops. These include rice, wheat, maize, barley, sorghum, jowar, bajra, sunflower, safflower, mustard, sesamum, jute, mesta, sugar beet, gram, pigeon pea, lentil, mungbean, soyabean, tomato, chilli, brinjal, cabbage, cauliflower, onion, orka, lettuce, carrot, petunia, nicotiana, antirrhinum, phlox and tithonia. In fact, there is no problem with relatively small and medium-sized orthodox seeds of most crop plants. The small seeded materials will do very well in the shorter plates (15 × 15 cm) but the longer ones (25 × 15cm) can also be employed. The shorter plates are, however, more convenient to handle. For seeds of bigger size, in which the seedlings are also larger, only the longer plates should be used.

The placement of the seeds on the plate may take a little more time than other methods, especially for the beginner, but there would be considerable saving of time in recording root and shoot growth.

The present method has, in general, the following advantages over the other commonly used methods of germination:

(i) Seedling growth is better in most of the materials, there is greater uniformity in growth as is evident from the low standard deviation values (except in a few cases) for root and shoot length.

(ii) All the seeds on the plate received equal amount of water; as water travels up by capillary movement there is no chance of waterlogging of the seed.

(iii) The effect of different chemicals on germination can be very conveniently studied by putting the solutions in the polythene packets.

(iv) Little attention is required; addition of water over the initial 20-30 ml is not necessary during the entire period of germination.

(v) The possibility of fungal contamination from one plate to another is completely eliminated.

(vi) For seeds which require light during germination, there is no problem of giving the necessary exposure; for seeds requiring wet prechilling(as in some lettuce cultivars), the plastic stands with the plates may be directly placed in the required shelf of the refrigerator for the recommended period.

(vii) Day-to-day visual observation of the progress of germination especially shoot and root growth is possible without disturbing the seedlings in any way.

(viii) There is little entangling of the root system as noted in the petri dish method.

(ix) At the end of the test, the plates may be stacked in the refrigerator or dried up and data may be recorded afterwards.

(x) The method involves the minimum of cost as compared to most other methods. The polythene packets, slit tubes, bands, *etc.* can be used many times after thorough washing with detergent and distilled water. No elaborate equipment is required and hence it can be followed under all laboratory conditions.

The technique has not, however, been successful in seeds of cotton in which the germinating seedlings tend to come out of the blotter strip disturbing the alignment of seeds on the plate. Perhaps, for such seeds further modifications of the technique would be required. Studies are presently being conducted to use the methodology for successful vigour rating of seeds of different germination potential and vigour status.

Basu, R.N. (1994). Seed Invigoration studies in the University of Calcutta; glimpses of the past and present* - Seed Tech. News, pp. 1-12.

Let me first express my deep sense of gratitude to the Indian Society of Seed Technology for the formal cognizance of our humble contributions to seed science and technology. I must say that the honour associated with this felicitation has to be shared with all my students, scholars and research collaborators, but for whom the ideas would have remained ungerminated.

I confess that I was a bit hesitant to come over here as I would prefer to remain away from such events. The remembrance of a particular day long back, about which I shall tell you shortly, made me change my mind and I am here to express our thanks to ISST for the support we have been receiving from the Society all the time. Another near-compulsive reason is the choice of the present venue by ISST – the city of Coimbatore, one of the finest centres of seed production in the country. Nearly a decade back, in 1985, the Tamil Nadu Agricultural University invited me to deliver a course special lecture on seed treatment. It was perhaps the first University to do so. Not only that, I found people here at Coimbatore brave enough to soak 10 metric tonnes of hybrid bajra seed for viability maintenance following our techniques. Further, I am so happy to find here my doctoral student and friend Dr. C. Dharmalingam as the Professor and Head of Seed Technology, Dr. K. Ramamoorthy who also obtained his doctorate degree working with me is now an Associate Professor of Seed Technology and good old friend Dr. Karivaratharaju is the Director of Research of the University. It is almost like a home coming for me.

Friends, unlike most of you assembled here this evening, I had little knowledge of seed science and technology when I ventured into this largely unknown but fascinating field of study. As a student of agriculture, and plant physiology in particular, I have always been aware of the importance of seed biology and its manifold implications. In my research on the physiology of regeneration of adventitious roots, especially for elucidating the mechanism of synergism between auxins and non-auxinic chemicals in regeneration, I had been routinely using good quality seeds of mung and French beans as I required uniformly grown seedlings for making standard cuttings. Often there were problems with germination but we had little time left to bother about that. We simply changed the seed lot.

My switch over from researches on the physiology of root regeneration on cuttings to seed physiology (thematically, however, both are propagules) was because our studies on synergism had virtually reached a dead end, at least from the fundamental standpoint. Of course, we could go on making effective formulations, consisting of auxins, auxin synergists and other accessory chemicals, for practical use in horticulture or in tissue culture; and in fact, some of our formulations have gone into commercial use through colleagues and friends. The infrastructure required for basic studies on synergism, especially the elucidation of the primary events in the dedifferentiation phase of regeneration, was beyond our reach. I was indeed looking forward to explore other avenues. At that juncture, the serious problem of rapid loss of viability of stored seeds, particularly of *boro* (winter) rice and wheat, in Bengal drew our attention. Dr. T.S. Ghosh the Director of Jute Agricultural Research Institute (ICAR) and Dr. S.L. Basak, Head of the Division of Genetics and Plant Breeding of the Institute emphasized the need to work on carried-over jute seed stored under warm-humid ambient conditions of the State. The Director of Agriculture, West Bengal, Shri B. Mandal, and Joint-Director of Agriculture Research, Shri D. Mukherjee, showed keen interest in studies on vigour and viability of stored seed. I responded to the situation and submitted research proposals, which were sanctioned in the due course.

The views of the eminent plant physiologist, Professor S.M. Sircar and his associates on dormancy and viability of rice seed attracted us. According to them, a supraoptimal concentration of auxin and other autotoxic metabolites and inhibitors like abscisic acid could be responsible for the loss of viability in storage. If such accumulations could be minimized, loss of viability could possibly be controlled. I remembered an important observation in our early root-regeneration studies wherein the dipping of bases of easy-to-root cuttings from girdled mango seedlings in dilute solutions of auxin and non-auxinic chemicals or even in pure distilled water resulted in complete failure of the cutting to regenerate roots due to leaching out of vital root regenerating factors accumulated in the girdled seedlings. May be the toxic metabolites and inhibitors accumulated in the dry stored seed could be leached out in the same way by soaking the seeds halfway during storage before actual loss of viability took place. Of course, the seed must be dried back before restorage to have a new lease of life. Fortunately for us, the seed used was medium vigour rice and carried-over low vigour jute seed. My two honorary lady scholars, Prativa Pal and Krishna Chattopadhyay were entrusted with the work on rice and jute, respectively. Had the seed lots used by them been from harvest-fresh high vigour stocks, soaking-drying would have adversely affected storability and perhaps we would have abandoned that line of approach. Another lucky coincidence was our selection of short soaking durations, otherwise germination advancement would have been incompatible with the drying back before storage. The redried seeds were stored at 98 per cent RH for 2-3 months. Later, we subjected the seed to accelerated ageing condition at 40-45°C and 100 per cent RH to cut down the time for ageing to weeks and days. The first germination test of the aged material showed a difference in germination of about 30 per cent in rice between respective untreated controls and water soaking-drying treatments. After three months of ageing, soaking-drying showed 50-60 per cent improvement in germination over control along with significant betterment of shoot and root growth of seedlings. The differences were so spectacular that we ourselves found it difficult to accept the results, what to talk of others, many of whom outrightly rejected the findings, attributing the same to gross experimental errors. We went on repeating the experiments over and over again till we were convinced to take the decision of publishing the results. While all other journals refused to oblige, Indian Agriculturist, the quarterly journal published by the Agricultural Society of India, whose editor was a friend of long standing and who himself saw the experiments, agreed to publish the results in the form of a short note in the first number of volume 17 of 1974, with the explicit understanding that the responsibility would be ours only.

The paper was subsequently abstracted in Abstracts of Tropical Agriculture (Amsterdam) in August, 1975 vol. *1*(8) Ref. no. 7503771 and Field Crops Abstracts (CAB) in March 1976 Ref.no. 2099.

Subsequently, the editorial board of the celebrated journal of the USA, "Science" found the contents of the paper too specialized although the reviewers recommended its publication. If I remember correctly, 'Nature' also was equally evasive. We found it really difficult to convince others about the efficacy of the soaking-drying treatments. After all, water for treating a dry-stored orthodox seed for storage is a taboo almost as old as the agrarian civilization.

Nevertheless, the news spread and I was invited by Dr. M.V. Rao of the ICAR to give a talk at he 14th All India Wheat Workers' workshop in August 1975 at the Bidhan Chandra Krishi Viswavidyalaya, West Bengal. The paper, "Prolonging the viability of wheat seed" - authored by Malavika Dasgupta (now Dr. Malavika Dadlani) and myself was presented by me after the routine workshop transactions and I believe I did pretty well. But then, in the question time, the-then Chairman of the Food Corporation of India made a scathing criticism of our paper and in unequivocal terms denounced the approach. Although, I was disheartened and discouraged at that time, looking back to the event after nearly two decades, I must admit that the learned member of the audience did a great service to our cause. On coming back to Calcutta, the very next day I told Malavika to send copies of our paper to Professor Eric Roberts of the University of Reading and Dr. Walter Heydecker of the University of Nottingham, two of the leading seed scientists of the world, for their valued comments. Almost by the return of post, Dr. Heydecker wrote to Malavika, "I must say they are amongst the most exciting pieces of literature, I have seen for quite a while. Who would have thought that simply soaking in water would have this effect on the viability and vigour of seeds?" Not only did Dr. Heydecker in his subsequent papers cite our work as 'outstanding', 'most important discovery' and so on, but he also sent me an invitation to present a paper in an international congress being held in the UK. Although the programme for the same had already been finalised and printed, Dr. Heydecker sought to accommodate me by squeezing the already invited speakers and wrote directly to the-then Vice-Chancellor of the Calcutta University so that I could make it. We could not, but that is another story. Professor Roberts also wrote back to Malavika after a lapse of almost three years (may be after confirmation of the findings!) with the comment that he found our work as 'most interesting' and suggested a few points on the mode of action of the treatments.

In the meantime, I was invited by Dr. P.K. Agrawal the organizer of the 3rd All India Seed Seminar held under the auspices of the Indian Society of Seed Technology at the Indian Agricultural Research Institute, New Delhi in February, 1976. The paper "Physico-chemical control of seed deterioration" (published in Seed Research volume 4 in the same year) summed up the findings on different crop seeds. I talked for more than an hour; the chairman of the session either forgot or intentionally did not stop me. The paper was widely acclaimed by the large gathering of seed scientists and when the ISST is felicitating us for our humble contributions, I remember that important day of February 17, 1976 when I believe we were finally accepted with confidence.

I may be permitted to recall another event which happened in Delhi a couple of years later. I was invited to an Indo-Soviet Cereal Improvement Seminar held at IARI. Immediately after I finished speaking, the leader of the Soviet delegation Academician N.V. Turbin rushed to the podium, shook hands with me and pinned on my coat lapel an insignia of Indo-Soviet friendship as a mark of his admiration for our work on seed invigoration. I was really overwhelmed by the spontaneous response from such a highly respected academician of the Soviet Academy of Sciences. I can still visualise the huge figure of Professor Turbin towering over my small physical stature. Academician Turbin and his team even made an unscheduled 2-hour visit to our laboratory in Calcutta to see for themselves what we had been doing. The inclined glass plate blotter method of germination test standardised by us (with Dr. Bina Punjabi as the main collaborator) so impressed them that they took some plates with seed placed on them for germination along with the plastic rack. Friends, those were really happy moments for us in the field of seed physiology. I know I have taken a lot of your time over personal reminiscences but I believe that to the younger generations of seed scientists these may give some provisions for contemplation.

I now come to the subject proper – the seed, a most wonderful creation of nature; and briefly discuss the ways and means to further improve its performance through invigoration.

By seed invigoration we mean an improvement in seed performance by any post-harvest physico-chemical treatment resulting in better storability, improved germination, and field performance over a wide range of edaphoclimatic conditions than the corresponding untreated seed. Although the efficacy of various treatments such as low dosages of irradiation, exposure to magnetic and electric fields and other treatments, as well as various post-sowing to pre-harvest *in situ* cultural treatments for improvement of seed quality, has been reported in the literature, we shall deal here only with post-harvest physiological treatment especially 'wet' and 'dry' treatments for prestorage, mid-storage and post-storage (pre-sowing) seed invigoration. An obvious reason is our own involvement in the subject for over two decades and also because of the author's insufficient background and lack of expertise in the other fields of specialisation on seeds.

Wet Seed Treatments

Wet treatments are suitable for pre-sowing and mid-storage invigoration of seeds. The pre-sowing treatments were developed in the erstwhile Soviet Union by Henckel and coworkers primarily for the purpose of imparting drought resistance and involved 2-3 cycles of 24 hour-imbibition-drying. The effects of such pre-sowing treatments have, however, not been very consistent; reports of beneficial effects in many crops have been countered by marginal effects or no effects at all. The reasons of such inconsistency have been analysed elsewhere.

The relatively short-term presoaking in water or dilute solutions of potassium and sodium phosphate (duration of soaking not exceeding 6-8 hours) would significantly improve germinability, uniformity of stand and subsequent crop performance.

For the maintenance of vigour and viability in storage, wet seed treatments are very effective when given about midway during ordinary uncontrolled storage to the seed which has gone down in vigour but not yet in viability. In the mid-term hydration-dehydration treatments, it is necessary to raise moisture content to 25-30 per cent (on wet-weight basis) for certain preventive and restorative intracellular reactions prior to drying back for restorage. Except in certain seeds with deep endogenous dormancy, orthodox seeds cannot be stored in the hydrated state. Hydration can be accomplished in various ways and the efficacy of the different methods of hydration depends on the initial seed vigour status. Thus, soaking for 1-6 hours in water (or dilute solutions of certain chemicals), depending upon the kind of seed, is effective in low and medium vigour stocks. In relatively high vigour seed, moisture equilibration with a water-saturated atmosphere for 24-48 hours (humidification) or just dipping the seed in water or dilute chemical solutions for 2-5 minutes, spraying or adding a fixed quantity of water to the seed (to attain seed m.c.of about 25-30 per cent) and then keeping the surface-wetted seed covered for 2-6 hours, would serve the purpose. In all cases, it is absolutely essential to dry the seed back to at least the original moisture content. For the farmer, we suggest that starting wih 10 kg of seed, the final weight after drying should be slightly less than 10 kg to account for some inadvertent loss during treatment.

For leguminous seeds such as the common pulses, peas and beans which are susceptible to soaking injury, direct immersion in water must be avoided. For such seeds, very slow hydration is necessary which is possible through osmoconditioning *via* moist sand employing osmotica like PEG (-5 bar) Jalshakti (a modified starch-cellulose preparation, 10 per cent solution in moist sand conditioning) potassium phosphate (0.3 M), *etc.*, [seed: sand ratio, 1:3, 1:4] followed by careful washing of the chemicals and short periods of imbibition for some kinds of seeds, moisture equilibration with water-saturated atmosphere or moist sand conditioning with or without subsequent short-term imbibition. Successful conditioning of seed with inert carriers such as expanded vermiculite, Micro-Cel E to improve seed performance has been done in the USA also. Incidentally, moist sand conditioning using the aforesaid osmotica but without drying; constant contact has shown considerable extension of seed viability of recalcitrant sal (*Shorea robusta*) and semirecalcitrant jackfruit (*Artocarpus integrifolia*) and tea (*Camellia assamica*) seeds.

Of the aforesaid slow hydration treatments, moist sand conditioning is the cheapest and most easily practicable. The dry seed is mixed with premoistened sand (5-8 per cent m.c. seed: sand ratio 1:3 or 1:4 for most kinds of seeds) and incubated in a covered container for 16 to 48 hours depending upon the material for a slow uptake of moisture. The moisture content of the sand should be so adjusted that final seed moisture content does not exceed 30 per cent on wet weight basis;

to avoid advancement of germination process to an extent not compatible with drying back. Almost total uptake of water by the hygroscopic seed will leave the sand virtually air-dry and simply sieving out the seed will enable re-use of the sand for further treatment.

While the effect of the mid-storage hydration treatment on immediate germinability of the treated seed is only marginal, highly significant differences in vigour and viability are noted after accelerated or natural ageing under ambient conditions. The invigoration effect is manifested upon all aspects of seed performance, namely, enhanced tolerance to ageing, heat, irradiation and other stresses in terms of germination percentage, seedling growth; and biochemically, on amylase and dehydrogenase enzyme activities, reduced lipid peroxidation and significantly lowered volatile aldehyde production (as determined by chemical estimation, and jute seedling bioassay – a method standardised by us with Dr. Kalpana Sur as the major contributor), and more importantly from the agricultural view point, improved field performance and productivity.

Dry Seed Treatments

Initially, I thought I would tell you only about dry seed treatments, as we have talked enough about wet seed treatments over the past two decades in different national and international symposia and seminars. However, in view of the ISST's decision to felicitate us for our contributions to seed science, I gave a second thought; after all, the greater part of our humble contribution has been in the field of wet seed treatment and I considered it my duty to tell you, particularly our young researchers, the developments in wet seed treatment and also share with you our experience on dry seed invigoration treatments.

The concept of 'dry' treatment originated from our studies on the elucidation of the mode of action of the wet physiological seed invigoration treatments. Our studies suggested that oxidation of the unsaturated lipid moiety of the lipoprotein membranes of the cell was one of the major reasons of loss of vigour and viability of dry-stored seeds. If the unsaturated moiety is made saturated by iodination, the risk of peroxidation, hence rapid leakage of ions and metabolites from the lipoprotein membranes, would be minimized. Iodine was used in the vapour form directly or through inert carriers like calcium carbonate and, as such there was no need of drying back. The problem of drying back, particularly in humid conditions (especially in the absence of artificial drying facilities), could be effectively eliminated. Further, unlike the wet treatments, the dry treatments, although less effective than the wet treatments, can even be given to high vigour harvest-fresh seed before bagging and storage which is distinctly advantageous to the seed producer.

A paper on the counteraction of lipid peroxidation of mammalian liposomal membrane by capsicum, the active ingredient of *Capsicum frutescene*, tempted us to use chilli powder for improving storability of seeds. We also used leaf powder of neem (*Melia azadirachta*), and rhizome powder of curcuma (*Curcuma domestica*), which are used as insect repellents and preservative-cum-colouring agents in cooking, respectively. Initially our experiments were confined to wheat but recently we find that these crude plant powders are effective in quite a good number of seeds. An added advantage of dry-dressing with plant preparations, besides applicability in high vigour seeds and elimination of drying, is the safe eco-friendly approach; the treated edible seeds can be consumed if necessary.

Although we initiated our studies on seed vigour and viability with the practical objective of maintaining good germinability at sowing and ensuring better field performance in the rather non-conducive seed storage environment of West Bengal, we realised that an understanding of the mode of action of the seed treatments was necessary to develop still more practicable methodologies for farmers and seedsmen. If I have been able to take you along the path traversed by us over the years, you would agree that we have just followed the dictates of the experimental findings. It is true that we have not been able to fully elucidate the mode of action of the physico-chemical treatments but through hypothesis testing and rejection-acceptance procedures, the focal area has been considerably narrowed down. Thus, our initial postulation of leaching out of toxic metabolites from the seed was ruled out in view of the high degree of effectiveness of humidification-drying where the question of leaching in liquid water does not arise. Advancement of germination, differential moisture uptake by the treated seed, antimicrobial effect of the treatments cannot be accepted as basic reasons of invigoration because of lack of response of high vigour seed, similar moisture absorption or desorption by treated and untreated seeds and effectiveness of the treatments under storage conditions not conducive to microbial invasion. Whether repair of vital bio-organelles during hydration is responsible for invigoration is still a matter of conjecture, since there is very little change in membrane functions or enzyme activities before ageing. Differences show up after ageing, implying a prophylatic rather than curative action of the treatments. Further the observation that humidification-drying is effective even in seeds pretreated with the cytoplasmic protein synthesis inhibitor cycloheximide, is indicative of non-involvement of *de novo* protein synthesis during hydration-dehydration treatment. Any repair of bio-organelles would thus be dependent on pre-existent enzyme proteins. The possibility of metabolic breakdown of toxic inhibitors is also not clear as our studies have not shown any activation of the enzyme superoxide dismutase, a scavenger of endogenous free radicals immediately after hydration. The role of other enzymes responsible for breakdown of toxic metabolic byproducts is yet to be studied. Three common observations namely, reduced post-ageing lipid peroxidation, low volatile aldehyde production and better dehydrogenase enzyme activity are suggestive of a significant protective action of treatments on lipids and proteins and particularly lipoprotein membrane structures. The leakage of ions and metabolites from the deteriorated seed is indeed one of the early symptoms of damage to the semi-permeable cell membrane. We have now enough evidences, though mostly indirect to suggest that lipid peroxidation reactions (catalyzed by free radicals), involving the unsaturated lipid moiety of lipoprotein membranes, is possibly a primary reason of deteriorative senescence of the stored seed. Free radical reactions also effect the structural integrity of proteins resulting in reduced activity of many vital enzymes. How hydration alone can quench free radical reactions is yet to be elucidated, but besides other possibilities, mutual annihilation of free radicals in the dry stored seed by short-term hydration could be a reasonable speculation. The effectiveness of the chemicals used in the soak-water could be attributed to their role as antioxidants or antioxidant-synergists. The enhanced beneficial activity of ascorbic acid in presence of the iron chelating agent, desferrioxamine B is a typical example of such synergism.

The dry invigoration treatments with diverse substances and methods possibly enter and react with seeds, in ways ultimately resulting in lower lipid peroxidation and reduced production of volatile aldehydes. In fact, these parameters are the attributes of all good quality seeds. Iodination or chlorination has been suggested to remove, to a certain extent, the degree of unsaturation in the lipid moiety of lipoprotein membranes and render the same less susceptible to peroxidative attack. The antioxidative or antioxidant-synergistic action of the chemicals used for dry-dressing would check the free radical reactions provided that they reach the target. As such, the volatiles have a distinct advantage in dry treatment. We can not, however, tell how certain volatile compounds, some of which are indeed products of anaerobiosis or lipid peroxidation, could improve storability at low concentrations. Could it be that their early presence would depress in a feedback or end product inhibition mechanism, the subsequent autocatalytic reactions? An answer to that would require critical biochemical studies at the cellular level.

Dry dressing with powdered plant preparations was accompanied with low post-storage volatile aldehyde production, suggesting a controlling effect on endogenous lipid peroxidation. However as pointed out earlier these could be mere secondary manifestations rather than the primary reasons for improved storability. Our very recent studies with a wide range of plant preparations including powdered tamarind, mango, colocasia and thevetia leaves have shown significant beneficial effects. Possibly there is some effective chemical(s) common in them, beside the more specific active ingredient(s) in the particular plant, which renders a protection to the stored seed. We also studied the effect of acetyl salicylate (aspirin) in view of its reported anti-ageing effect. The entry of sugar molecules between collagen fibres of the retinal proteins, leading to diabetic cataract, could be checked by early intake of aspirin, the acetyl group of which preoccupies the position, thus effectively preventing the entry of the sugar molecule. A systematic approach may be made towards understanding the protective effect of such chemicals on lipids and proteins in improving post-storage germinability.

Friends, to some of you, our approach may appear unconventional and arbitrary, but throughout in our studies there has been an attempt to prevent free-radical induced lipid peroxidation and its consequent damage to proteins, other vital biomolecules and organelles. Perhaps some of you may discover better methods and take up appropriate experiments for the purpose. More and more critical experiments are not merely necessary but imperative. Regarding experiments, I would like to conclude with the wise saying of the great artist-cum-experimentalist, Leonardo da Vinci, amongst whose many superb creations is the immortalization of the mystic smile on the lips of Mona Lisa Giocondo. I quote, "Experiments never deceive, it is our judgement that sometimes deceives itself because it expects results that experiment refuses. We must consult experiment, changing the circumstances until we have deduced reliable rules".

Once again, I on behalf of all my past and present research associates, express our gratitude to the Indian Society of Seed Technology for this felicitation and sincerely thank all of you for your kind attention.

**Special invited lecture at the Tamil Nadu Agricultural University, organized by Indian Society for Seed Technology.*

Bhattacharya, A.K., Rahman, F. and Basu, R.N. (1994). Preservation of Tea (*Camellia sinensis* L.) seed viability. *Seed Research*, 22(2), 108-111, 1994.

Clonal propagation is practiced for commercial cultivation of tea and seed obtained from clonal varieties is considered an important planting material due to its wider adaptability in respect of varied agro-climatic conditions, pest incidence, nutrient uptake, soil heterogeneity and other cultural practices (Barua, D.N. 1989). The availability of tea seed is restricted only for a short period ranging from October to December which is the normal harvesting time in the North Eastern Region of India, the largest tea producing area of the Indian subcontinent (Bezbarua, H.P. 1975 and Barua, D.N. 1961). Tea seed exhibits recalcitrant behaviour with respect to loss of viability which is rapid (King, M.W. and E.H. Roberts 1980) on exposure to ambient condition due to loss of moisture and associated deteriorative senescence and microbial attack (Harrison, J.C. 1926, Rashid, A., B. Alam and H. Chakraborty 1973, Verma, M.M. and S.K. Banerjee 1983, Barman, T.S. 1988). So, for raising seedlings, the common practice is to sow the seeds in the nursery immediately after harvest. However, seedlings thus obtained have to pass through the subsequent dry spell in the summer months which necessitates extra investment for nursery irrigation. Storage of tea seeds under cold conditions with or without prestorage seed treatment with fungicide have been reported by several workers (Barman, T.S. 1988, Hume, P.F. 1955, Sebastiampillai, A.R. and T.I. Anandappa 1979, Arma, S. and A. Watanabe 1983, Barman, T.S. and C.M. Sarma 1987, Rahman, F. 1987). However, this method is not cost-effective for commercial (bulk) seed storage. The present study aims (i) to determine the safe limits of seed moisture content at which it would remain viable for prolonged period without germination, and (ii) to standardise a suitable treatment for the maintenance of vigour and viability of seeds under ambient conditions for more than 6 months.

Freshly harvested biclonal tea seeds of TS-450 were collected from the Research and Development Department of Macneill and Magor Limited, Distt.

Sobitpur, Assam in the 2nd week of November, 1989. After sinker-floated test, only the sinkers were selected for the present experiment.

To determine relationship between the moisture content and the viability, seeds were exposed to a current of air at a temperature of 30 ± 1°C in a forced air circulation cabinet. Seeds were divided into 8 lots, each containing 50 seeds and were placed inside the cabinet. One lot each was taken out after 8, 16, 24, 36, 48, 72 and 96 h for determining moisture content and germinability. From each lot 25 seeds were used for moisture determination and the rest 25 seeds were placed for germination test. Moisture content was determined by oven drying at 105°C for 16 h. For germination test seeds were soaked overnight in distilled water before sowing in sand at 25°C. Germination (per cent) including non-germinated viable, cracked and sprouted seeds, root and shoot length were recorded after 35 days of sowing. Sand was kept sufficiently moist by watering uniformly at regular intervals.

***Table 5.37: Effect of Moist-Sand Pre-conditioning Treatments on the Germinability and Vigour of Tea Seeds after 7 Months of Storage**

	Treatments	*Germinability (Per cent)*		*Root Length (mm)*	*Shoot Length (mm)*	*Total Seedling Length (mm)*
1.	Control	0	(.0.0)	-	-	-
2.	Water	5	(12.9)	-	5	5
3.	0.25 per cent Dithane M-45	65	(53.7)	131	53	184
4.	0.05 per cent Mercuric chloride	30	(33.2)	-	8	8
5.	0.2M Pot. dihydrogenphosphate	30	(33.2)	60	56	116
6.	Tr.3 + Tr.5	35	(36.2)	48	63	111
7.	Tr.4 + Tr.5	40	(39.2)	95	77	172
8.	Polyethylene glycol 6000 (-5 bar)	60	(50.7)	117	72	189
9.	Tr.3 + Tr.8	65	(53.7)	130	84	214
10.	Tr.4 + Tr.8	70	(56.7)	113	58	171
11.	Control at 10°C	40	(39.2)	119	45	164
C.D. (P = 0.05)		(2.69)		3.49	3.96	6.32

Values in parentheses denote transformed angular values; Treatments No.2 to 10 with moist sand conditioning, Seed: Sand = 1:3, air-dry sand moistened to 5 per cent (W/W) with water or respective chemical solution in water.

*Table 1 in original publication.

For storage experiment, medium grained sand was thoroughly washed in running tap water, sun-dried and then dried overnight in the oven at 100°C followed by cooling outside. In different treatments (Table 5.37) water or chemical solutions @5 per cent W/W were thoroughly mixed with the sand and seeds were uniformly covered with this pre-moistened sand in the ratio of 1 part seed and 3 parts sand (weight/weight) and then stored at room temperature in loosely bound polythene packet for easy exchange of gases. After 7 months of storage germination was recorded in three replicates of 50 seeds following the method mentioned before.

Viability of tea seed was found to be critically influenced by its moisture content. Freshly harvested seeds having 43 per cent moisture content recorded 96 per cent germination (Table 5.38). But when the moisture content of the seed dropped below 29 per cent, viability started declining and at 26 per cent m.c. seeds became virtually non-viable. Moisture content is an important factor for storage of recalcitrant seeds, because when moisture content drops below a critical level seeds start to deteriorate but when the moisture content exceeds a supraoptimal level, germination starts. It is rather difficult to determine the critical moisture content at whichthe seed would remain viable for a longer period without germinating (King, M.W. and E.H. Roberts 1980 and Chin, H.F., Y.L.Hor and M.B. Mohd. Lassim 1984). The present results are in general conformity with the earlier reports of Sebastiampillai and Anandappa (Sebastiampillai, A.R. and T.I. Anandappa 1979) and Barman (Barman, T.S. 1988). Besides germination percentage, seedling vigour as determined by the growth of the seedlings, also declined progressively with the loss of seed moisture content. The present result thereby confirms the recalcitrant nature of tea seed.

***Table 5.38: Effect of Moisture Content on Germinability of Tea Seed**

Duration of Forced Air Drying (h)	*Moisture Content after Forced Air Drying (Per cent wet weight)*	*Germination (Per cent)*	*35 Days after Sowing*		
			Root Length (mm)	*Shoot Length (mm)*	*Total Seedling Length (mm)*
0	42.71	96	76	68	144
8	41.67	93	75	66	141
16	39.52	88	69	62	131
24	38.38	85	64	58	122
36	37.22	83	61	53	114
48	29.59	74	52	46	98
72	26.82	46	34	23	47
96	23.41	15	16	12	28

*Table 2 in original publication.

Among the various treatments polyethylene glycol (-5 bar) with or without fungicides (Tr. 8, 9 and 10) showed significant extension of seed viability (Table

5.37). The other treatments, such as, moist sand conditioning (MSC) with dithane M-45 (Tr. 3) was also quite effective in extending tea seed storability. Beneficial treatment effects were also manifested in better growth of root and shoot. The maintenance of optimum seed moisture content by sand, premoistened with water or osmotic solution, is a critical step in the retention of tea seed viability. As such, if the initial seed moisture is high, germination may take place during storage itself, and if too low, the final seed moisture in a particular treatment may fall below the desirable moisture range resulting in the loss of viability. The germinability and subsequent growth of the cracked and sprouted seeds was, however, as good as the non-germinated viable seeds. The beneficial effect of fungicides during storage (Barman, T.S. 1988, Barman, T.S. and C.M. Sarma 1987 and Rahman, F. 1987) especially in moist seeds was clearly demonstrated in this study.

From the above study it is concluded that improved storability of tea seed under ambient condition is possible by maintaining a proper seed moisture content and controlling microbial spoilage. Further studies on the standardization of moisture content using other chemicals and media for storing seeds without germination/sprouting during storage is under progress.

Pan, D. and Basu, R.N. (1985). Absence of *de novo* protein synthesis in moisture equilibration-drying treatment for maintenance of lettuce (*Lactuca sativa* L.) seed viability. *Indian Journal of Experimental Biology*, 23, 375-379.

Maintenance of seed viability by short-term soaking-drying treatment of stored seed was first reported from this laboratory. Hydration of stored lettuce seed by moisture equilibration with a water-saturated atmosphere followed by drying back proved very effective in the maintenance of vigour and viability. But the mode of action of hydration-dehydration treatments is still not clear.

The greater storage potential of moist-stored (dormant) lettuce seed than corresponding dry-stored seed might be attributed to the ability of the former to undertake repair of damage to bioorganelles in the imbibed state; enzymatic repair of biochemical lesions and membrane synthesis would not occur in the dry seed in which the ageing damage would accumulate leading to loss of viability.

As protein synthesis should normally be necessary for membrane turnover and for repair of bioorganelles, we thought it worthwhile to verify whether the beneficial effects of the hydration-dehydration treatments would also be obtained in the presence of cycloheximide which is an effective inhibitor of cytoplasmic protein synthesis.

Medium vigour (MV) lettuce seeds of the cultivar 'Suttons A' which requires a wet prechilling (5°C for 48 hr to break dormancy) for germination around 28°C have been used in this study. The seeds can be kept in dark at 30°C and above in the fully imbibed conditions for long periods without germination.

Two sets of experiments were carried out to study the role of protein synthesis in viability maintenance by hydration-dehydration (H-D) treatments.

Firstly, cycloheximide was introduced into the seed by the dry permeation technique employing acetone as the solvent. The soaking duration was 8 hr and the concentrations of cycloheximide were 3×10^{-3}, 5×10^{-3} and 6×10^{-3} M. We could not give longer soaking duration for dry permeation employing acetone because soaking in acetone for prolonged period reduced storability. In the second experiment, the seed was soaked in different aqueous solutions (10^{-4}, 2×10^{-4} and 5×10^{-4} M) of cycloheximide for 2 hr and then the soaked seeds of different treatments were kept in an incubator at 30°C in the imbibed state in petri dishes on wet inhibitor-soaked blotters for another 70 hr. The same concentration of inhibitors could not be used in dry and wet permeation sets because of much greater inhibitory effect of cycloheximide in aqueous solution. The long period of imbibition was adopted to ensure entry of cycloheximide inside the seed. As this material required prechilling to break dormancy, advancement of germination was prevented by keeping at 30°C. Thereafter, the seeds were dried and subjected to ageing at 90 per cent RH and 40°C for 18 days (basic ageing) to make them receptive to further moisture equilibration-drying treatments. In both the sets of experiments, one-half of the dried seeds of different treatments, including the untreated controls, were equilibrated with a water-saturated atmosphere (approx. 100 per cent RH at 26°C) for 24 hr and then dried back. The moisture equilibrated-dried and non-equilibrated seeds were subjected to accelerated and natural ageing before germination test. The soaking in aqueous solution was done in view of some doubts regarding the efficacy of organic solvents to introduce chemicals into the embryo as would be case with aqueous solutions.

The effects of seed treatment (hydration-dehydration) were analysed statistically and treatment means within a column have been scored with different letters using Duncan's multiple range test. The interaction of cycloheximide and moisture equilibration treatments on germinability was calculated by analysis of variance after angular transformation of germination percentage data.

Germination test immediately after dry permeation of cycloheximide into the seed showed significant reduction in root length of seedlings by the chemical but germination percentage and shoot length were not affected (Table 5.39). No beneficial effect of moisture equilibration-drying was noted in the before ageing set but the trend of inhibition of root growth by cycloheximide at 5×10^{-3}M was more pronounced than that noted in the non-equilibrated seed.

After accelerated and natural ageing, however, moisture equilibrated-dried seeds showed highly significant beneficial effect on germinability over non-equilibrated seeds. Under accelerated ageing at 90 per cent RH and 40°C, higher concentration of cycloheximide (5×10^{-3} and 6×10^{-3}M) caused significant reduction in germinability of seeds but the moisture equilibrated seeds of different cycloheximide pretreatments showed greater vigour and viability than the corresponding non-equilibrated seeds (Table 5.39). The germination percentage and seedling vigour were inhibited especially by 5×10^{-3} and 6×10^{-3}*M*

***Table 5.39: Effect of Moisture Equilibration-Drying (ME-D) on Germination Percentage and Mean Root Lengths of Cycloheximide-Pretreated (Acetone solution) Lettuce Seeds without Ageing and with Accelerated Ageing at 90 per cent RH and 40°C for 14 Days and Natural Ageing for 120 Days (av. RH 77.4 ± 3.1 per cent; av. Temp. 29.4°± 0.5°C) (Values within a column under a particular ageing, scored by the same letter, using Duncan's multiple range test, are not significantly different at $p = 0.05$)**

Cycloheximide Pre-treatments (Molar conc.)		*Germination (Per cent)*		*ME-D Advantage (Per cent)*	*Root Length (mm)*		*ME-D Advantage (Per cent)*
		Post-CHI Treatments			*Post-CHI Treatments*		
		NME	*ME-D*		*NME*	*ME-D*	
I	Without Ageing Control (untreated) Cycloheximide (*M*)	98.1[a]	98.6[a]	0.5	71[a]	72[a]	1.4
	0 (acetone)	88.0[a]	98.4[a]	0.4	70[ab]	71[a]	1.4
	3×10^{-3}	98.3[a]	98.1[a]	-0.2	68[ab]	70[a]	2.9
	5×10^{-3}	97.0[a]	98.1[a]	1.1	67[b]	65[b]	-3.0
	6×10^{-3}	96.2[a]	96.7[a]	0.5	66[b]	64[b]	-3.0
II	Accelerated Ageing Control (untreated) Cycloheximide (*M*)	24.4[a]	73.8[a]	202.4	25[a]	44[a]	76.0
	0 (acetone)	16.6[b]	64.4[b]	287.9	22[ab]	43[a]	95.4
	3×10^{-3}	17.3[b]	60.2[b]	247.9	24[a]	45[a]	87.5
	5×10^{-3}	10.0[c]	47.0[c]	370.0	19[b]	37[b]	94.7
	6×10^{-3}	6.0[d]	40.6[d]	576.6	15[c]	33[c]	120.0
III	Natural Ageing Control (untreated) Cycloheximide (*M*)	29.0[a]	83.5[a]	187.9	32[a]	52[a]	62.5
	0 (acetone)	23,1[b]	81.0[a]	250.6	31[a]	48[a]	54.8
	3×10^{-3}	22.5[b]	78.2[a]	247.5	31[a]	50[b]	61.3
	5×10^{-3}	15.0[c]	68.8[b]	358.6	24[b]	39[b]	62.5
	6×10^{-3}	14.0[c]	62.1[c]	346.6	20[c]	32[c]	60.0

NME, no moisture equilibration; ME-D, moisture equilibration-drying, CHI, cycloheximide.

*Table 1 in original publication.

concentrations of cycloheximide. Cycloheximide also inhibited the germinability of seeds subjected to natural ageing under ambient conditions for 120 days (Table 5.39). Although storability of seeds was reduced in both moisture equilibrated and non-equilibrated seeds at high cycloheximide concentrations the beneficial effect of moisture equilibration was clearly evident in all cases. As such, the difference between moisture equilibrated and non-equilibrated seeds could be noted over the entire range of cycloheximide concentrations used in the experiment.

In case of aqueous cycloheximide pretreatment, the germination percentage and seedling vigour of non-equilibrated and moisture-equilibrated cycloheximide-pretreated seeds, before ageing, was significantly reduced at higher concentration ($5 \times 10^{-4} M$) of cycloheximide (Table 5.40). That would confirm a significantly greater entry of chemicals into seed when applied in aqueous solutions than through dry permeation employing organic solvents. Before ageing germination test showed only slight beneficial effect of moisture equilibration-drying treatment but moisture equilibration greatly slowed down the deterioration of seeds subjected to accelerated and natural ageing conditions after treatment. The high degree of effectiveness of hydration by moisture equilibration with a saturated atmosphere followed by drying back would eliminate the possibility of leaching of toxic metabolites or inhibitors from the seed as a reason of the beneficial effect of seed treatment. Under accelerated ageing at 90 per cent RH and 40°C, moisture equilibration of cycloheximide pretreated seeds showed highly significant beneficial effect on germination and seedling growth over the corresponding non-equilibrated seeds (Table 5.40). As would be expected, in medium vigour seeds, water soaking-drying alone showed significant beneficial effect on storability. After accelerated ageing the germination percentage and seedling vigour showed a gradual reduction with increasing cycloheximide concentrations in both equilibrated and non-equilibrated seeds (Table 5.40). However, the difference in germinability between non-equilibrated and equilibrated seeds remained similar over the entire range of cycloheximide concentrations irrespective of the extent of inhibition caused by the chemical.

Similar results were noted when the seeds were subjected to natural ageing under ambient conditions for 120 days (Table 5.40). Here also with increasing cycloheximide concentrations, the germination percentage and seedling vigour were progressively reduced. As under accelerated ageing, the inhibition by cycloheximide was clearly evident at the concentration of $2 \times 10^{-4} M$. Although cycloheximide pretrestment adversely affected germinability, moisture equilibration of such seeds showed much better storability than corresponding non-equilibrated seeds.

The statistical non-significance of the interaction between cycloheximide and moisture equilibration treatments on germinability (Table 5.41) is of great interest from the physiological standpoint. It shows that the effect of moisture equilibration on germinability was not influenced by cycloheximide at least at the

***Table 5.40: Effect of Moisture Equilibration-Drying (ME-D) on Germination Percentage and Mean Root Lengths of Cycloheximide-Pretreated (Aqueous solution) Lettuce Seeds without and after Accelerated Ageing at 90 per cent RH and 40°C for 9 Days and Natural Ageing for 115 Days (av. RH 77.4 ± 3.1 per cent; av. Temp. 29.4°± 0.5°C) (Values within a column, under a particular ageing, scored by the same letter, using Duncan's multiple range test, are not significantly different at $p = 0.05$)**

	Cycloheximide Pre-treatments (Molar conc.)	*Germination (Per cent)*		*ME-D Advantage (Per cent)*	*Root Length (mm)*		*ME-D Advantage (Per cent)*
		Post-CHI Treatments			*Post-CHI Treatments*		
		NME	*ME-D*		*NME*	*ME-D*	
I	Without Ageing Control (untreated) Cycloheximide (*M*)	97.5[a]	98.1[a]	0.6	62[a]	61[a]	-1.6
	0 (water)	96.8[a]	98.1[a]	1.3	62[a]	62[a]	0.0
	10^{-4}	95.3[a]	97.8[a]	2.6	61[a]	58[ab]	-4.9
	2×10^{-4}	88.8[b]	87.7[b]	-0.7	60[a]	57[b]	-5.0
	5×10^{-4}	45.0[c]	54.0[c]	20.0	41[b]	39[c]	-4.9
II	Accelerated Ageing Control (untreated) Cycloheximide (*M*)	40.0[d]	90.5[b]	126.2	38[b]	49[ab]	28.9
	0 (water)	67.0[a]	98.1[a]	46.4	46[a]	53[a]	15.2
	10^{-4}	50.9[b]	96.6[a]	89.8	39[b]	50[ab]	28.2
	2×10^{-4}	45.8[c]	67.3[c]	46.9	36[b]	46[b]	27.8
	5×10^{-4}	9.6[a]	34.1[d]	255.2	32[c]	39[c]	21.9
III	Natural Ageing Control (untreated) Cycloheximide (*M*)	35.6[d]	86.3[a]	142.4	32[b]	48[a]	50.0
	0 (water)	65.8[a]	89.6[a]	36.2	40[a]	50[a]	25.0
	10^{-4}	55.3[b]	88.1[a]	59.3	32[b]	46[a]	43.7
	2×10^{-4}	41.5[c]	71.1[b]	71.3	28[b]	40[b]	42.8
	5×10^{-4}	9.0[c]	19.2[c]	113.3	14[c]	22[c]	57.1

NME, no moisture equilibration; ME-D, moisture equilibration-drying, CHI, cycloheximide.

*Table 2 in original publication.

concentrations used in this study. The interaction effects, calculated for root and shoot length of seedlings, were also found to be statistically non-significant. Thus, notwithstanding the inhibitory effect of cycloheximide, in both acetone permeation and water permeation sets, moisture equilibration showed highly beneficial effects clearly implying that the moisture equilibration-drying treatment gives its beneficial

***Table 5.41: Effect of Moisture Equilibration-Drying on Germinability (Germination Angles) of Cycloheximide-Pretreated (Dry permeated and Wet permeated) Lettuce Seeds after Accelerated Ageing at 90 per cent RH and 40°C for 12 Days**

Treatments	*Germination Angles*		*Mean*
	No Moisture Equilibration	*Moisture Equilibration*	
(a) Dry permeation set			
Control Cycloheximide (*conc. M*)	34.7	64.7	49.7
0 (acetone)	31.1	54.3	47.7
3×10^{-3}	31.6	54.8	43/2
5×10^{-3}	25.7	49.9	37.8
6×10^{-3}	22.8	46.3	34.6
Mean	29.2	54.0	
L.S.D. for Main and Interaction Effects	*Probability Levels*		
	P = 0.05	*P = 0.01*	
Pretreatment	13.8	NS	
Moisture equilibration	8.7	12.3	
Interaction	NS	NS	
(b) Water permeation set			
Control Cycloheximide (*conc. M*)	35.7	67.6	51.7
0 (water)	47.3	74.2	60.8
10^{-4}	42.3	72.5	57.4
2×10^{-4}	39.9	52.9	46.4
5×10^{-4}	14.4	31.1	22.8
Mean	36.0	59.7	
L.S.D. for Main and Interaction Effects	*Probability Levels*		
	P = 0.05	*P = 0.01*	
Pretreatment	10.4	14.7	
Moisture equilibration	6.6	9.3	
Interaction	NS	NS	

NS – Not significant.

*Table 3 in original publication.

effect on vigour and viability even if cytoplasmic protein synthesis is inhibited by cycloheximide. In fact, the relative advantages of moisture equilibration-drying were even more pronounced at higher cycloheximide concentrations when the values were expressed as the respective per cent increases over the corresponding non-moisture equilibrated seeds. Although the absolute differences in germinability were similar, the values become magnified when expressed as percentages. Usually, the per cent increases by hydration-dehydration treatments become more prominent as the ageing is advanced and control values become lower than the hydration-dehydration treatments.

Based on the present observations, it is concluded that *de novo* protein synthesis in the cytoplasm, which is usually inhibited by cycloheximide, is not involved in viability maintenance by the moisture equilibration-drying treatment. Otherwise, in the presence of inhibitory concentrations of cycloheximide inside the seed, the beneficial effect of moisture equilibration on vigour and viability would not show up.

Chapter 6

Quantification of Seed Vigour

Studies in the University of Calcutta have led to the development of easily practicable seed invigoration treatments which would not only slow down seed deterioration but would also improve the productivity of the crop raised from such seeds. The treatments have been successfully used for preservation of seeds of a large number of cereals, pulses, oilseeds, fibre crops and also vegetable and flower seeds. In addition to the physico-chemical hydration-dehydration treatments, pre-storage 'dry' treatments with certain chemicals (using organic solvents) or natural plant constituents and also chemicals in vapour form have effectively controlled loss of vigour and viability of seeds stored under ambient conditions.

In view of the significant positive effects of a wide range of seed invigoration substances their general role in seed quality improvement *vis-à-vis* seed deterioration was investigated through extensive physiological and biochemical studies on control and invigorated seeds especially after accelerated and natural ageing. These studies unequivocally confirmed the involvement of lipid peroxidation and emanation of volatile aldehydes in seed deterioration and their effective counteraction by seed invigoration treatments in all crop seeds without any exception. Along with chemical assay employing MBTH (3-methyl-2-benzothiazolinone hydrazone), we developed a simple bioassay method for quantifying seed vigour (Sur and Basu, 1990).

Bioassay of Seed Vigour

The underlying principle of this seed vigour bioassay is the concentration dependent seedling growth inhibition of a sensitive bioassay material in a quantitative manner upon exposure to gaseous emanations of growth inhibiting volatile substances by germinating stock seeds in suitable air-tight containers as loss of vigour of seeds is clearly related to increasing emanations of volatile aldehydes and associated gases.

In this method high vigour jute seeds (*Corchorus olitorius* L.), just sprouted, are grown in vitro along with stock seeds (materials of unknown vigour status) in suitable air-tight containers. The volatile inhibitory substances emanated by the stock seeds, depending on the concentration, would inhibit the growth of the jute seeds (bioassay material).This, when expressed as the percentage of the growth of jute seedlings in the absence of stock material would serve as the jute seedling bioassay score(JSBS)of the material being tested. The method has been described in details and the results obtained are discussed in Annexure No A–1. The mathematical-statistical approaches towards quantification of seed vigour status have been described in details in our papers on "Approaches to Quantification of seed Vigour (R.N. Basu, 2004) and "A Holistic Approach to Quantification of Seed Vigour through Integration of Moorphological and Physiological-Biochemical Parameters" (R.N. Basu, 2007). The first paper in reproduced below along with a rejoinder and the second paper in Annexure No A–2; also see abstracts in chapter 5.

APPROACHES TO QUANTIFICATION OF SEED VIGOUR

Reproduced from Indian Biologist, Basu, R.N. (2004) **36** (2), 1-23 with kind permission of the Indian Association of Biological Sciences (Abstract in chapter 4).

Background

Seed vigour is a vital qualitative character of a seed lot and is of great importance for successful establishment of plants in the field especially under relatively stressful growing conditions. The issue of seed performance *vis-à-vis* seed deterioration and loss of vigour in storage and its assessment has been receiving a lot of attention from researchers over a long time as would be evident from published literature on the subject (Heydecker[11], Heydecker and Coolbear[12], Coolbear[5], McDonald[15,16,17], Hampton[8,9], Hampton and TeKrony[10], Powell[23,24], Wilson[31], Wilson and McDonald[32,33], Dornbos[7], and others), but despite sustained individual, institutional and organizational efforts no vigour test has official sanction in terms of inclusion in the seed testing certificate as an important parameter of seed quality. Only two vigour tests, namely the electrical conductivity of the seed-steep water and the accelerated ageing test are considered to be of sufficient merit by the International Seed Testing Association (ISTA) to be recommended as official but optional tests (Hampton and TeKrony[10]). Perhaps, in future more such tests may have to be seriously considered for acceptance. But more than the tests themselves, the issue regarding quantification of test results to serve as vigour scores (vigour rating) needs critical adjudication in view of the fact that quite often the tests results are masked by undue deviations and variances due to sampling errors, uncontrolled test conditions, *etc.* Many of the important vigour test parameters are negatively correlated with seed vigour while the morphological vigour test parameters, such as seedling length, dry weight, *etc.* are positively correlated. Thus, the most acceptable vigour test, electrical conductance, which is one of the easiest vigour testing methods, is negatively correlated, along with tests based on leaching of sugars (Matthews and Bradnock[14], Takayanagi and Murakami[29]) and amino acids (Basu and Rudrapal[2], Rudrapal and Basu[25,26]), from seeds dipped in water. The conductivity and leaching tests are based on sound physiological considerations in that they measure the integrity of semipermeable cell membranes and the loss of integrity is most probably one of the earliest indications of seed deterioration leading to loss of seed vigour, and ultimately viability. If done properly with due consideration to variability in moisture content of seed lots (often requiring moisture stabilization and moderately controlled laboratory conditions), the test results would be free from personal errors. Unfortunately, however, variations in test results have been noted that may be attributed to uncontrollable seed lot differences and often due to confounding of germination percentage and vigour status of those seeds just surviving (many of which may not be included as normal seedlings in a standard germination test) and those with relatively high vigour indices, and commercial seed lots may be heterogenous enough to have variable proportions of such seeds. A large number of samples drawn from a seed lot in conductivity as well as in other tests may partly overcome the problem. Other important physiological and biochemical tests, besides the membrane permeability tests that are negatively correlated with seed vigour include lipid peroxidation (possibly the reason behind loss of membrane integrity), volatile compounds emanation (again, most likely lipid peroxidation byproducts, *vide* Wilson and McDonald[33], Sur and Basu[28], Sung[27]) which have shown considerable promise as effective parameters for vigour tests.

The positively correlated biochemical parameters that are used for certain specific purposes include dehydrogenase enzyme assay (the basis of topographical tetrazolium, TZ test as the enzyme turns TZ into red coloured formazan enabling identification of dormant viable seeds, with the additional advantage of eluting the colour from excised incubated embryos with solvents like methyl cellosolve and assay the enzyme colorimetrically, as discussed later), and amylolytic enzyme activity (widely used in the laboratories of brewing industry requiring highly germinable cereal seeds), and bioassay vigour tests (based on the emanation of growth inhibitory volatile substances produced by germinating low vigour seeds).

A major problem of quantification would be forming a suitable scale (preferably in the range of 0 to 100 as in case of germination) and then, if so required, combining positively as well as negatively correlated scores into a unified score for a seed lot. A simple approach towards such quantification could be the use of correlation coefficients (r values) obtained from bivariate analysis, and ignoring the + or – signs, multiplying the r value by 100 to get a positive percentage value. Even better would be use of r^2 (coefficient of determination) because that is always positive and gives a better estimate of the coverage of the observed variances by the two variates). The coefficients of determination expressed as per cent values (on multiplication by 100) may be amalgamated with other parameters to get a tentative vigour score. However, there are problems such as the question of viability differences (similar correlations may be noted at different levels of viability) and of undesirable deviations that have been discussed later. Prior transformation of the data to improve the correlations and minimize the deviations of the observed variates from those predicted by the relevant regression equations would be necessary for the purpose. With the aforesaid background, a range of vigour related data on membrane functions, notably electrical conductance, sugar and amino acid leaching, lipid peroxidation, volatile inhibitory substances emanation, bioassay vigour scores, dehydrogenase enzyme activity, amylolytic enzyme activity have been variously transformed and scored for subsequent correlation-regression analysis with a view to determine the efficacy and predictability of the different approaches. The data for the study have been obtained from a number of outside sources (published materials in international journals) as specifically mentioned in the tables as well as those originating in the laboratory of the author in the University of Calcutta (mostly published in national journals).

Data Collection, Processing, Transformation and Statistical Analysis

Data from several kinds of seeds have been used to have a generalized view on the issue and those included are: crimson clover, cotton, soybean, wheat, mustard, rice, sunflower and onion. The particulars on methods, units, *etc.* of different vigour parameters are given in the respective publications cited in the tables and in the section on references. Emphasis has been given on electrical conductivity and data for the same have been available for each of the materials. For the other seeds, the available data on parameters such as lipid peroxidation, volatile inhibitor (mostly aldehydes) production, dehydrogenase and amylase activity have been analyzed. Further, an attempt to combine individual vigour scores has been made only for soybean.

The methodology for transformation of data is outlined below for the different parameters starting with electrical conductivity.

Electrical Conductivity (EC)

Usually conductance is measured in μmhos; for the purpose of transformation, the value is first converted to millimhos (mmhos). The conductance in mmhos is then deducted from 1 and the remainder is multiplied by 100 to get a percentage value which is named electrical conductivity resistance score (ECRS).

so, ECRS = (1 – 0. millimho) × 100

Thus for the first value in Table 1, namely 70 μmhos the transformed value is:

ECRS = (1 – 0.070) × 100 = 93 per cent

The other transformations of electrical conductivity are: EC reciprocal (abbreviated as EC reci), obtained simply by dividing one (1) by the EC μmhos value and is actually the resistance in ohms. Another transformation of the EC data is division or multiplication by germination factor (GF), which is equal to germination percentage divided by 100, and is 1 for 100 per cent germination and always less than 1 when germination is lower. As would be evident (and

Table 1: Correlation-Regression Analysis of Electrical Conductance Data of Crimson Clover Seed after Various Transformations for Vigour Rating (Data from Ching and Schoolcraft[3], as quoted by Heydecker[11])

A. Data of 10-year-old viable seed lots along with their transformations

Seed Lot No.	*Germination[a]*		*EC (µmho per g)*	*ECRS[b] Per cent*	*EC Reciprocal[c] (ohm × 10⁶)*	*EC µmho ÷ GF (µmho)*	*EC Reciprocal × GF (ohm × 10⁶)*	*ECRS (Per cent) × GF*
	G per cent	*GF*						
2	98	(0.98)	70	93.0	14285	71.4	13999	91.1
3	98	(0.98)	89	91.1	11235	90.8	11010	89.2
4	98	(0.98)	44	95.6	22727	44.9	22272	93.7
5	99	(0.99)	51	94.9	19607	51.5	19411	93.9
6	94	(0.94)	49	95.1	20408	52.1	19183	89.4
7	86	(0.86)	78	92.2	12820	90.7	11025	79.3
8	89	(0.89)	57	94.3	17543	64.0	15613	83.9
Correlation with germination (per cent)			-0.1778	0.1778	0.2492	-0.3891	0.4106	0.9532**

(a) Germination factor (GF, germination percentage divided by 100) in parenthesis alongside germination percentage. (b) ECRS: electrical conductivity resistance score calculated as: (1–0.millimhos) × 100; ECRS × GF: same multiplied by germination factor. (c) EC reciprocal: reciprocal of EC µmho *i.e.*, 1 ÷ EC µmho, it would give the value of resistance in ohm; EC reci × GF: same multiplied by germination factor.

B. Summary of correlation-regression analysis with germination per cent as the x variate and EC values as the y variate (See Table 1C for detailed analysis of three sets of variates)

y Variate	*Correlation Coefficient r*	*Coefficient of Determination r^2*	*Mean of x Variate $\overline{x}$*	*Mean of y Variate $\overline{y}$*	*Sample Co-variance SC (n=7)*	*Regression Equation y = A + Bx*		*Sum of Least Squares*		*Mean Deviation*	
						A	*B*	$\Sigma(x-\hat{x})^2$	$\Sigma(y-\hat{y})^2$	$(x-\hat{x})$	$(y-\hat{y})$
EC µmho	-0.178	0.032	94.6	62.6	-15.38	117.2	-0.578	4489.2 (27.94)	1632.4 (24.41)	23.88 (25.25)	13.80 (22.05)
EC reci	0.249	0.062	94.6	16946	5493.7	-2571.4	206.38	2412.5 (19.63)	103065296 (22.64)	16.88 (17.83)	3474 (22.64)
ECRS	0.178	0.032	94.6	93.7	2.01	88.28	0.058	4889.2 (27.94)	16.32 (1.63)	23.88 (25.25)	1.38 (1.47)
EC µmho ÷ GF	-0.389	0.151	94.6	66.5	-37.61	200.1	1.413	895.1 (11.96)	2299.1 (27.26)	10.29 (10.88)	16.21 (24.38)
EC reci × GF	0.411	0.169	94.6	16073	9270.1	-16861	348.2	778.1 (11.15)	85887636 (21.79)	9.25 (9.78)	3311 (20.60)
ECRS × GF	0.953	0.909	94.6	88.6	26.17	-4.34	0.983	16.06 (1.60)	15.53 (1.68)	1.35 (1.43)	1.34 (1.51)

C. Detailed correlation-regression analysis of three sets of variates (*vide* Table 1A); data of Ching and Schoolcraft

i) Germination (per cent) versus EC µmho

			G %	*EC µmho*				
$(x-\hat{x})^2$	$(x-\hat{x})$	$\hat{x}$	(x)	(y)	$\hat{y}$	$(y-\hat{y})$`	$(y-\hat{y})^2$	
265.195	+16.284	81.715	98	**70**	60.590	+9.409	88.541	r = -0.1778
2417.411	+49.167	48.832	**98**	**89**	60.590	+28.410	807.128	r^2=0.0316
824.383	-28.712	126.712	**98**	**44**	60.590	-16.59	275.228	$\overline{x}$ =94.57
243.266	-15.597	114.597	**99**	**51**	60.012	-9.012	81.225	$\overline{y}$ =62.57
578.826	-24.058	118.060	**94**	**49**	62.90	-13.901	193.254	Sample covariance (n=7) = -15.380
328.696	+18.130	67.870	**86**	**78**	67.524	+10.475	109.743	
231.454	-15.213	104.213	**89**	**57**	65.790	-8.791	77.276	
$\Sigma(x-\hat{x})^2$ = 4889.23 (27.94 per cent of $\overline{x}$)	Σdeviation ÷ n = 23.88 (25.25 per cent of $\overline{x}$)					Σdeviation ÷ n = 13.8 (22.05 per cent of $\overline{y}$)	$\Sigma(y-\hat{y})^2$ = 1632.39 (24.41 per cent of $\overline{y}$)	Regression coeff. A = 117.216 B = -0.578

ii) Germination (per cent) versus ECRS

			G per cent	*ECRS*				
$(x-\hat{x})^2$	*$(x-\hat{x})$*	*$\hat{x}$*	*(x)*	*(y)*	*$\hat{y}$*	*$(y-\hat{y})$`*	*$(y-\hat{y})^2$*	
265.195	+16.284	81.715	98	**93.0**	93.941	-0.940	0.885	r =0.1778
2417.411	+49.167	48.832	**98**	**91.1**	93.941	-2.840	8.071	r^2=0.0316
824.383	-28.712	126.712	**98**	**95.6**	93.941	+1.659	2.752	$\overline{x}$ =94.57
243.266	-15.597	114.597	**99**	**94.9**	93.999	+0.901	0.812	$\overline{y}$ =93.74
578.826	-24.058	118.060	**94**	**95.1**	93.709	+1.390	1.932	Sample covariance (n=7) = 2.0095
328.696	+18.130	67.870	**86**	**92.2**	93.247	-1.047	1.097	
231.454	-15.213	104.213	**89**	**94.3**	93.421	+0.879	0.772	
$\Sigma(x-\hat{x})^2$ = 4889.23 (27.94 per cent of $\overline{x}$)	Σdeviation ÷ n = 23.88 (25.25 per cent of $\overline{x}$)					Σdeviation ÷ n = 1.378 (1.47 per cent of $\overline{y}$)	$\Sigma(y-\hat{y})^2$ = 16.322 (2.44 per cent of $\overline{y}$)	Regression coeff. A = 88.278 B = 0.0578

iii) Germination (per cent) versus ECRS × GF

			G per cent	*ECRS × GF*				
$(x-\hat{x})^2$	*$(x-\hat{x})$*	*$\hat{x}$*	*(x)*	*(y)*	*$\hat{y}$*	*$(y-\hat{y})$`*	*$(y-\hat{y})^2$*	
0.863	+0.929	97.070	98	**91.1**	92.013	-0.913	0.834	r =0.9532
8.190	+2.861	95.138	**98**	**89.2**	92.013	-2.813	7.917	r^2=0.9086
2.941	-1.715	99.715	**98**	**93.7**	92.013	+1.686	2.843	$\overline{x}$ =94.57
0.843	-0.918	99.918	**99**	**93.9**	92.997	+0.903	0.815	$\overline{y}$ =88.64
1.799	-1.341	95.341	**94**	**89.4**	88.081	+1.318	1.739	Sample covariance (n=7) = 26.17 Regression coeff. A = -4.338 B =0.9832
0.867	+0.931	85.068	**86**	**79.3**	80.215	-0.915	0.838	
0.558	-0.747	89.747	**89**	**83.9**	83.165	+0.734	0.540	
$\Sigma(x-\hat{x})^2$ = 16.06 (1.60 per cent of $\overline{x}$)	Σ deviation ÷ n = 1.348 (1.43 per cent of $\overline{x}$)					Σdeviation ÷ n = 1.337 (1.51 per cent of $\overline{y}$)	$\Sigma(y-\hat{y})^2$ = 15.528 (1.68 per cent of $\overline{y}$)	

Note: Values in parentheses indicate average deviations of the individual variates from the estimates, $\hat{x}$ and $\hat{y}$, expressed as per cent of respective means ($\overline{x}$ or $\overline{y}$); In case of simple deviation (+ or –) the summations, irrespective of + or – signs (otherwise the same would be around zero), are divided by n and expressed as the respective percentages of $\overline{x}$ or $\overline{y}$. In case of squared deviations (all positive values), after division by n, the average values are square rooted and expressed as respective percentages of $\overline{x}$ or $\overline{y}$. Thus, taking the example of data given in Table 1C(iii) $\Sigma(x-\hat{x})^2 = 16.06$ (n=7), the procedure followed has been: $\Sigma(x-\hat{x}) = 16.06 \div 7 = 2.2943 \rightarrow \sqrt{2.2943} = 1.5147 \rightarrow \div \overline{x}$ (= 94.57) × 100 = 1.601 per cent; the same procedure being followed for the y variate.

expected as well) the inclusion of GF (multiplication in case of positive correlations and division for negative correlations) would considerably improve the value of correlation coefficients and minimize the deviations of the observed variates from those predicted by the regression equations. For EC readings in μmhos, the transformations are thus: ECRS, EC reci (in ohms), EC μmho ÷ GF (value in μmhos), ECRS × GF and EC reci × GF.

Leaching of Sugars and Amino Acids

As mentioned earlier leaching of metabolites such as sugars and amino acids is attributed to loss of membrane integrity (the leaching of metabolites is, however, slower than electrolytes such as K^+ ions).

Sugar leaching is calculated by measuring the optical density (OD) of the anthrone colour reaction product. The transformation is done by deducting from 1 the optical density value and then multiplying the remainder by 100. Thus sugar leaching resistance score (SLRS) for the first item of Table 8(B) for sunflower, namely 0.205 OD would be equal to 79.5 per cent as shown below:

SLRS = (1– 0.OD) × 100 = (1 - 0.205) × 100 = 79.5 per cent

Although the optical density values for the purpose should be theoretically less than 1, optical density values should preferably lie between 0.5 OD to 0.05 OD, corresponding to transmission per cent (per cent T) approximately from 30 to 80 (that indicates the avoidance of 1, corresponding to log 10 per cent T, as the upper limit of usual OD readings (logarithmic) in the scale of 0 – 100 per cent transmission). Incidentally, the direct transmission reading (that are positively correlated with seed vigour, higher per cent T values showing lower concentrations of a solute in solution), can be used directly as a parameter for negatively correlated traits as has been tested in case of lipid peroxidation and volatiles emanation.

Amino acid leaching resistance score (ALRS) can be calculated in the same way as sugar leaching.

Thus, ALRS = (1–0.OD) × 100; for the first item 8 (B) namely, 0.238 OD for ninhydrin colour reaction it is equal to 76.2 per cent.

Sometimes the amount of sugar or amino acid leached out is given in μg per g of seed. In that case the value has to be converted to mg (upto three decimal places) and deducted from 1 and then the remainder multiplied by 100 as shown below:

SLRS/ALRS = (1 – 0. mg) × 100

The aforesaid transformation would be useful for conversion of violatile aldehyde production data given as μg aldehyde (as formaldehyde) per g seed (or per 100 seeds) as done by Wilson and McDonald[32].

Lipid Peroxidation (LP)

The optical density of the thiobarbituric acid (TBA) colour reaction product is deducted from 1 and then the remainder multiplied by 100 and expressed as LPRS (lipid peroxidation resistance score). Thus the value of LP OD for mustard given as OD × 10^3 = 220, *i.e.*, 0.220 (Table 5, first item) when transformed, has given a score of 78 per cent. Incidentally, LP per cent T values in wheat (recalculated from OD values in Table 6) also give positively correlated vigour indices in percentages that are mathematically different from LPRS values.

Volatiles (Aldehydes) Production

Although a range of volatile substances would be emanated from deteriorating seeds, the emphasis is given on low molecular weight aldehydes (presumably products of lipid peroxidation) that are preferentially absorbed by the reagent Methylbenzothiazolinone hydrazone (MBTH) and the optical density of the coloured reaction product is read at 635 nm on a colorimeter. This optical density value is deducted from 1 in the same manner as above, multiplied by 100 and expressed as volatiles emanation resistance score (VERS), the term aldehyde has been excluded to keep uniformity as a 4-letter abbreviation (also partly due to the fact that MBTH colour reaction is not strictly specific and certain other volatile substances may be absorbed). The VERS value for the first lot (T_1) of soybean derived from a VE OD value of 0.30 (given in Table 7 as OD × 10^2 = 30) would be as follows:

$$\text{VERS} = (1 - 0.\text{OD}) \times 100 = 0.70 \times 100 = 70 \text{ per cent}$$

When the volatiles are expressed as μg per g or per 100 seeds, conversion of μg to 0.mg would be necessary for VERS transformation, as mentioned earlier. The other transformations of VE OD data are: VE OD ÷ GF, VE reci, VE reci × GF, VE per cent T, VE per cent T × GF, all of which, except the first one, are positively correlated with germinability.

Bioassay Vigour Score

The jute seedling bioassay score (JSBS) is based on the growth of very high vigour jute seedlings (other similar small seeded, if possible just sprouted, materials characterized by rapid and uniform growth, *e.g.*, lettuce, amaranthus, *etc.* may also be used depending on ambient temperatures especially when temperature control is not possible) grown for around 48 to 72 hours (Sur and Basu[28]) along with germinating stock materials (seeds being tested for vigour rating) in small air-tight containers, finally expressed as percentage of the value in the blank set (air-tight containers with only the bioassay material, no stock seeds). The number of stock seeds (usually from 15-100, depending on seed size) has to be so adjusted that the score value for a very-high vigour reference stock seed lot is nearly 100. The physiological basis of the bioassay method is very strong and the growth inhibition is attributable to emanation of volatile inhibitory substances, mostly aldehydes (and some alcohols) that are presumably byproducts of lipid peroxidation (Wilson and McDonald[33]). The growth of the bioassay material is negatively correlated with such emanation, and, as such, positively correlated with the vigour status of the stock seed. The jute seedling bioassay score is expressed as JSBS per cent and its further transformation on multiplication by the germination factor (*i.e.* JSBS ×

GF) greatly improves the magnitude of the positive correlation and minimizes the deviations between observed values and statistical estimates.

Dehydrogenase and Amylolytic Enzyme Scores

The activity of the dehydrogenases and amylases are positively correlated with germinability. The topographical tetrazolium test (TZ tests, Moore[18]), performed in many seed laboratories, has been modified to assay the net dehydrogenase activity (Kittock and Law[13]). Excised embryos are imbibed in dilute solution (usually around 1 per cent) of 2,3,5-triphenyl tetrazolium chloride (or its other salts) which is reduced by the enzymes of the dehydrogenase group of enzymes to a red coloured non-diffusible substance called formazan (in cells and tissues lacking dehydrogenases, as in dead areas of non-viable seeds, formazan is not formed) that can be eluted with organic solvents like methyl cellosolve and the optical density is read at 470 nm. The value of the optical density is divided by the experimentally determined optical density of a well preserved very high vigour reference material of the same seed and then multiplied by 100 to get dehydrogenase enzyme activity score (DEAS) expressed as a percentage value, as shown for rice and onion in Table 8.

If no well preserved reference material is available, an estimate of the benchmark enzyme activity may be indirectly obtained by using a set of JSBS (jute seedling bioassay score) data along with corresponding enzyme activity readings of the seed lots being studied and estimating through regression analysis the enzyme activity at 100 per cent JSBS value.

The amylotic activity of usually α- and β- amylases is determined by extracting crude enzymes from duly imbibed and incubated seeds. The extent of hydrolysis of a standard starch solution by the crude enzyme extract is obtained by assaying the amount of unhydrolyzed residual starch using starch-iodine colour (blue) reaction. The amount of starch hydrolyzed, is converted to a percentage value by dividing with the value obtained parallelly from a very high vigour reference material followed by multiplication with 100. The amylolytic enzyme activity score (AEAS) shown for rice in Table 8 can be transformed to improve its vigour rating efficiency by multiplication with GF (*i.e.* AEAS × GF; the same is true for the dehydrogenase enzyme activity score, DEAS).

It needs to be mentioned here that all vigour rating values for originally negatively correlated characters (*vis-à-vis* original/basic data) which have become positively correlated on transformation, such as ECRS, EC reci, SLRS, ALRS, LPRS, LP reci, LP per cent T (recalculated from LP OD), VERS, VE reci, VE per cent T (recalculated from OD), each with or without GF incorporation (multiplication by GF), would actually give an estimate of physiological resistance of the seed to the occurrences of the untoward processes and events leading to the loss of vigour and viability and, as such, would measure the intrinsic vigour of the seed. The use of the term 'resistance' (with 'R' as its abbreviation) must not be mixed up with electrical resistance which has a specific connotation in being the reciprocal of "mho" (*i.e.*

Table 2: Transformation of Electrical Conductivity Data of Seeds of Acala 44 Cotton and their Correlation Regression Analysis for Seed Vigour Rating (Data of Andersen, Hart and French[1])

Seed Lot No.	*Germination (Per cent)*	*Conductivity Resistance CR, ohm × 10⁶*	*EC µmho*	*EC Milli-mho*	*ECRS (Per cent)*	*CR × GF*	*EC µmho ÷ GF*	*ECRS (per cent) × GF*
1	54.5	5580	179	0.179	82.1	3041	328	44.7
4	57.5	6108	163	0.163	83.7	3512	283	48.1
8	60.0	6744	148	0.148	85.2	4046	247	51.1
21	78.0	7450	134	0.134	86.6	5811	172	67.5
13	72.5	7630	131	0.131	86.9	5532	181	63.0
20	90.0	7830	128	0.128	87.2	7047	142	78.5
18	70.0	8420	119	0.119	88.1	5894	170	61.7
9	75.5	8820	113	0.113	88.7	6659	150	67.0
16	80.0	9840	102	0.102	89.8	7872	127	71.8

For abbreviation and details see Table 1 and text; original data of Andersen *et al*,[1] in conductivity resistance (CR, in ohms × 10^6). For statistical notations also see Table 1 and text.

Table 2 Contd....: Summary of Correlation-Regression Analysis with Germination per cent as the x Variate

y Variate	*r*	r^2	$\overline{x}$	$\overline{y}$	*SC (n=9)*	*A*	*B*	$\Sigma(x-\hat{x})^2$	$\Sigma(y-\hat{y})^2$	$(x-\hat{x})$	$(y-\hat{y})$
Conductivity Resistance (CR)	0.730	0.532	70.9	7602	11385	1678	83.57	956.61	6681548	8.64	722.6
								(14.54)	(11.33)	(12.20)	(9.50)
EC µmho	-0.776	0.603	70.9	135.2	-221.6	250.5	-1.63	718.13	1899.5	7.80	12.73
								(12.60)	(10.74)	(11.05)	(9.42)
ECRS	0.776	0.603	70.9	86.5	22.16	74.95	0.163	718.13	18.99	7.80	1.27
								(12.60)	(1.68)	(11.05)	(1.47)
CR × GF	0.914	0.836	70.9	5530.4	16910	-3268	124.12	213.37	3287394	4.01	497.9
								(6.89)	(10.93)	(5.66)	(9.0)
EC µmho ÷ GF	-0.912	0.833	70.9	200	-741.0	585.6	-5.44	200.36	6273.5	4.31	22.59
								(6.66)	(13.30)	(6.08)	(11.29)
ECRS × GF	0.996	0.992	70.9	61.48	132.12	-7.26	0.970	9.04	6.94	0.861	0.873
								(1.41)	(1.43)	(1.21)	(1.41)

ohm); the term resistance (meaning physiological resistance) has been preferred to other synonyms such as deterrence, hindrance, opposition, *etc.* because of its easy understandability and general but more meaningful appeal in the present context. The CR value in Table 2 (Andersen *et al.*[1]) that stands for conductivity resistance would be equivalent to EC reci values in ohms and ECRS is a totally different mathematical entity. The scores, namely ECRS, LPRS, VERS, JSBS, DEAS, *etc.* and all such values in the present discourse with their abbreviations ending with a capital 'S' (S for score), with or without GF, are actually percentage values strictly within the scale ranging from 0-100 (the upper limit will not be exceeded even on GF incorporation as the maximum value of the latter is 1). As such, the term "score" in the context of this paper has a specific connotation that singles it out from the general term "parameter" of vigour which stands for all measurements and indices of vigour including scores.

Statistical Calculations

Correlation-regression analysis of the original data obtained from the different sources, and of their various transformations, have been done following standard methods (Panse and Sukhatme[22]) with especial emphasis on bivariate analysis. The regression equation, $y = A + Bx$ (where y is the dependent variate, x is the independent variate, A and B are the two regression coefficients) would yield information on the deviations of observed values from those predicted by regression. A perusal of data in Table 1 dealing with the original data of Ching and Schoolcraft[3] and their transformations and correlation-regression analyses would be of help in understanding the approach. Table 1(A) shows the original data and their transformation ready for statistical analysis, Table 1(C) gives examples of three sets of data with detailed correlation-regression analysis that has been presented only in case of the first table (only the summarized versions of the analyses are shown in Tables 2–7), Table 1(B) gives the summary of the correlation-regression analysis for critical comparative study and as mentioned earlier, the same is available for all the materials covered in Tables 1–7. The statistical terms, notations and abbreviations mentioned in the tables and in different places in the text are as follows (full name in parentheses): r (coefficient of correlation), r^2 (coefficient of determination), $\overline{x}$ (mean of x variate), $\overline{y}$ (mean of y variate), SC (sample covariance), Reg. Eqn. (regression equation), Reg. coeff. A (regression coefficient A), Reg. coeff. B (regression coefficient B), $\hat{x}$ (estimated value of x), $\hat{y}$ (estimated value of y), $x-\hat{x}$ (deviation of a x variate from the estimate), $y-\hat{y}$ (deviation of a y variate from the estimate), $\Sigma(x-\hat{x})^2$, $\Sigma(y-\hat{y})^2$ (sum of least squares for the x and y variates, respectively). Below the columns of sum of least squares and also of mean deviations, certain values have been given in parentheses which indicate the extent of deviation as per cent of $\overline{x}$ and $\overline{y}$. Those below the sum of least squares column were obtained as follows: $\Sigma(x-\hat{x})^2 \div n$, the mean sum of squares is square–rooted ($\sqrt{\ }$) and then expressed as per cent of $\overline{x}$, *i.e.* (mean deviation ÷ $\overline{x}$) × 100; the procedure is exactly the same in case of $\Sigma(y-\hat{y})^2$. The values under the simple deviation column $\Sigma(y-\hat{x})$ or $\Sigma(y-\hat{y})$ have been obtained from mean deviations (summing up the mean deviations ignoring the positive or negative signs and dividing by n, as otherwise the summation would give zero) and expressed as per cent of $\bar{x}$ or $\bar{y}$ for the respective variates; however, for obvious mathematical reasons, the latter estimates are always lower than those obtained from the sum of least squares; they nevertheless, serve as useful checks. All these per cent deviation values in parentheses have been calculated for easy understanding of general seed scientists regarding the effect of data transformations on the deviations of observed values of the variates from those predicted by the regression equations especially when the $\bar{x}$ and $\bar{y}$ values are very different.

In bivariate correlation-regression analysis, if all the data in either x or in y set of variates are exactly the same, the calculation will be subjected to mathematical error. Sometimes that may happen in very high vigour seed lots, all showing 100 per cent germination; in that case r may be calculated taking the seedling length as the x variate (which may even be converted to per cent values against a reference or benchmark value) or by taking corresponding JSBS values as the x variate.

The regression analysis for most of the data has been done keeping values upto 3-4 places of decimals for accuracy but for presentation in the tables, those have often been rounded off to a fewer places to save space and, as such, may show minor differences on recalculation.

The Findings and Implications

The data analyzed in this paper literally speak for themselves. Beginning with the data of Ching and Schoolcraft[3], the need for transformation and analysis has been clearly brought about in all the materials considered in Table 1–9. The salient findings are very briefly narrated hereunder for a general overview of their relevance in seed vigour rating:

(1) The original data of Ching and Schoolcraft[3] on electrical conductance in EC µmhos have shown negative correlation with germination per cent but the r value is very low and statistically nonsignificant. When the conductivity data are reciprocally converted to ohms (resistance) there is a marginal improvement in r value which becomes positive but still far below the level of statistical significance (Table 1A). ECRS transformation does not show any change in r value but it greatly reduces the deviations of the observed values of dependent y variates from the statistical estimates provided by the regression equation (Table 1C) and with the incorporation of the germination factor (GF), the correlation with germination in ECRS × GF versus germination per cent has reached the level of statistical significance at 0.01P. The inclusion of GF has improved the r value in EC reci × GF and EC µmho ÷ GF approaches as well but those are still statistically nonsignificant (Table 1B). Further, ECRS × GF transformation has greatly minimized the deviations of observed and statistically estimated values of both x and y variates that show a

fall from around 25 per cent in the original data to less than 2 per cent demonstrating a vast improvement in predictivity (Tables 1B and 1C).

(2) In cotton, the original data of Andersen, Hart and French[1] have been provided in ohms as a conductivity resistance value (Table 2). There has been no change in the value of r with ECRS or EC µmho (the sign becomes negative) transformations but the former (ECRS) considerably minimizes the deviations between observed and estimated values of y variates. Following incorporation of GF, the correlation has improved and the ECRS × GF transformation has reduced the deviations between the observed values of the two variates (x and y) and the respective predicted values to less than 1.5 per cent of $\overline{x}$ and $\overline{y}$.

(3) In case of soybean, the data from McDonald's group (Tables 3, A, B) have shown similar improvements in correlation and predictive abilities. The data of Wilson and McDonald[32] on ECRS transformation show reduced deviation of y variate to around 0.5 per cent and on inclusion of GF (*i.e.* ECRS × GF), to less than 0.4 per cent for both x and y variates. The r value as well as the coefficient of determination has nearly touched the theoretical maximum of 1. The deviations in the y variate side of the original data of Van Toai, McDonald and Staby[30] were less than in the former study but on ECRS × GF transformation there has been further overall improvement in correlation and predictive efficiency

(4) The data of Rudrapal and Basu[25] on electrical conductivity of wheat seeds *vis-à-vis* germination per cent fully confirm the benefit of transformation, and in ECRS × GF, the value of r and r^2 (coefficient of determination), similar to that of Wilson and McDonald[32], have almost reached the theoretical limit of 1 along with minimization of deviations between observed values and statistical estimates (0.48 – 0.62 per cent of $\overline{x}$ and $\overline{y}$). Unfortunately, in the case of onion, the original data of Choudhuri and Basu[4] show very high deviations between observed values and statistical estimates. ECRS × GF transformation has not minimized the deviations to the level seen in the other seed materials (only to half in case of x variate) though correlation coefficient has gone up to +0.943 from the original -0.742 and the coefficient of determination (r^2) from 55 per cent to about 90 per cent (both in percentages on multiplication by 100). A close scrutiny of the data would reveal certain shortcomings of the data for the purpose of transformation for correlation-regression analysis. The very wide range of germination (22 – 84 per cent) and very high EC µmho value could not be duly taken into account by the present ECRS × GF transformation. Perhaps, a relatively narrow range of germination and lower EC readings (with greater dilution of the leachates) would have surmounted the problem. That would, however, not be in conformity with the objectives of the study of Choudhuri and Basu[4] the source from which the data have been collected.

Table 3: Electrical Conductivity Data of Soybean and their Transformations for Correlation-Regression Analysis

A. Data of Wilson and McDonald[32]

Lot No.	*Germination (Per cent)*	*EC μmho*	*EC mmho*	*ECRS (Per cent)*	*ECRS (Per cent) × GF*
2	93	27	0.027	97.3	90.49
4	79	28	0.028	97.2	76.79
3	92	29	0.029	97.1	89.33
1	71	31	0.031	96.9	68.80
5	62	49	0.049	95.1	58.96

Summary of Correlation-Regression Analysis with Germination per cent as the x Variate

y variate	*r*	r^2	$\overline{x}$	$\overline{y}$	*SC (n=5)*	*A*	*B*	$\Sigma(x-\hat{x})^2$	$\Sigma(y-\hat{y})^2$	$(x-\hat{x})$	$(y-\hat{y})$
EC μmho	-0.797	0.635	79.4	32.8	-97.9	76.15	-0.546	412.5	122.97	8.35	4.56
								(11.44)	(15.12)	(10.52)	(13.91)
ECRS	0.797	0.635	79.4	96.7	9.79	92.38	0.055	412.5	1.23	8.35	0.46
								(11.44)	(0.51)	(10.52)	(0.47)
ECRS × GF	0.999	0.999	79.4	76.9	180.42	-3.02	1.006	0.453	0.459	0.28	0.28
								(0.38)	(0.39)	(0.35)	(0.36)

B. Data of Van Toai, McDonald and Staby[30]

Lot No.	*Germination (Per cent)*	*EC µmho*	*EC mmho*	*ECRS (Per cent)*	*ECRS (Per cent) × GF*
Amsoy 71	78	187	0.187	81.3	63.41
Pella	85	158	0.158	84.2	71.57
Williams 79	88	153	0.153	84.7	74.54
Cumberland	85	120	0.120	88.0	74.80
Cutler 71	90	107	0.107	89.3	80.37

Summary of Correlation-Regression Analysis with Germination per cent as the x Variate

y variate	*r*	r^2	$\overline{x}$	$\overline{y}$	*SC (n=5)*	*A*	*B*	$\Sigma(x-\hat{x})^2$	$\Sigma(y-\hat{y})^2$	$(x-\hat{x})$	$(y-\hat{y})$
EC µmho	-0.793	0.628	85.2	145	-115	618.3	-5.55	48.94	1510.4	2.76	14.97
								(3.67)	(11.99)	(3.16)	(10.32)
ECRS	0.793	0.628	85.2	85.5	11.5	38.17	0.555	48.94	15.10	2.76	1.50
								(3.67)	(2.03)	(3.16)	(1.75)
ECRS × GF	0.962	0.926	85.2	72.9	27.15	-38.83	1.312	6.59	11.34	0.99	1.30
								(1.35)	(2.06)	(1.16)	(1.79)

See Table 1 for abbreviations and transformations, and also for statistical notations.

Table 4: Transformations of Electrical Conductivity Data of Wheat and Onion and their Correlation-Regression Analysis

Wheat[a]								*Onion*[b]				
Seed Lot	*Germination (Per cent)*	*EC µmho*	*EC Reci*	*ECRS (Per cent)*	*EC µmho ÷ GF*	*EC Reci × GF*	*ECRS (Per cent) × GF*	*Seed Lot No.*	*Germination (Per cent)*	*EC µmho*	*ECRS (Per cent)*	*ECRS (Per cent) × GF*
T_1	71	96	10416	90.4	135.2	7396	64.2	T_1	22	597	40.3	8.9
T_2	80	89	11236	91.1	111.2	8989	72.9	T_2	55	566	43.4	23.9
T_3	85	86	11628	91.4	101.2	9884	77.7	T_3	68	557	44.3	30.1
T_4	88	65	15385	93.5	73.9	13538	82.3	T_4	77	533	46.7	36.0
T_5	100	58	17241	94.2	58.0	17241	94.2	T_5	79	420	58.0	45.8
								T_6	80	450	55.0	44.0
								T_7	84	357	64.3	54.0

(a) Data of Rudrapal and Basu[25]; (b) Data of Choudhuri and Basu[4].

Summary of Correlation-Regression Analysis with Germination per cent as the x Variate

	y variate	*r*	r^2	$\overline{x}$	$\overline{y}$	*SC (n=5)*	*A*	*B*	$\Sigma(x-\hat{x})^2$	$\Sigma(y-\hat{y})^2$	$(x-\hat{x})$	$(y-\hat{y})$
Wheat	EC µmho	-0.923	0.852	84.8	78.8	-161.3	199.1	-0.142	79.25 (4.69)	159.5 (7.17)	3.28 (3.87)	4.65 (5.91)
	EC reci	0.917	0.840	84.8	13181	28987	-8438	254.9	86.24 (4.90)	5621569 (8.04)	3.65 (4.30)	930 (7.06)
	ECRS	0.923	0.852	84.8	92.1	16.13	80.1	0.142	79.25 (4.69)	1.59 (0.61)	3.28 (3.87)	0.46 (0.50)
	EC µmho ÷ GF	-0.969	0.938	84.8	95.9	-315.4	331.4	-2.77	29.96 (2.89)	230.6 (7.08)	1.90 (2.24)	5.26 (5.49)
	EC reci × GF	0.959	0.920	84.8	11410	40.54	-18824	356.5	39.49 (3.31)	4280366 (8.11)	2.59 (3.05)	922 (8.08)
	ECRS × GF	0.999	0.998	84.8	78.3	118.78	-10.32	1.045	1.074 (0.55)	1.17 (0.62)	0.40 (0.48)	0.42 (0.54)
Onion	EC µmho	-0.742	0.550	66.4	497.1	SC(n=7) -1344.7	692.5	-2.811	2345 (27.56)	18540 (10.18)	15.29 (23.02)	42.97 (8.50)
	ECRS	0.742	0.550	66.4	50.3	134.47	30.75	0.281	2345 (27.56)	185.4 (10.41)	15.29 (23.02)	4.30 (8.69)
	ECRS × GF	0.943	0.889	66.4	34.7	303.67	-8.18	0.635	359.3 (10.78)	144.84 (13.37)	6.32 (9.03)	3.81 (11.20)

Note: With a wide range of germination in case of onion (22-84 per cent) the deviations $(x-\hat{x})$ or $(y-\hat{y})$ though considerably reduced in ECRS × GF still remain quite high indicating reduced predictability; this has since been attributed not to viability but to unusually high EC µmho readings in the original data set which should better be around 100-150 or so for the efficacy of data transformation and GF integration.

(5) The transformation and analysis of electrical conductance data of mustard (Rudrapal and Basu[26]) once more categorically bring out the benefits of transformation and incorporation of germination factor (Table 5).

(6) In mustard, an important oil seed crop of the country, the role of lipid peroxidation in seed deterioration was studied with the view to control the process by exposing the seeds to low concentrations of iodine vapour to stabilize unsaturated bonds in lipid fractions with great success (Basu and Rudrapal[2]), the data from that project, as well as those of Rudrapal and Basu[25] as presented in Table 5, very clearly show that lipid peroxidation can be used as an important indicator of seed vigour and lipid peroxidation resistance score (LPRS) is very highly correlated with germination (r and r^2 values nearly reaching the theoretical limit of 1); with the incorporation of GF (*i.e.* LPRS × GF), the deviations between observed values and statistical estimates have been further minimized (Table 5). As would be statistically and logically expected the correlation between LPRS × GF and ECRS × GF is highly significant along with concomitant lowering of deviations between observations and estimates. From the physiological viewpoint, this correlation would most likely indicate a cause-effect relation between lipid peroxidation and loss of membrane integrity. The highly significant correlation between LPRS × GF and ECRS × GF in wheat also would indicate such a possibility (Table 6).

(7) Further confirmation of the benefits of transformation of LP data prior to regression-correlation analysis is brought out by the findings in wheat (Rudrapal and Basu[25]). The transformation of LP OD values to the lipid peroxidation resistance score has not shown any effect on r value except changing its sign from negative to positive but has greatly reduced the deviation in the y variate. However, LPRS × GF has vastly improved the correlation with concomitant minimization of deviations in both x and y variates. LP per cent T × GF transformation is also almost equally good.

As in case of mustard (Table 5) and soybean (Table 7), the possibility of using LPRS × GF for vigour rating of wheat seed is very strong at least for specific purposes.

(8) The data presented in Table 7 originating from a paper of Mukhopadhyay, Pal and Basu[20] on soybean bring out the benefits of transforming volatiles emanation (VE) data in the same manner as in case of lipid peroxidation (LP) data. The VERS × GF score giving the best correlation amongst the seven approaches made on the original VE OD data. It needs to be emphasized here that in all correlation-regression analyses on soybean, the r values have always been over 0.9 and it is the minimization of deviation of observed values of variates from the statistical estimates that has been the major advantage of the transformations. Incidentally,

Table 5: Transformations of Electrical Conductivity and Lipid Peroxidation Data of Mustard and their Correlation-Regression Analysis (Data of Rudrapal and Basu[25])

Seed Lot No.	*Germination Per cent*	*EC µmho*	*ECRS (Per cent)*	*EC µmho ÷ GF*	*ECRS (Per cent) × GF*	*LP OD × 10³*	*LPRS (Per cent)*	*LPRS (Per cent) × GF*
1	47	203	79.7	432	37.45	220	78	36.66
2	52	190	81.0	365	42.12	208	79.2	41.18
3	60	179	82.1	298	49.26	200	80	48.0
4	68	150	85.0	220	57.8	180	82	55.76
5	75	138	86.2	184	64.65	170	83	62.25
6	65	135	86.5	208	56.225	176	82.4	53.56

Details of transformations and abbreviations same as in Tables 1 and 6; for further details see text.

Summary of Correlation-Regression Analysis with Germination per cent as the x Variate

y Variate	*r*	r^2	$\overline{x}$	$\overline{y}$	*SC (n=6)*	*A*	*B*	$\Sigma(x-\hat{x})^2$	$\Sigma(y-\hat{y})^2$	$(x-\hat{x})$	$(y-\hat{y})$
EC µmho	-0.923	0.852	61.2	165.8	-274.97	321.90	-2.551	93.23	606.9	2.75	7.02
								(6.44)	(1.64)	(4.50)	(4.23)
ECRS	0.923	0.852	61.2	83.4	27.50	67.81	0.255	93.23	6.07	2.75	0.702
								(6.44)	(1.21)	(4.50)	(0.84)
EC µmho ÷ GF	-0.971	0.944	61.2	284.5	-993.5	848.4	-9.219	20.83	2736.16	1.69	15.55
								(3.05)	(7.51)	(2.76)	(5.47)
EC µmho × GF	0.998	0.996	61.2	51.2	106.12	-8.978	0.985	2.34	2.26	0.46	0.45
								(1.02)	(1.20)	(0.75)	(0.88)
LP OD × 10^3	-0.9667	0.9345	61.2	192.3	-200.07	305.89	-1.856	37.79	130.23	1.90	3.52
								(4.10)	(2.42)	(3.10)	(1.83)
LPRS	0.9667	0.9345	61.2	80.8	20.007	69.411	0.186	37.79	1.30	1.90	0.35
								(4.10)	(0.58)	(3.10)	(0.44)
LPRS × GF	0.999	0.999	61.2	49.7	98.926	-6.581	0.918	0.57	0.56	0.22	0.20
								(0.50)	(0.62)	(0.35)	(0.40)
LP OD versus EC µmho	0.989	0.978	192.3	165.8	565.67	-107.89	1.423	44.24	89.6	2.29	3.26
								(1.41)	(2.33)	(1.19)	(1.97)
LPRS × GF versus ECRS × GF	0.999	0.999	49.7	51.2	99.467	-3.806	1.107	0.3	0.37	0.19	0.21
								(0.45)	(0.49)	(0.38)	(.0.42)

Table 6: Transformations of Lipid Peroxidation Data of Wheat and their Correlation-Regression Analysis for Vigour Rating (Data of Rudrapal and Basu[25])

Seed Lot No.	*Germination (Per cent)*	*LP OD × 10^3*	*LPRS (Per cent)*	*LP Reci × 10^3*	*LP Per cent T*	*LP OD ÷ GF*	*LPRS (Per cent) × GF*	*LP Reci × GF*	*LP Per cent T × GF*
T_1	71	155	84.5	6.45	69.98	218.3	59.99	4.58	49.69
T_2	80	144	85.6	6.94	71.78	180.0	68.48	5.55	57.42
T_3	85	142	85.8	7.04	72.11	167.1	72.93	5.98	61.29
T_4	88	148	85.2	6.75	71.12	168.2	74.98	5.95	62.59
T_5	100	137	86.3	7.29	72.95	137.0	86.3	7.30	72.95

Abbreviations: LP, lipid peroxidation, LPRS, lipid peroxidation resistance score; LP reci., reciprocal of LP OD; LP per cent T, Lipid peroxidation per cent transmission (per cent T) recalculated from LP OD; the rest are GF (germination factor) incorporated. For further details on transformations see text.

Summary of Correlations and Regressions with Germination (per cent) as the x Variate

y Variate	*r*	r^2	$\overline{x}$	$\overline{y}$	*SC (n=5)*	*Value of A*	*Value of B*	$\Sigma(x-\hat{x})^2$	$\Sigma(y-\hat{y})^2$	$(x-\hat{x})$	$(y-\hat{y})$
LP OD × 10^3	-0.852	0.727	84.8	145.2	-61.45	-0.540	-61.45	171.0	49.9	5.10	2.75
								(6.90)	(2.18)	(6.01)	(1.90)
LPRS	0.852	0.727	84.8	85.5	6.14	80.89	0.054	171.0	0.499	5.10	0.27
								(6.90)	(0.37)	(6.01)	(0.32)
LP reci × 10^3	0.862	0.743	84.8	6.90	2.92	4.720	0.026	156.9	0.103	4.89	0.12
								(6.60)	(2.08)	(5.76)	(1.82)
LP per cent T	0.853	0.728	84.8	71.6	10.12	64.04	0.089	169.7	1.343	5.07	0.45
								(6.90)	(0.72)	(5.98)	(0.63)
LP OD ÷ GF	-0.977	0.954	84.8	174.0	-304.75	401.29	-2.68	22.09	158.71	1.99	5.33
								(2.48)	(3.24)	(2.34)	(3.06)
LPRS × GF	0.999	0.999	84.8	72.5	102.41	-3.84	0.901	0.41	0.33	0.24	0.21
								(0.34)	(0.35)	(0.29)	(0.29)
LP reci × GF	0.991	0.982	84.8	5.87	10.34	-1.84	0.091	8.47	0.07	1.06	0.10
								(1.53)	(2.014)	(1.30)	(1.70)
LP per cent T × GF	0.998	0.997	84.8	60.8	90	-6.33	0.791	1.43	0.89	0.46	0.36
								(0.63)	(0.70)	(0.54)	(0.60)
LPRS × GF ver-sus ECRS × GF	0.998	0.997	72.4	78.2	105.9	-5.78	1.160	1.09	1.46	0.34	0.40
								(0.64)	(0.69)	(0.47)	(0.51)

Table 7: Transformations of Data on Volatiles Release from Soybean Seeds for Correlation-Regression Analysis along with Data on Jute Seedling Bioassay Score for Seed Vigour Rating (Data of Mukhopadhyay, Pal and Basu[20])

Seed Lot No.	*Germination (Per cent)*	*VE OD × 10^2*	*VERS (Per cent)*	*VE Reci × 10^2*	*VE Per cent T*	*VE OD ÷ GF (× 10^2)*	*VERS (Per cent) × GF*	*VE Reci × GF (×10^2)*	*VE Per cent T × GF*	*JSBS (Per cent)*	*JSBS (Per cent) × GF*
T_1	40	30	70.0	3.333	50.118	75	28	1.333	20.05	74	29.6
T_2	69	18	82.0	5.555	66.069	26.087	56.58	3.833	45.59	83	57.3
T_3	82	12	88.0	8.333	75.857	14.634	72.16	6.833	62.2	90	73.8
T_4	77	14	86.0	7.143	72.443	18.181	66.22	5.5	55.78	88	67.8
T_5	68	17	83.0	5.882	67.608	25	56.44	4	45.97	86	58.5
T_6	73	16	84.0	6.25	69.183	21.918	61.32	4.562	50.5	88	64.2
T_7	47	25	75.0	4	56.234	53.191	35.23	1.88	26.43	81	38.1
T_8	90	9	91	11.111	81.283	10	81.9	9.999	73.15	93	83.7

Abbreviations: VE, volatiles emanation, in OD × 10^2; VERS, volatiles emanation resistance score (per cent); VE reci × (10^2), reciprocal transformation of VE OD × (10^2); VE per cent T, volatiles emanation in per cent transmission; GF, germination factor for division or multiplication; JSBS, jute seedling bioassay score.

Summary of Correlations and Regressions with Germination (per cent) as the x Variate

y Variate	*r*	r^2	$\overline{x}$	$\overline{y}$	*SC (n=8)*	*A*	*B*	$\Sigma(x-\hat{x})^2$	$\Sigma(y-\hat{y})^2$	$(x-\hat{x})$	$(y-\hat{y})$
VE OD × 10^2	-0.995	0.989	68.2	17.62	-115.75	45.11	-0.403	21.56	3.50	1.30	0.52
								(2.40)	(3.75)	(1.91)	(2.97)
VERS	0.995	0.989	68.2	82.4	115.7	54.88	0.403	21.56	3.50	1.31	0.52
								(2.40)	(0.80)	(1.91)	(0.64)
VE reci	0.929	0.862	68.2	6.541	38.85	-2.78	0.135	320.56	5.86	5.3	0.72
								(9.27)	(13.27)	(7.76)	(11.1)
VE per cent T	0.995	0.989	68.2	67.3	170.75	26.79	0.594	21.95	7.75	1.37	0.81
								(2.43)	(1.46)	(2.01)	(1.21)
VE OD ÷ GF	-0.970	0.942	68.2	30.5	-364.8	117.15	-1.269	123.99	199.86	3.41	4.33
								(5.77)	(16.39)	(5.00)	(14.21)
VERS × GF	0.998	0.997	68.2	57.2	304.70	-15.14	1.060	6.10	6.86	0.76	0.80
								(1.28)	(1.62)	(1.11)	(1.40)
VE reci × GF	0.929	0.862	68.2	4.74	43.64	-5.62	0.152	321.1	7.41	5.27	0.80
								(9.28)	(20.29)	(7.72)	(16.87)
VE per cent T × GF	0.994	0.987	68.2	47.5	295.55	-22.74	1.028	25.65	27.13	1.55	1.60
								(2.62)	(3.88)	(2.28)	(3.37)
JSBS	0.959	0.920	68.2	85.3	96.75	62.40	0.33	175.13	19.85	3.57	1.20
								(6.85)	(1.84)	(5.24)	(1.41)
JSBS × GF	0.997	0.995	68.2	59.1	302.28	-12.67	1.052	10.8	7.01	0.83	0.87
								(1.70)	(1.80)	(1.22)	(1.48)

For other details on statistical notations and abbreviations see Table 1 and text.

a testing of the efficiency of transformation has been done in case of active absorption by MBTH of volatile aldehyde released by soybean seeds (Wilson and McDonald[32]) during 27-48 hr versus field emergence (per cent) after VERS × GF transformation of the same (data in μg converted to 0.mg, deducted from 1 and multiplied by 100 and then by GF of 7-day germination test); the transformation has shown a r value of +0.878** (against the original correlation of -0.69 that was statistically nonsignificant). The jute seedling bioassay score (JSBS) is by itself a percentage value; it has shown very good correlation with germination and demonstrates still higher correlation with the incorporation of GF (Table 7) with requisite minimization of deviations in both x and y variates. When laboratory conditions are inadequate for other biochemical tests, with a little practice this test can be done to get a very reliable vigour score. Techniques being developed in the laboratory of the author suggest that even autobioassay (the same seed acting as a test material) would give an effective vigour score.

(9) Taking advantage of the transformation techniques, data on several vigour parameters of rice, sunflower and onion have been subjected to bivariate correlation analysis with the view to evaluate the relative efficacy of several vigour scores employing the coefficient of determination (r^2) as a measure for the purpose. For the sake of easy understanding of the comparative efficiency of the different scores, the r^2 values have been converted to percentage values by multiplying with 100 or with mean germination or field emergence per cent values (Table 8,9). It is revealed that the transformed EC data are much more reliable as vigour parameters of rice than dehydrogenase or amylase-based indices. In sunflower also, the ECRS × GF and ALRS × GF, both concerned with membrane integrity, would prove useful. In onion, however, LPRS × GF and DEAS × GF have shown good efficiency as vigour scores. The vigour scores for soybean (Table 9) have demonstrated good correlations with germination singly as well as in combinations of two, three and four scores. But more important is the high r and r^2 values when correlations are worked out with field emergence. The percentage values of coefficient of determination ($r^2 \times 100$) have shown 89-98 per cent efficiency for most of the calculated scores. The present statistical treatment would not, however, allow the final determination of superiority of any particular vigour score or a combination of two or more vigour scores over other scores for which there could be several options. More detailed analysis with appropriate computer-based software packages, rather than ordinary scientific calculators, would be required for the purpose. For the present, however, EC data with required transformation, namely ECRS × GF is recommended for practical use in seed and crop production oriented practices, and if necessary in seed trade and commerce. In due course, there would be a possibility of combining ECRS × GF score with VERS × GF or JSBS × GF score for even more reliable vigour rating of crop seeds.

Table 8: Correlations between Germination and some Vigour Parameters of Rice, Sunflower and Onion Seeds to Evaluate the Use of Coefficient of Determination as Possible Vigour Scores

A. Rice: Data of Pal and Basu[21]

Seed Lot No.	*Germination Per cent*	*EC µmho*	*ECRS (Per cent)*	*ECRS (Per cent) × GF*	*AEA µg Starch Hydro-lyzed*	*AEAS (Per cent)*	*AEAS (Per cent) × GF*	*DEA OD*	*DEAS (Per cent)*	*DEAS (Per cent) × GF*
T_1	67	117	88.3	59	82	54.7	36.8	0.091	31.4	21.0
T_2	91	59	94.1	85	100	66.7	61.0	0.140	48.2	43.9
T_3	93	52	94.8	88	108	72	67.0	0.148	51.0	47.5
T_4	96	56	94.4	91	116	77.3	73.9	0.161	55.5	53.3
T_5	91	59	94.1	86	98	65.3	59.1	0.161	55.5	50.5
T_6	98	58	94.2	93	150	100	98.0	0.290	100	98.0
(Refrigerated material)										
Correlation coefficient (r)		-0.967	0.967	0.999	0.743	0.736	0.851	0.692	0.692	0.747
Coeff. determination (r^2)		0.936	0.936	0.999	0.552	0.542	0.724	0.479	0.479	0.559
$r^2 \times 100$		93.6	93.6	99.9	55.2	54.2	72.4	47.9	47.9	55.9
$r^2 \times \bar{G}$ per cent ($CSVS_1$)		89.3	89.3	89.3	49.3	48.4	64.7	42.8	42.8	49.4

Abbreviations: AEAS, amylase enzyme activity score; DEAS, dehydrogenase enzyme activity score, also see Table 1 and text.

B. Sunflower: Data of Dey and Basu[6]

Seed Lot No.	*Germination Per cent*	*EC µmho*	*ECRS (Per cent)*	*ECRS (Per cent) × GF*	*Sugar Leaching OD 580*	*SLRS (Per cent)*	*SLRS (Per cent) × GF*	*Amino Acid Leaching OD 610*	*ALRS (Per cent)*	*ALRS (Per cent) × GF*
T_1	48	381	61.9	29.7	0.205	79.5	38.2	0.238	76.2	36.6
T_2	59	354	64.6	38.1	0.201	79.9	47.1	0.225	77.5	45.7
T_3	68	342	65.8	44.7	0.148	85.2	57.9	0.215	78.5	53.4
T_4	79	302	69.8	55.1	0.136	86.4	68.2	0.198	80.2	63.3
T_5	75	299	70.1	52.6	0.155	84.5	63.4	0.218	78.2	58.6
T_6	65	338	66.2	43	0.171	82.9	53.9	0.220	78	50.7
T_7	100	290	71.0	71.0	0.090	91.0	91.0	0.095	90.5	90.5
Correlation coefficient (r)		-0.920	0.920	0.998	-0.967	0.967	0.998	-0.902	0.902	0.990
Coeff. determination (r^2)		0.847	0.847	0.997	-0.936	0.936	0.996	0.813	0.813	0.980
$r^2 \times 100$		84.7	84.7	99.7	93.6	93.6	99.6	81.3	81.3	98.0
$r^2 \times \bar{G}$ per cent ($CSVS_1$)		59.8	59.8	70.3	66.1	66.1	70.3	57.4	57.4	69.2

Abbreviations: SLRS, sugar leaching resistance score; ALRS, amino acid leaching resistance score; also see Table 1 and text; $CSVS_1$, correlative seed vigour score, calculated by multiplying r^2 with mean germination per cent ($\bar{G}$ per cent), also see note below Table 9.

C. Onion: Data of Choudhuri and Basu[4]

Seed Lot No.	*Germination Per cent*	*LP OD*	*LPRS (Per cent)*	*LPRS (Per cent) × GF*	*DEA OD*	*DEAS[(a)] (Per cent)*	*DEAS (Per cent) × GF*
T_1	97	0.170	83.0	80.5	0.242	99.2	96.2
T_2	96	0.155	84.5	81.1	0.235	96.3	92.4
T_3	95	0.166	83.4	79.2	0.240	98.4	93.5
T_4	99	0.142	85.8	84.9	0.240	98.4	97.4
T_5	71	0.243	75.7	53.7	0.154	63.1	44.8
T_6	89	0.173	82.7	73.6	0.222	91.0	81.0
T_7	92	0.193	80.7	74.2	0.215	88.1	81.0
T_8	93	0.155	84.5	78.6	0.228	93.4	86.9
Correlation coefficient (r)		-0.926	0.926	0.993	0.980	0.979	0.989
Coeff. determination (r^2)		0.857	0.857	0.986	0.960	0.960	0.979
$r^2 \times 100$		85.7	85.7	98.6	96.0	96.0	97.9
$r^2 \times \bar{G}$ per cent ($CSVS_1$)		78.4	78.4	90.2	87.8	87.8	89.6

Abbreviations: LPRS, lipid peroxidation resistance score, others same as in Table 6 and Tables 8A and Table 8B. (a) A refrigerated (non-aged) material of the same stock giving an optical density of 0.244 at 470 nm was used as a reference (DEAS value=100).

Table 9: Comparison of different Vigour Scores of Soybean after Data Transformations (Basic data for transformation from Mukhopadhyay, 2001)

A. Transformations of basic data

Seed lot No.	*Germination Per cent*	*Field Emergence (Per cent)*	*EC µmho*	*ECRS (Per cent)*	*ECRS (Per cent) × GF (a)*	*LP OD × 10^3*	*LPRS (Per cent)*	*LPRS (Per cent) × GF (b)*	*VE OD × 10^3*	*VERS (Per cent)*	*VERS (per cent) × GF (c)*	*JSBS (Per cent)*	*JSBS (Per cent) × GF (d)*
T_1	84	64	155	84.5	71.0	120	88.0	73.9	96	90.4	75.9	75	63.0
T_2	89	81	84	91.6	81.5	48	95.2	84.7	32	96.8	86.1	88	78.3
T_3	86	75	99	90.1	77.5	72	92.8	79.8	45	95.5	82.1	82	70.5
T_4	86	76	113	88.7	76.3	60	94.0	80.8	43	95.7	82.3	86	74.0
T_5	87	71	125	87.5	76.1	96	90.4	78.6	42	95.8	83.3	79	68.7
Correlations of vigour parameters with germination: (germination per cent versus vigour parameters)													
r	-	0.850	-0.832	0.832	0.941	-0.754	0.754	0.909	-0.854	0.854	0.956	0.761	0.86
r^2	-	0.722	0.692	0.692	0.885	0.569	0.569	0.827	0.730	0.730	0.914	0.580	0.74
$r^2 \times 100$	-	72.2	69.2	69.2	88.5	56.9	56.9	82.7	73.0	73.0	91.4	58.0	74.0
$r^2 \times \bar{G}$ per cent ($CSVS_1$)	-	-	59.8	59.8	76.5	49.2	49.2	71.4	63.1	63.1	79.0	50.1	63.9
Correlations of vigour parameters with field emergence: (field emergence per cent versus vigour parameters)													
r	0.85	-	-0.972	0.972	0.961	-0.984	0.984	0.990	-0.896	0.896	0.909	0.969	0.983
r^2	0.722	-	0.943	0.943	0.923	0.969	0.969	0.980	0.803	0.803	0.826	0.938	0.966
$r^2 \times 100$	-	-	94.3	94.3	92.3	96.9	96.9	98.0	80.3	80.3	82.6	93.8	96.6
$r^2 \times \overline{FE}$ per cent ($CSVS_2$)	-	-	69.2	69.2	67.7	71.1	71.1	71.9	58.9	58.9	60.6	68.8	70.9

B. Different combinations of GF-incorporated vigour scores and their correlations with field emergence

(a) ECRS × GF, (b) LPRS × GF, (c) VERS × GF, (d) JSBS × GF

Different combinations of vigour scores, a, b, c, d (see respective columns in Table 9A).

Lot No.	*abcd*	*abc*	*abd*	*adc*	*bcd*	*ab*	*ac*	*ad*	*bc*	*bd*	*cd*
T_1	70.9	73.6	69.3	70.0	70.9	72.4	73.5	67.0	74.9	68.5	69.5
T_2	82.7	84.1	81.5	82.0	83.1	83.1	83.8	79.9	85.4	81.5	82.2
T_3	77.5	79.8	75.9	76.7	77.5	78.6	79.8	74.0	81.0	75.2	76.3
T_4	78.3	79.8	77.0	77.5	79.0	78.6	79.3	75.1	81.6	77.4	78.1
T_5	76.7	79.4	74.5	76.1	76.9	77.4	79.7	72.4	81.0	73.7	76.0
Correlations with field emergence: (field emergence per cent versus vigour scores in combination)											
r	0.983	0.968	0.990	0.981	0.983	0.982	0.945	0.990	0.965	0.988	0.979
r^2	0.967	0.937	0.981	0.961	0.967	0.964	0.894	0.981	0.931	0.977	0.958
$r^2 \times 100$	96.7	93.7	98.1	96.1	96.7	96.4	89.4	98.1	93.1	97.7	95.8
$r^2 \times \overline{FE}$ per cent ($CSVS_2$)	71	68.8	72.0	70.5	71.0	70.5	65.6	72.0	68.3	71.7	72.3

Abbreviations and other details same as in Tables 1, 5, 6 and 7.

Note: The two tentative correlation based vigour scores, provisionally named as 'correlative seed vigour scores', *viz.* $CSVS_1$ and $CSVS_2$ obtained by multiplying r^2 with mean germination per cent $\bar{G}$ and mean field emergence per cent ($\overline{FE}$ per cent), respectively are found useful for comparing the individual vigour scores and their combinations (the number of pairs of variates, n, the degrees of freedom should be same in all cases for such comparisons) with the potential to predict final field performance in terms of agricultural productivity.

Lastly, this paper highlights the very important role of germination percentage in vigour rating; if germination factor is not incorporated in vigour scores, the predictions would be unreliable thereby vindicating the emphasis given to standard germination tests by the two major international organizations deeply concerned with research and development in seed science and technology– the ISTA and AOSA.

Although the present paper deals specifically on the issue of seed vigour rating, it is hoped that some of the data transformation techniques may be useful to other fields of biology in which masking of data due to various factors has been a long standing problem in correlation related studies.

Acknowledgements

The author gratefully acknowledges the use of data of many authors from their published works, including his own former students and collaborators. Thanks are due to Professor T. M. Das for inspiration to write this paper and for critical review of the manuscript and to Dr. S. L. Basak for fruitful discussion. The use of some data in one of the tables from the Ph.D. thesis of my former student, Dr. Anuradha Mukhopadhyay (Associate Supervisor, Dr. Prativa Pal) is gratefully acknowledged.

Note: *The individual vigour scores have, in general, successfully brought out the differences between seed lots (in case of materials from the author's laboratory attributable to seed invigoration and ageing conditions). However, the details of issues concerning differences between kinds of seeds, differences in seed lots within a kind and the statistical treatments of interactions between vigour scores and kinds of seeds and seed lots, have been taken up separately, elsewhere. – Author.*

REFERENCES

1. Andersen, A. M., Hart, J. R. and French, R. C. *Proc. Int. Seed Testing Ass.*,**29** (1), 81-96 (1970).
2. Basu, R. N. and Rudrapal, A. B. *Indian J. Exp. Biol.*, **18**, 492-494 (1980).
3. Ching, T. M. and Schoolcraft, I. *Crop. Sci.*, **8**, 407-409 (1968).
4. Choudhuri, N. and Basu, R. N. *Seed Sci. Technol.*, **16**, 51-61 (1988).
5. Coolbear, P. In: *Seed Quality, Basic Mechanisms and Agricultural Implications* (ed. A. S. Basra), Food Products Press, NY, USA, pp. 223-277 (1995).
6. Dey, G. and Basu, R. N. *Indian J. Plant Physiol.*, **25**, 87-97 (1982).
7. Dornbos, D. L. (Jr.) In: Seed *Quality, Basic Mechanisms and Agricultural Implications* (ed. A. S. Basra), Food Products Press, NY, USA pp. 45-80, (1995).
8. Hampton, J. G. Ibid pp. 81-118 (1995).
9. Hampton, J. G. In: *Seed Vigour Testing Seminar* (ed. H. A. van de Venter), ISTA, Zurich, Switzerland, pp. 10-28 (1995).
10. Hampton, J. G. and Tekrony, D. M. (eds) *Handbook of Vigour Test Methods.* ISTA, Zurich, Switzerland (1995).

11. Heydecker, W. In: *Viability of Seeds* (ed. EH. Roberts), Chapman and Hall, London, pp. 209-252 (1972).
12. Heydecker, W. and Coolbear, P. *Seed Sci. Technol.*, **5**, 353-425 (1977).
13. Kittock, D. L. and Law, A. G. Agron. J., **60**, 286-288 (1968).
14. Mathews, S. and Bradnock, W. T. *Proc. Int. Seed Testing Ass.*, 553-563 (1967).
15. McDonald, M. B. (Jr.). J. *Seed Technol.*, **93**-101 (1993).
16. McDonald, M. B. (Jr.). In: *Seed Vigour Tesing Seminar* (ed. H. A. van de Venter), ISTA, Zurich, Switzerland, pp. 88-97 (1995).
17. McDonald, M. B. (Jr.) Seed Sci. Technol., **27**, 177-232 (1999).
18. Moore, R. P. In: Viability of Seeds (ed. E. H. Roberts), Chapman and Hall, London, pp. 94-113 (1972).
19. Mukhopadhyay, A. Ph. D. *Thesis, University of Calcutta*, India (2001).
20. Mukhopadhyay, A., Pal, P. and Basu, R. N. *Indian Agric.*, **41**, 79-85 (1997).
21. Pal, P. and Basu, R. N. In: *Rice in West Bengal* **vol. II.** A Govt. of West Bengal Publication, W. B., India, pp. 159-166 (1982).
22. Panse, V. G. and Sukhatme, P. V. *Statistical Methods for Agricultural Workers.* (2nd edition) ICAR, New Delhi, India, pp. 112-121 (1967).
23. Powell, A. A. *Adv. Res. Technol. Seeds,* **11**, 29-80 (1988).
24. Powell, A. A. In: *Seed Vigour Testing Seminar* (ed. H. A. van de Venter), ISTA, Zurich, Switzerland, pp. 73-87 (1995).
25. Rudrapal, A. B. and Basu, R. N. *Indian J. Exp. Biol.*, **17**, 768-771 (1979).
26. Rudrapal, A. B. and Basu, R. N. *Ibid,* **20**, 465-470 (1982).
27. Sung, J. M. *Physiologia Plant.*, **97**, 85-89 (1996).
28. Sur, K. and Basu, R. N. *Seed Sci. Technol.*, **18**, 661-671 (1990).
29. Takayanagi, K. and Murakami, K. *Nature*, Lond.,**218**, 493-494 (1968).
30. Van Toai, T. T., McDonald, M. B. (Jr.) and Staby, G. L. *Seed Sci. Technol.*, **14**, 301-312 (1986).
31. Wilson, D. J. (Jr.). In: Seed Quality: *Basic Mechanisms and Agricultural Implications* (ed. A. S. Basra), Food Products Press, NY. USA, pp. 173-207 (1995).
32. Wilson, D. J. (Jr.) and McDonald, M.B. (Jr.). *Seed Sci. Technol.*, **14**, 259-268 (1986a).
33. Wilson, D. J. (Jr.) and McDonald, M.B. (Jr.). *Ibid*, **14**, 269-300 (1986b).

This paper is most respectfully dedicated to my esteemed teacher the late Professor R. D. Asana of the Indian Agricultural Research Institute, New Delhi, who led us to the firm conviction that authentic experimental data, if properly approached and rationally treated, would be silently but explicitly communicative.

REJOINDER TO "APPROACHES TO QUANTIFICATION OF SEED VIGOUR"

Vigour Related Factors and their Mutual Relationships

The vigour factors, GF (germination factor), SF (seedling factor) and their integration through multiplication giving GSF (germination cum seedling factor) are considered as the three important seed vigour and viability determining mathematical expressions or factors, each with values ranging from 0 to 1, obtained from germination percentage values and experimentally derived seedling evaluation percentage values. These two basic parameters are simply divided by 100 to get the two primary factors, GF and SF (germination per cent ÷ 100 = GF, if germination per cent is 80 then GF = 80 ÷ 100 = 0.80; if seedling evaluation per cent is 60, SF = 60 ÷ 100 = 0.60); the derivative of the two basic factors GF and SF namely GSF is obtained through integration by multiplication of GF and SF values (thus, if GF = 0.80, and SF = 0.60, GSF = 0.48). GSF on multiplication by 100 would give germination-cum-seedling evaluation per cent value; however, GSF as a factor is of much greater utility for the improvement of the predictive (regressional) efficiency of the experimental data requiring statistical correlation-regression analysis. Division of different factor integrated score or parameter values by the respective factors (GF, SF or GSF) would readily release the original score and parameter values from either or both of the two variates (namely x and y).

It is necessary to point out here that in some of our earlier publications we have been too strict with the benchmark value for seedling evaluation per cent calculation (see the next paragraph) by adding the full range value of the data set on (y) variate giving a higher reference value for the estimation of the 'divisor' for getting corresponding percentage values of y at 100 per cent germination (x). That would result in conspicuous lowering of resultant seedling evaluation percentage with consequent reduction of SF and GSF values. While the same would not affect the correlations (which are not influenced by scale values), low SF and GSF values may not be readily acceptable to the producers and the buyers of seeds. As such, we have now reduced the reference or benchmark value (used as the divisor to convert the different y variates to respective percentage values) as far as possible so that the regressionally derived value of y at 100 per cent G would be close to germinability. In fact such values (transformed y variates in percentages in respect of most physiological-biochemical vigour scores would be quite close to the corresponding x variates (germination per cent).

Seedling Growth Evaluation *in vitro* in the Laboratory

Seedling evaluation per cent of different seed lots is determined by germinating the seeds *in vitro* under laboratory conditions suitable for standard germination tests. The inclined glass plate blotter method for germination is very suitable as there is uniformity in growth and periodic observation on the progress of growth of the seedlings is possible from outside without disturbing the set up; other suitable methods may also be followed. Seedling evaluation may be based on extension growth of shoot and root or total growth of seedlings, fresh and dry weights of seedling, *etc.*; in our present studies we have preferred total seedling length (TSL) and have transformed the same to TSL per cent, after division by the regressionally estimated benchmark value for total seedling length at 100 per cent germination (often in the data set itself; TSL value at 100 per cent germination may be readily available, in case it occurs more than once, the highest value should be taken as the divisor which is applicable to regressionally calculated value at 100 per cent germination also (the divisor should be such that no other value would exceed that). The aforesaid conversions do not influence the coefficient of correlation r or determination (r^2). The mean germination per cent of the sample on division by 100 would give the mean GF value of the sample and division of TSL per cent value of the sample by 100 would give the seedling growth factor or seedling factor (SF), in short. GSF is derived through integration by multiplication of the values of the two basic factors GF and SF, as stated in the first paragraph.

Dependence of Seedling Factor (SF) on Germination Factor (GF) and Related Issues

Among the germinable seeds, as distinguished from non-germinable non-viable seeds (as broadly covered by germination per cent in germination tests, and GF values), there may be significant differences in vigour as reflected in early seedling growth (preautotrophic phase) that may be quantified to seedling evaluation per cent and SF values. GF and SF are intimately related with a high positive correlation; with SF physiologically fully dependent on GF. Seedling evaluation is possible only when a nondormant viable seed germinates and grows into a seedling. Germination per cent and GF in their terminological and mathematical meanings and perceptions would not directly cover the factual differences in vigour levels of fast and slow germinating seeds in a commercial seed lot, the majority of which would come under "germination per cent " of the officially certified seed.

As mentioned earlier, in our discussion germinability and viability are considered synonymous. GF incorporated into positive vigour scores and parameters would eliminate the number-wise (numerical) contribution of all non-viable seeds from positive vigour scores or parameters assuming equal contribution per seed from non-viable as well as viable seeds of different vigour status. However, for many physiological-biochemical parameters there could be significant qualitative and quantitative differences in the relative contribution of the viable and non-viable seeds in a seed lot. Simple GF incorporation would ignore such genuine possibilities and facts.

Exponential Loss of Seed Vigour during Seed Deterioration

It has been observed that reductions in seedling evaluation values would increase at a faster rate with fall in germinability and lowering of GF values suggesting more and more rapid reductions in vigour levels of seedlings with time. Progressively lower GF values therefore are indicative of the fact that the surviving germinable seeds are of still lower vigour status. Apparently, some sort of exponential fall in vigour level with a gradual decrease of germinability is clearly indicative of the fact. The same may be captured by raising the power of exponents, such as squaring the GF values or raising the power to cubes (GF^2, GF^3 and so on), and multiplying the positive vigour scores or parameters by such transformed GF values to effectively improve the subsequent correlations and regressions. It needs to be emphasized that integration of GF, GF^2 or GF^3, *etc.*, with positive vigour scores and parameters would always progressively reduce the value of the resultant integrated score or parameter values as the GF values range from 0 to 1 and, in between the aforesaid range, are always less than one, only the highest value is 1 for which there would be no exponential increase or decrease, the same is true for zero. On mathematical grounds, in the above mentioned calculations the powers (exponents) should be of positive value and above 1 and the bases below 1, as with all GF values (range 0 to 1) in between 0 and 1.

It needs to be highlighted again that among germinated seeds, as broadly covered by germination per cent and GF (*i.e.*, G per cent ÷ 100), there may be significant differences in seedling vigour that is reflected in early (preautotrophic) seedling growth phase in the usual laboratory germination tests. Even in seed lots of the same cultivar, all showing 100 per cent germination there may be statistically significant difference in growth of seedlings. The same would hold good for lower levels of germination as well (90 per cent, 80 per cent or still lower germination) suggesting the necessity of seedling evaluation for a more comprehensive understanding of the seedling vigour status of seed lots at similar germination levels.

Slowly germinating laggards and weaker seeds (*vis-à-vis* fast germinating ones) would very often fulfil the germination per cent criterion of seed certification but in terms of seedling evaluation and positive physiological-biochemical characters such 'officially' germinated weaker seeds would betray their relatively subdued vigour status.

GSF as an overall vigour factor therefore assumes great importance not only for a better understanding of the fundamental aspects of the seed vigour concept but also for its practical use in seed science and technology through crop performance improvement by different seed invigoration treatments. Hundred per cent germination-cum-seedling evaluation value (*i.e.*, GSF=1) should be the ideal reference value for seed vigour quantification. When the same cannot be readily worked out, the other reference value, y at 100 per cent G, or the highest value of '*y*' variate, whichever is higher, would be the best alternative as the 'divisor' to convert the parameters to respective percentage values or vigour scores.

For more details see Basu, R.N. 2007. A Holistic Approach to Quantification of Seed Vigour through Integration of Morphological and Physiological-biochemical Parameters. *Indian Biologist*, Vol.39(1), 1-44 (Annexure No A-2).

Chapter 7

Supplement to Thematic and Allied Issues

Some Important Observations and Issues

The dry-stored orthodox seed, a rudimentary and dormant embryonic plant in the dry state but capable of developing into a seedling under proper conditions, is a remarkable creation of nature that can be stored in a virtually air-dry state over a long period of time at a high level of vigour and viability. A rise in seed moisture and ambient temperature during the storage period would, however, be detrimental to retention of vigour and viability under ordinary uncontrolled storage; very often in hot and humid regions loss of vigour and viability could be substantial. Seed invigoration treatments would enable effective retention of vigour and viability in storage with due manifestation on field performance and productivity of a wide range of agricultural and horticultural crops of commercial importance.

The choice of a particular kind of seed invigoration treatment (presowing soaking-drying, a 'wet' treatment, or midstorage hydration-dehydration treatment, also a 'wet' treatment employing different methods of hydration or prestorage 'dry' treatment – nonaqueous, includes use of organic solvents, vapour phase treatment, dry-dressing, *etc.*) would depend on a number of important practical issues involving the nature ('dry' or 'moist' stored) and morphological-physiological characteristics of the seed being treated. Whether the seed is orthodox (dry-stored as in case of most agricultural field crops) or recalcitrant (moist-stored, as for a number of fleshy fruit and plantation crops in which drying would be positively injurious to seed viability and seed invigoration could be done only under moist storage) has been an important consideration requiring new approaches and practical solutions. Between a leguminous seed and a nonleguminous seed the former would be susceptible to soaking injury and hydration without soaking injury for legumes could be accomplished by moisture equilibration with a saturated atmosphere (humidification at 100 per cent RH) or moist sand conditioning (sand at 5 per cent moisture content, sand:seed ratio 3:1), followed by drying back. The vigour status of the seed at the time of invigoration is also of great importance in choosing the most suitable treatment for a particular seed lot as there exists a significant interaction between the initial vigour status of the seed and the kind of invigoration treatment (prestorage 'dry' treatments or 'wet' midstorage hydration-dehydration (H-D) treatments). Thus, midstorage H-D treatments are suitable for high-medium vigour (HMV), medium vigour (MV) and low vigour (LV) seeds, with appropriate (slow or fast) hydration for low-medium vigour (LMV) and low vigour (LV) seeds followed in all cases by drying back. Soaking injury prone legume seeds as well as many relatively high vigour non-leguminous crop seeds would always prefer slow humidification-drying or moist sand conditioning-drying for effective invigoration for the maintenance of vigour, viability and yield potential.

The conventional presowing 'hardening' treatments which are given to stored seeds, using relatively long duration soaking followed by light air-drying, may also invigorate them giving improved field performance but the effects are rather inconsistent. In our modified presowing treatments, short duration soaking or slow hydration in various ways (similar to midstorage 'wet' treatments) followed by light air-drying, would give consistent beneficial effects on field performance and productivity. We strongly recommend the same for our small and marginal farmers.

Prestorage 'dry' seed invigoration treatments should be given to harvest-fresh high vigour (HV) seeds of virtually all kinds of dry-stored orthodox seeds of cereals, millets, pulses, oilseeds, ornamental flower seeds, preferably at the time of postharvest bagging and storage. Such 'dry' physiological invigoration treatments are not effective in low vigour stored seeds (in which 'wet' treatment would show much better effect). The 'wet' (aqueous) treatments should be avoided in case of harvest-fresh HV seeds which may adversely affect vigour, viability and yield potential; the situation would be still more serious for soaking injury susceptible harvest-fresh (HV) leguminous seeds.

During the last four decades, besides the successful retention of seed vigour and viability the researchers in the Calcutta University have repeatedly demonstrated highly significant invigoration effects on field performance and productivity of a wide range of agricultural and horticultural crops. In fact, since the first report on the maintenance of seed vigour and viability of rice and jute (Basu, Chattopadhyay and Pal, 1974), the same was duly followed up with two publications on field performance and productivity (Basu and Pal, 1978 on rice and Basu *et al.*, 1978 on jute). In rice cv. Ratna, post-ageing grain yield was 4.6 t/ha in untreated control against 6.1 t/ha in water S-D and 6.7 t/ha in S-D with sodium phosphate (dibasic, 10^{-4}M). In jute cv. JRC 321, seed invigoration treatments were given to carry-over seed. Post-ageing field performance was significantly improved with water soaking-drying recording 5.8 g dry weight of bast fibre per plant against 3.9 g per plant in the untreated control. For most orthodox crops of agricultural and commercial importance, we have confirmed significant beneficial effects of invigoration treatments on field performance and productivity as would be evident from the list of selected abstracts of papers presented in Chapter 4.

Although, the studies have been conducted mostly with dry-stored orthodox seeds, a few attempts have been made to improve the storability of moist-stored recalcitrant seeds. Two such examples include the significant increase in storage life of jackfruit (*Artocarpus heterophyllous* Lam.) seeds which are large and fleshy, employing a modified hydration system, moist sand conditioning (seed sand ratio 1:3, 5 per cent sand moisture content, continuous incubation in loosely bound polythene packets, no drying) that maintained around 90 per cent viability for 9 months against 35 per cent viability in control under indoor ambient conditions in Kolkata (Calcutta). Similar moist storage retained viability of recalcitrant tea (*Camellia sinensis* L.) seeds for 7 months giving 65 per cent germination (with the aforesaid moist sand conditioning along with 0.25 per cent Dithane M-45 as a fungicide) against total loss of germinability in the control set (see relevant data in Chapter 5). However, with recalcitrant seeds, no field performance studies could be undertaken.

It was realized quite early that an understanding of the mode of action of the seed invigoration treatments was necessary not on theoretical considerations only but also for their more effective use in diverse crops in different situations. Various physiological and biochemical studies were taken up in the laboratories duly equipped for the purpose. Based on systematic studies on the radioprotective role of seed invigoration treatments and the observed broad parallelism between age- and irradiation (x-ray and *y*-ray) induced adverse effects on germinability (coupled with magnification of radiation damage by ageing known as biomagnification) and also the parallelism between the role of free radical reaction controlling agents in ageing as well as radioprotection (opposite effect with radiosensitizing agents and free radical reaction promoters, which increase damage by irradiation and ageing, natural as well as accelerated ageing), the involvement of free radical reactions in deteriorative senescence of seeds was confirmed. The same was already convincingly established and reported by eminent radiobiologists like Haber and Randolph, 1967, Kesavan, 1973 and others in respect of radiation induced damage to seeds. We have demonstrated the role of free radical controlling agents in lipid peroxidation of deteriorating seeds and its counteraction by seed invigoration treatments in our physiological and biochemical studies on a range of crop seeds (see Chapter 5 for details). It has been concluded that free radical reactions involving peroxidation of membrane lipids are the primary reasons of cellular membrane damage and loss of semipermeability (as manifested by leaching of electrolytes and metabolites) and also production of bioinhibitory (toxic) volatile aldehydes by senescing seeds besides adversely affecting the activity of a range of enzymes and other vital cellular constituents. Seed invigoration treatments would significantly reduce the adverse effects of ageing on vital cellular microconstituents and the resultant invigoration effect would be manifested in improved field performance and crop productivity.

The ups and downs in crop performance in the field may be partly attributed to the rigours and uncertainties of the difficult-to-control outdoor edaphoclimatic conditions but despite that 2-3 repetitions with adequate replications would certainly bring out the desired invigoration effect. The significant positive correlations between viability-cum-vigour parameters on one hand and field performance-cum-productivity parameters on the other would remain largely unaffected although the absolute values (contributions) of such parameters might show changes in magnitude as a healthier and stronger invigorated plant would be expected to outperform the weaker control plant in diverse field conditions; of course, the performance of both would be poorer under adverse field conditions compared to normal (optimum) conditions. In the chapter on thematic issues (Chapter 5) we have therefore concluded that if properly carried out there should not be any doubt about the positive response of seed invigoration on productivity. Vigour, viability, field performance and productivity are all positively correlated characters – that is the basic axiom, things cannot be otherwise under normal conditions.

According to the Intergovernmental Panel on Climate Change (IPCC), the irregularity of climate systems would be a major problem for the successful implementation of the existing cropping schedules. Besides, optimum temperature regimes, adequate rainfall and its timely distribution over the cropping season for a rainfed crop are the basic requirements for the success of arid zone agriculture. Inadequate rainfall causing drought and too much rainfall within a short period causing waterlogging would be undesirable. However, erraticism in the distribution of rainfall is increasing over the years and it is perhaps an indirect but expected outcome of global climate change.

Cultivators may choose less water requiring drought tolerant crop which may suffer from standing water as it might be inherently susceptible to waterlogging. Employing conventional breeding methods, development of cultivars equally tolerant to drought and waterlogging situations would be a very difficult proposition. The same would be true for simultaneous tolerance to opposing characters such as acidic as well as alkaline soil conditions, *etc.*

Preliminary studies in the faculty of agriculture in the University of Calcutta have shown that seed invigoration treatments would enable invigorated seedlings to tide over to a considerable extent, the adverse effect of partial drought, semisubmergence and medium levels of salinity under simulated laboratory conditions simply because the invigorated seedlings are morphologically and physiologically stronger and as such, capable of overcoming the short-term stress situations than the noninvigorated less vigorous counterparts (Pal, Mandal and Basu, 2012; Govt. of West Bengal Project Report; unpublished). Inexpensive seed invigoration treatments would be of considerable advantage for normal as well as untoward environmental and edaphoclimatic conditions.

The aforesaid issues are discussed in greater details later in this chapter under eco-sustainable and organic agriculture with seed invigoration as a complementary technical tool. Prior to that we would very briefly highlight several other important findings that include seed size and its relevance to seed vigour and seed invigoration treatments in wheat, decade long germplasm conservation in rice, the role of age- and irradiation-induced chromosomal aberrations in wheat and the counteraction of the same by seed invigoration treatments. The relative role of seed invigoration treatments on embryo and endosperm of rice also needs to be highlighted. The salient points of our experimental and mathematical-statistical approaches to quantification of seed vigour also needs to be conveyed very briefly to all those interested in the subject; of course the detailed information is given in the relevant chapters and Annexures for critical appraisal. We would thereafter take up eco-sustainable agriculture (with particular emphasis on organic agriculture) and discuss the possible complementary role of seed invigoration treatments for still more remunerative and nature-friendly agriculture that would ensure seed security as well as food security to millions of small, marginal and landless farmers in the developing countries like India.

Seed Size, Seed Vigour and Seed Invigoration Treatments

Size grading of harvest-fresh high vigour (HV) wheat seed into big, medium and small ones, using appropriate sieves would show that in wheat big seeds are always better than medium sized seeds which in turn are better than small seeds in respect of germinability, early seedling growth, field performance and productivity. 'Wet' as well as 'dry' seed invigoration treatments would show similar beneficial effects in terms of percentage of increase for big, medium and small seeds against their respective untreated (noninvigorated) controls. There has been no statistically significant interaction between different seed sizes and different ('wet' and 'dry') seed invigoration treatments which is contradictory to the significant differential response (interaction effect) of seed lots with different vigour levels (HV, MV or LV as obtained with accelerated or natural ageing of harvest-fresh seed lots). As such, big, medium and small seeds are not physiologically equivalent to HV, MV and LV seeds respectively; the size-graded seeds derived from the same harvest-fresh seed lot are of the same vigour status (HV) which on differential ageing would give MV or LV seeds and interact differently with 'wet' and 'dry' invigoration treatments. Perhaps, the greater nutritional reserve in the endosperm of the bigger seed would give better seedling growth in the preautotrophic (seed reserve-dependent) phase giving added benefit (additive effect). See Chapter 5 on Thematic Issues and papers by Pal *et al.,* 2001; De *et al.,* 2003 on wheat.

Decade Long Germplasm Conservation of Rice Seed under Ambient Conditions

Basu and Pal (1978) reported that hydration-dehydration treatments were highly effective in maintaining vigour, viability and productivity of rice varieties so far studied (that included Ratna, Jaya, IR8, Pusa 2-21, Mahsuri, Rupsail, Patnai 23, Sitabhog, Dular and Dhairal) and emphasised the need to maintain germplasm collections under ordinary storage conditions for a considerable period and eliminate the need to regrow the genetic stocks each year.

Soaking-drying treatment of 7-month-old rice seed (cv. Ratna) stored in air-tight containers under ambient conditions retained good seed viability for more than a decade (Pal and Basu, 1989). At the end of 12 years the S-D treatment retained 85 per cent seed viability against 12 per cent in the control, with good seedling growth in the former. It has been noted that after thorough moisture stabilization in glass desiccators containing fused calcium chloride, the seeds can be transferred to rubber-stoppered (wax-sealed) glass bottles for long-term preservation without the need for temperature control. It is also possible to use moisture equilibration-drying (ME-D) or moist sand conditioning-drying for the purpose (Basu and Pal, 1980). In fact, Rao, Roberts and Ellis (1987) working in the University of Reading, UK have recommended this method (ME-D, also known as humidification-drying, originally developed in the University of Calcutta for the maintenance of seed viability and vigour for long-term preservation of plant genetic resources at the international level; they reported significant reduction of chromosomal aberrations in lettuce by humidification-drying (the same as moisture equilibration-drying) confirming our earlier observations as reported hereunder.

Counteraction of Ageing and Irradiation Induced Chromosomal Aberrations in Wheat by Seed Invigoration

Dasgupta *et al.* (1977) working on wheat and jute showed a parallelism between age- and X-irradiation induced seed deterioration and the counteraction of the same by hydration-dehydration seed invigoration treatments. Cytological studies taken up on wheat very clearly demonstrated a significant protection offered by post-irradiation soaking-drying treatments with water and several radioprotective chemicals on chromosomal abberrations such as fragmentation, deletion, rearrangement, *etc.*, during mitosis in wheat root tip cells. The occurrences of such aberrations (number of breaks per cell) were as follows: X-irradiated (30 kR) untreated control 11.6, S-D with water 6.3, S-D with disodium phosphate 10^{-4}M 4.1, S-D with EDTA (Na-salt) 10^{-4}M 5.5, S-D with caffeine 10^{-4}M 11.9, against non-irradiated control, 0.008. Hydration (S-D) itself was very effective but sodium phosphate and EDTA further reduced chromosomal abberrations. Caffeine, a radiosensitizing chemical completely erased the large radioprotective action of hydration with water alone. (Table 5.13)

Regarding the possible *modus operandi* of the different treatments, we are in full agreement with the suggestion by earlier researchers that the same is basically physico-chemical in nature (Haber and Randolph, 1967; Kesavan, 1979 and others). Using radioprotective chemicals like cysteine, *p*-aminobenzoic acid, potassium iodide, *etc.* in mustard and also in lettuce employing acetone as an organic solvent (for eliminating the hydration effect), highly significant (0.05 P) radioprotective action as well as counteraction (prevention) of deteriorative senescence by ageing could be consistently confirmed (Pathak and Basu, 1982 in mustard, Punjabi and Basu, 1988 in lettuce).

In rice, age-and irradiation-induced seed deterioration was effectively controlled by hydration-dehydration treatments including S-D and ME-D (moisture equilibration-drying) (Sur and Basu, 1986). In fact, for both pre- and post-irradiation damage control by seed invigoration, ME-D was much more effective than S-D in high vigour (HV) rice seeds than in medium vigour (MV) or low vigour (LV) seeds; in the latter (LV) in particular, ME-D was ineffective while S-D significantly prevented ageing as well as irradiation damage of LV seeds. It needs to be pointed out here that the cellular repair mechanism (enzymatic) in imbibed lettuce seed giving prolonged longevity (*vide* Villiers and Edgecumbe, 1975) would be rather inoperative in short-term S-D or ME-D (Pan, Punjabi and Basu, 1981).

Reciprocal Embryo-Endosperm Transplantations Involving Hydration-Dehydration Seed Invigoration Treatments in Rice

Embryo-endosperm transplantation experiments involving the placement of invigorated or noninvigorated embryos on noninvigorated or invigorated endosperms, with all possible combinations, have shown that hydration-dehydration treatment besides benefitting the embryo, would effectively prevent a decline in the post-ageing functional activity of the aleurone cells of the endosperm. The same was evident from increased amylolytic activity causing greater solubilization of starch reserves for better nourishment of the embryo transplanted on it (Sur and Basu, 1985). It was subsequently shown that in rice not only ageing but also the adverse effects of irradiation and heat stress of the endosperm could be greatly reduced by pretreatment of the endosperm with hydration-dehydration invigoration treatment. When both the embryo and endosperm were invigorated, the mobilization from endosperm to embryo and utilization of the endospermic reserves by the growing embryo were promoted, even under various stress situations such as ageing, irradiation and heat stress (Sur and Basu, 1990).

Issues on Quantification of Seed Vigour

Bioassay of Seed Vigour

The development of the bioassay method has been of great advantage in experimentally quantifying vigour status in the percentage scale of 0-100.

In this method high vigour jute (*Corchorus olitorius* L.) seeds just sprouted prior to use as bioassay material, are grown *in vitro* along with stock seeds (materials of unknown vigour status) in air-tight containers. The volatile growth inhibitory substances emanated by the stock seeds, depending on concentrations, would inhibit the growth of the bioassay material which when expressed as the percentage of growth of jute seedlings in the absence of stock materials would serve as the JSBS (jute seedling-bioassay score) of the material being tested. The bioassay method has been discussed in greater details in Annexure No A-2. Quantification of vigour status of different seed lots and seed invigoration treatments using a range of vigour related morphological and physiological-biochemical parameters such as germination and early seedling growth, leaching of electrolytes, sugars and amino-acids from seeds as indicators of membrane permeability, lipid peroxidation, activity of selected enzymes associated with vigour and viability, *etc.*, are described in Annexure No A-2.

Mathematical-Statistical Approaches to Quantification of Vigour

In our rather simple mathematical-statistical approaches to quantification of seed vigour we have transformed the basic vigour parameter data so that the same would be positively correlated with germination and the vigour rating data would range between 0 and 100, the same as with germinability values and would be positively correlated with germination per cent.

Thus for electrical conductivity of a set of seed leachate readings (negatively correlated with germination) the data need to be duly converted for positive correlation. For example a value, say 101 μmho is converted to electrical conductivity resistance score (ECRS) by first changing the value to millimho (=0.101mmho), then deducting the same from 1 and multiplying the remainder by 100 to convert to a percentage value as follows: $(1 - 0.101) \times 100 = 89.9$ per cent ECRS. In case of colorimetric or spectrophotometric data, the optical density (OD) values are deducted from 1 and multiplied by 100 to get the required vigour scores with positive connotation. Thus for lipid peroxidation (LPO, negatively correlated with germination) we would get lipid peroxidation resistance score (LPRS); volatile products emanation in OD to volatiles emanation resistance score (VERS) and so on. For vigour parameters already positively correlated with germinability such as dehydrogenise enzyme activity score (DEAS) and many others, no such conversions would be necessary; only such values (scores) are expressed as the percentage of non-aged (refrigerator stored) harvest-fresh high vigour (HV) seed or, by regressionally estimating the enzyme activity at 100 per cent germination (benchmark value) with a set of appropriate data generated from the seed lot being studied.

The integration of the germination factor, GF (obtained by dividing germination per cent by 100, range 0 to 1) through multiplication of vigour parameters with positive connotations such as many morphological, physiological and biochemical characters, would eliminate the role of non-viable seeds in a seed lot for all positive vigour parameters, including the transformed values like ECRS, LPRS, VERS, *etc.* Thus in ECRS × GF, LPRS × GF, VERS × GF, compared to data before transformation and GF incorporation, there has been a great improvement in the correlation coefficients (r values) between germination per cent (x variate) and transformed GF integrated vigour parameters (y variate).

With the correlations usually reaching the level of statistical significance at 0.01P, often *r* values as high as 0.98 – 0.99, the mean deviations of statistical estimates and experimental observation of both x and y variates; *i.e.* ($\hat{x}$ -x) or ($\hat{y}$ -y) would go down to 1-2 per cent or even less of the respective means $\hat{x}$ and indicating thereby a great improvement in the predictive efficiency of the revised regression equations.

It is worthwhile to note in this connection that interchange of x and y variates (*i.e.*, vigour parameter as independent variate and germination as the dependent variate) would have little influence on mean deviations at such higher *r* values attributable to data conversion and GF integration.

We have further shown that SF (seedling factor or seedling growth factor), which is a measure of vigour of seedlings that may or may not be associated with germination differences, as for example vigour differences of seedlings with all samples showing 100 per cent germination or at still lower levels (say vigour differences of seedlings with all samples showing 80 per cent germination level), and GSF (germination-cum-seedling factor on integration of GF and SF) that would stand for the overall vigour status of a seed lot (see Chapter 6 for details and also paper by Basu, R.N., 2007 on 'A Holistic Approach on Quantification of Seed Vigour through Integration of Morphological and Physiological-Biochemical Parameters', Annexure No A-2).

Chapter 8

Seed Priming: A General Review on Priming and Invigoration

Genesis of Seed Priming for Improved Germinability

The problem of seed germinability and attempts to improve seed performance would date back to the pre-Christian times. As early as 620 BC, the Aryan settlers in India expressed their genuine concern on the germinability of seeds as documented in prayers like "May the seeds be viable" in the Yajurveda (one of the four Vedas, the sacred religious texts).

The Greek Philosopher and naturalist Theophrastus (371-287 BC) observed that germination of cucumber seeds soaked in water was faster than untreated seeds. Pliny the Elder (the popular name of the Roman naturalist, Gaius Plinius Secundus, AD 23-79) made a reference to the importance of presowing soaking of cucumber seeds in honey diluted with water (Paparella *et al.,* 2015).

The effectiveness of presowing soaking of wheat, rye, *etc.* in manure water for two days followed by drying back in shade, a practice undertaken by farmers was described by the French agronomist and botanist Olivier de Serres (1539-1619). That was similar to the experiment by J.E. Ells in 1963 on the improvement of emergence rate and seedling growth of tomato by seed treatment with nutrient solutions.

It would also be of great interest (of course not surprising!) to record the studies of Charles Darwin, the father of organic evolution, on the improvement of germination of lettuce and cress seeds through submergence in salty sea-water (in principle similar to the present-day osmo-halo priming treatments). Darwin's four papers in 1855 published in four issues of Gardeners Chronicle and Agricultural

Gazette on vitality of seeds, longevity of seeds, effect of salt water on germination of seeds and another, "Does sea water kill seeds?" are not only of historical importance to seed priming but also a testimony to the diversity and depth of an exceptionally brilliant naturalist's intellectual pursuits.

Present-day presowing priming treatments owe greatly to the extensive research work on seed hardening treatments conducted in the erstwhile USSR by P.A. Henckel (syn. Gekel) and coworkers around 1950 onwards with the publications mostly in the Russian language. The same was brought to the notice of seed researchers outside Russia as "An appraisal of the contribution by P.A. Henckel" in 1962, in English language by May, Milthorpe and Milthorpe in the *Field Crop Abstracts* volume 15 (pp. 94-98). The impact was immediate and a good number of research papers on the subject were brought out from different countries including India (see Table 1 and also the review by Heydecker and Coolbear 1977 for details).

The voluminous work of Heydecker and coresearchers over the years on seed presoaking (hydropriming, osmopriming as seed invigoration treatments) might be looked upon as further advanced forms of seed priming and stress amelioration treatments for improved crop performance (Heydecker, 1974, 1975; Heydecker and Hendy 1975; Heydecker, Higgins and Turner, 1975).

Beneficial effects of presowing seed hardening treatments have been reported by Austin, Longden and Hutchinson (1969) on carrot, Hutchinson (1969) on sweetcorn and Berrie and Drennan (1971) on oats and tomatoes. However, Evenari (1964) working with sorghum did not find any advantage when seeds were allowed to imbibe water upto 50 per cent of their dry weight and then dried back, repeating the same three times. The situation could possibly be attributed to overpriming or desiccation intolerance.

Heydecker *et al.* (1973) could avoid too rapid imbibition of water by successfully using polyethylene glycol (PEG) of very high moleculer weight (6000-8000) which is, viscous but nontoxic, and does not enter the cells. Therefore, unlike salt solutions, PEG acts as a safer and effective osmopriming agent. However, curtailment of oxygen availability (as such, requiring aeration) and difficulty in separating PEG from seeds have been practical handicaps, besides high costs.

In due course of time, osmotic priming with PEG or salts has been replaced by solid matrix priming (SMP) as a more practicable and viable alternative. The usefulness of SMP was experimentally verified with tomato, carrot and onion seeds by Taylor, Klein and Whitlow (1988) who concluded that solid matrix priming would provide an effective presowing treatment for improving seedling establishment of small seeded vegetables. Their study clearly indicated that seeds sown from the SMP treatments were superior, or equal to seeds from treatments of conventional liquid priming with PEG or inorganic salts. 'Agro-lig', a SMP proprietary material of the American Colloid Company was statistically equal to or better than conventional priming treatments for improving seedling emergence, decreasing T_{50} time for 50 per cent seedling emergence and increasing plant dry weight.

Besides the valuable review by Heydecker and Coolbear (1977), the reviews of Hegarty (1970), Evenari (1984), Bradford (1986), Parera and Cantliffe (1994) and others have effectively brought out the development in the preplanting seed priming treatments. In a review on seed invigoration for improved crop performance (Basu,1994) involving appraisal of 'wet' and 'dry' seed invigoration treatments, high vigour (harvest fresh) and stored medium and low vigour seeds, the vigour level at the time of treatment has been shown to be the decisive factor in the choice of treatment protocols for assured improvement of field emergence and ultimate crop performance for productivity. Very high vigour seeds of all crops studied in the University of Calcutta were responsive only to dry treatments and low vigour seeds only to wet treatments with the medium vigour seeds falling in between; further most high vigour seeds were susceptible to soaking injury.

Some of the important reviews on presowing seed priming from 1995 onwards include the following : Pill (1995), McDonald (1999, 2000), Ashraf and Foolad (2005), Paparella *et al.* (2015), Araujo *et al.* (2016), Oliveira and Gomes-Filho (2016). We shall deal with the information provided therein besides our own observations based on the reviews (Basu, 2004, 2005 and Basu *et al.,* 2010, *etc.*) as rejoinders on the issues involved.

According to Pill (1995) priming would operate to enhance seed quality by a combination of processes that may include cellular repair and improved membrane integrity and decreased seed exudation, enhanced mobilization of seed reserves as a result of activation of relevant enzymes and osmotic adjustments (a secondary effect in case of halopriming with salts in particular) and advancement of embryo development (short of radicle emergence) and increased potential for metabolic functions during seedling establishment upon planting (Pill, 1995) with all observations duly supported by relevant references.

Crop Responses to Henckel's "Seed Hardening" Treatments. (vide Heydecker and Coolbear, 1977)

Table 1 : Effect of Presowing Seed Tempering against Drought (Hardening) on Crop Yield of Plants in Field Experiments (Trambovsky, Province, 1963); (Henckel, 1968)

Crop	*Cultivar*	*Management*	*Yield of Untreated Seeds (Control) t/ha*	*Yield Increase by Hardening (Per cent over control)*
Barley	Kazansk 6/4	(a)	2.10	11.0
"	"	(b)	1.84	8.7
Sunflower	Perebodchik	(b)	0.46	114.0
Maize	VIR – 25	(b)	16.95	23.7
Maize	Bukovinskaya	(b)	12.40	29.0
Sugar beet	Ramonskaya 06	(b)	20.46	11.4

Crop	*Cultivar*	*Management*	*Yield of Untreated Seeds (Control) t/ha*	*Yield Increase by Hardening (Per cent over control)*
"	" 15-37	(c)	15.06	8.1
Table beet	Bordo	(d)	12.50	7.2
Carrot	Nantskaya.4	(d)	6.05	11.7
"	".4	(e)	5.10	23.5
Tomato	Kolkhoznui 34	(f)	21.35	87.0
"	Tambovskiy (early)	(f)	10.0	32.0

Note: a) Farm 'Lenin' Soshovski region; b) Trambovsk Agri Experimental station; c) Farm 'Novo Pokrovsky' region; d) Farm '"Roshcha' Michurinsky region; e) Farm 'Tambovsky' T. region; f) Farm 'Lenin' Tambovsky region

Comments: The beneficial effects of the 'hardening treatments' on yield were recorded in all the crops but species and varietal effects in case of maize and tomato, particularly in the latter, wide differences between the two varieties were recorded. These are consistent with present-day priming treatments showing significant differences attributable to location, species, varieties as well as initial physiological status of seed at the time of treatment.

Some Important Observations on the Effectiveness of Hydration-Based Priming (*vide* Pill, 1995)

Observations

While PEG (polyethylene glycol, H-(O-CH_2-CH_2)$_n$-OH is inert and nonphytotoxic it is sufficiently viscous to hinder aeration; it is costly.	Maxal *et al.* (1975)
The term 'overpriming' is usually used for the greater water content of seeds primed with salt solutions (*e.g.* KNO_3) than with PEG.	Alvardo and Bradford (1988)
The nitrate (NO_3^-) salts may be absorbed preferably to lower the internal osmotic potential and thereby encourage water influx which would explain alleviation of 'underpriming' (less water content of seeds) by inclusion of salts like KNO_3.	Haigh and Burlow (1987)
Synthetic sea water (a patented product) at about one-third full strength was as effective as PEG as a priming agent for tomato seeds.	Fret *et al.* (1991)
In tertiary phosphate (eg.K_3PO_4) priming solutions of the same osmotic potential, pH had a major effect in preventing germination and pH6 promoting germination.	Suzuki *et al.* (1990)
Biopriming is an integration of the principles of hydropriming with microbial organisms or biological materials as effective crop protectants.	Callan *et al.* (1991)

It has been observed that maize seeds coated with 1.5 per cent methylcellulose suspension of *Pseudomonas fluoroscens* then allowed to hydrate for 20 hours at room temperature(23°C) were protected from pre-emergence damping off caused by *Pythium ultimatum* to the same level achieved by the fungicide metalaxyl.	Callan *et al.* (1991)
Matric priming involving the use of solid or semi-solid carriers has been found to be very effective for seed conditioning for improved emergence and seedling growth rate and field performance. Solid matrix priming (SMP) (using synthetic silicate matrix) showed greater percentage and rate of emergence and higher shoot fresh weight than those primed with PEG.	Bennet and Wafers 1984, 1987 Taylor *et al.*, 1988 (SMP) Khan *et al.*, 1990 Khan, 1992 (comprehensive review on priming)
As regards post priming planting the seeds may be transferred directly from the priming medium to the ultimate planting medium (in the greenhouse, field, etc.) without drying with the yet to emerge radicle tip ready for emergence leading to seedling establishment. The situation is almost similar to sowing pre-germinated seeds. (drying back, as conventionally done, would require additional time for reimbibition essential to enter the growth phase).	Bradford, 1986 (phasic development) also see 'Hydrotime concept' of Bradford and Still, 2004 (*vide* Paparella *et al.*, 2015)
There are conflicting reports on storability of primed seeds particularly for long-term storage if not properly dried. Primed tomato seeds stored at 4°C retained viability for a long period but at 30°C loss of viability, as such the researchers concluded primed tomato seed, vigorous but reduced shelf-life.	Argerich *et al.*, 1989

Some Important Observations from the Reviews of McDonald (1999 and 2000)

A. Physiological Events during Seed Deterioration

The physiological events involved in seed deterioration included activities of a range of enzymes, changes in sugars, proteins and nucleic acids, membrane permeability, fatty acid contents, respiratory competence, lipid peroxidation, evolution of volatiles, evidences for repair, *etc.* It would be of great interest to note that as important indicators of deteriorative senescence of seeds such as increased membrane permeability, decreased respiratory competence, increased lipid peroxidation and evolution of volatiles have always been associated with seed deterioration in all the materials covered in the review.

Although the review has covered only the literature from 1984 onwards till 1997 or so, the data, before 1984 or after 1998-99 onwards, would be in tune with the observations of the aforesaid indicators of seed deterioration.

Understanding Seed Deterioration (*vide* McDonald, 1999)

McDonald (1999) has outlined the difficulties towards common interpretations of the physiology of seed deterioration based on several reasons.

(I) Firstly, short-term seed deterioration in the field would be a different phenomenon than the long-term deterioration in storage. Mechanical damage or lesions would influence seed performance in a different manner than uniform physiological deterioration during seed storage. (II) Secondly, accelerated ageing under controlled high temperature and high humidity, though more reproducible, may be different from natural ageing under ambient conditions. For example, Priestley and Leopold (1983) and Priestley *et al.* (1985) reported little increase in free radical levels of naturally aged soybean seeds but a doubling of free radicals in seeds subjected to accelerated ageing. Powell and Harman (1985) noted the absence of a consistent association of changes in membranal lipids with accelarated ageing and natural ageing of pea seeds. On the other hand, in a sharply contrasting view, Likhlatchev *et al.* (1984) concluded that physiological changes in seeds under accelerated ageing were the same as natural ageing, the only difference between different ageing methods would be the rate of ageing at which they would occur. In this connection, it is worth mentioning that our studies in the University of Calcutta over the last four decades, covering almost all the commonly cultivated agricultural and horticultural crop seeds would strongly support the conclusions of Likhlatchev and co-researchers (the matter has been dealt with in greater details elsewhere). (III) Thirdly, the rate of seed deterioration would be influenced by confounding of environmental and biological factors. The interaction of those factors may far exceed the sum of their individual effects. (IV) Seed treatment with fungicides, in particular, may indirectly influence seed deterioration. However, the effect may be transitory, and would need caution if repeated. Fumigation with methyl bromide, an insecticide resulted in delays in germination or even loss of germinability (Hanson *et al.,* 1987). (Seed treatments for maintenance of vigour and viability would be a different issue and the same has been covered under seed invigoration treatments). (V) Most seed deterioration studies are conducted using whole seeds but studies with vital constituents such as embryos would be more informative and useful. (VI) Lastly, seed deterioration is an individual event, but most studies are done with seed samples containing viable as well as non-viable seeds making interpretations difficult for distinguishing the cause and effect of seed deterioration in a population of seeds in a commercial seed lot. (This particular issue has been discussed more critically in two papers on quantification of seed vigour by Basu in 2004 and 2007 in which the role of non-viable seeds could be effectively eliminated through integration of germination factor (GF) which is obtained by dividing germination percentage by 100, (*i.e.* germination per cent divided by 100 = GF). Physiological, biochemical and morphological growth parameters or scores (also field performance and productivity parameters) can be multiplied by the germination factor (GF) for eliminating the role of non-viable seeds that would lead to a considerable improvement of regressional predictive efficiency (see Chapter 4 on selected abstracts of papers, and Chapter 5 on theme papers and thematic issues).

B. Presowing Seed Priming Treatments for Improved Crop Performance

The physiological events during the deteriorative senescence of seeds have been discussed by McDonald (1999) as a prelude to seed priming for improved crop performance. As seed quality is a very important factor in the determination of pregerminative metabolism leading to the point upto pre-radicle emergence stage, an understanding of physiological events occurring in seeds of different qualities would be helpful. Crossing the limits of desiccation tolerance would be undesirable for storability of primed seeds. It is considered necessary to stop priming at the optimum point between underpriming and overpriming for successful seed invigoration, which to a considerable extent would depend on physiological quality of the seed at the time of treatment besides other factors related to priming protocols.

The salient observations in the reviews of McDonald (1999, 2000) are tabulated hereunder citing the relevant references.

Observations

(a) Physiological quality of seed as a factor in priming	
Initial seed quality (seed vigour in particular) has a bearing on the success of priming.	Trawatha *et al.*, 1990
In pepper, osmo-conditioning improved the performance of moderate but not of poor quality seed lots.	Passam *et al.*, 1997
Hydration-dehydration treatments improved the germination of medium quality but not high quality seed lots of tomato subjected to accelerated ageing before treatment.	Penazola and Eira, 1993
Priming imposed additional damage on already deteriorated cauliflower seed primarily due to increased abnormal seedlings with damaged radicle tip.	Fujikura and Karssen, 1992
In cereals, seed germination and vigour were improved by priming high quality seeds but much more improvement was noted in low quality seeds.	Aschermann-Koch *et al.*, 1992
The success of priming was dependent on seed maturity in musk-melon.	Wellbaum and Bradford, 1991
Priming enhanced the vigour of immature mechanically harvested seeds of broccoli.	Jett and Wellbaum, 1996
(b) Suitable priming technique, optimum moisture, aeration, *etc.* for successful priming.	
No difference in sugar beet seed response to the use of water or salts as long as the osmotic potential was in the range of-10 to -20 bars.	Durrant *et al.*, 1983
Priming response of pepper seed was dependent on the duration of treatment and osmotic potential of the solution rather than a specific salt.	Smith and Cobb, 1991

The use of inorganic salts compared to PEG was deleterious to tomato and asparagus seeds and caused membrane damage and enzymatic changes that affected subsequent growth.	Fret *et al.*, 1991
The use of salts in priming was toxic to *Sorghum* seeds.	Haigh and Burlow, 1987
On the other hand priming with salts was superior to PEG for three herbage grasses.	Mauromicale and Cavallaro, 1996
Imbibition injury to very dry seeds (legumes in particular could be avoided by pre-humidification (moisture equilibration with a water saturated atmosphere) to avoid soaking injury.	Saha and Basu, 1984 (in soybean) Ellis *et al.*1990 (in peas)
Prolonged soaking and undue advancement in germination may lead to loss of desiccation tolerance.	Armstrong and McDonald, 1992
Storability of conventionally primed (long duration hydration period) seed is generally poor.	Bradford and Tarquis, 1989.
Priming of tomato seeds enhanced germination but lowered seed resistance to deterioration. It was concluded by the authors that the primed seeds should be considered vigorous with a reduced storage life. Similar reduced shelf life has been reported by many other researchers.	Also see Tarquis and Bradford 1992 for lettuce; Dearman *et al.* 1987 for leek and carrot; Singh *et al.*, 1985 with celery; Lanteri *et al.*, 1997 For pepper; Maude *et al.*, 1994 for leek seeds.
Osmotically primed asparagus and tomato seeds should be stored at 4°C rather than 25°C for better post-storage vigour and viability. Similar low temperature controlled storage has been noted by other researchers.	Owen and Pill, 1994
(c) Biochemical mechanisms of priming.	
Free radical scavenging enzymes such as superoxide dismutases, catalase and peroxidase and glyoxysome enzymes such as isocitric lyase were increased by seed priming.	Jeng and Sung, 1994
Priming effectively favoured membrane repair in watermelon seeds which was attributed to the stimulation of peroxidase scavenging enzymes.	Chiu *et al.*, 1995
Osmopriming increased the RNA content of leek and tomato seeds.	Bray *et al.*, 1989 (in leek seed) Coolbear *et al.*, 1990 (in tomato seeds)
It has been reported that 70-80 per cent of the DNA synthesis which continued into the post-priming period culminated in significant increase in plastid and mitochondrial DNA in primed leek seeds.	Ashraf and Bray, 1993 (the avoidance of mitochondrial dysfunction by priming could be a key factor in quality enhancement in primed seeds).

Comments on the Observations of McDonald (1999, 2000)

The difference between short-term hydration dehydration seed invigoration treatments as originally developed in the University of Calcutta for the maintenance of seed viability and vigour and the long-term presowing seed priming is apparently the duration of seed hydration (that has been the probable reason of inclusion by McDonald (1999) of Saha and Basu (1984), Saha *et al.* (1990) as references under hydropriming and matripriming respectively. The consequential effect of that difference in duration of hydration is large enough to require further elucidation.

The short-term soaking-drying treatments are usually given midway during the ambient storage period with soaking duration not exceeding 2-3 hours is common, even dip-dry (5 minutes dipping then drying back within an hour) or spraying-drying would delay loss of seed viability in several kinds of seed. Hydration can be accomplished by moisture equilibration (humidification) of seed with a water saturated atmosphere for 16-24 hours followed by drying back. Interestingly, unlike conventionally primed seed, the aforesaid hydration-dehydration treatments would show little or no increase in germination percentage or seedling growth immediately after treatment. Significant beneficial effects of the treatments on seed viability and vigour would be seen after natural ageing for long periods or accelerated ageing suggesting a preventive or prophylactic effect of the treatments on the maintenance of seed vigour and viability. The significant beneficial effects have been consistently associated with reduced lipid peroxidation, retention of cell membrane permeability and integrity and reduced emanation of volatile (gaseous) emanations from treated seeds along with increase in the activity of dehydrogenase enzymes (suggesting respiratory competence) and various other characters associated with seed vigour and viability.

The important point of contention in this connection is the fact that all the afore-said developments would take place in the very early initial phase of conventional seed priming (not exceeding a couple of hours), but the subsequent events involving the operation of the cellular repair mechanisms and undue advancement of germinative metabolic reactions would not be compatible with drying back resulting in desiccation intolerance which would reduce the longevity of the primed seed in storage. In the prolonged conventionally primed seed, despite all the early beneficial prophylactic reactions, the desiccation intolerance attributable to undue advancement of pregerminative metabolism would certainly reduce the shelf life of the primed seeds. Storage of conventionally primed seed should be avoided.

However, a suitable compromise needs to be worked out in view of the fact that sufficient pregerminative metabolism is an essential feature of the conventional seed priming protocols aimed towards alleviation of varied edaphoclimatic stresses.

On practical considerations, determination of optimum time, (may be a bit approximate) to stop priming (which is as yet a trial and error process) need to be streamlined not only upto crop species and cultivar levels but also for quality status (particularly, initial seed vigour levels, high, medium, low, *etc.* of the same cultivar at the time of priming treatment).

Studies have been taken up in the University of Calcutta with moist sand conditioning (MSC-D) treatment, a kind of solid matrix priming (SMP), aimed towards benefitting small and marginal farmers through inexpensive seed priming treatments for improved crop productivity. It is proposed to prepare one easy-to-follow vernacular language (Bengali) pamphlet for the farmers who are willing to take up seed priming as a part of their farming schedule. Necessary training, *etc.* would be provided by a reputed social service organisation for the purpose.

Some Observations from the Review on Seed Priming by Ashraf and Foolad (2005)

In an extensive review on seed priming very appropriately termed as 'A shotgun approach' (which as per dictionary meaning would imply 'covering a wide field with hit and miss effectiveness', Ashraf and Foolad (2005) deliberated on the major findings in the field of conventional water-based seed pretreatments especially for salt-stressed situations. The review has been based on salient research findings of different priming protocols(over a dozen priming techniques) including hydro-priming, osmopriming, halopriming, solid matrix priming, thermopriming, priming with growth hormones, biopriming with bacterial control agents, drum priming, as well as seed pelleting, fluid drilling, *etc.* The effects of the priming treatments on germination, seedling emergence, physiological and biochemical changes, *etc.* as reported in the relevant research publications have been duly cited as references.

The reviewers noted that individual priming protocols would show advantages and quite often disadvantages as well.The extent of beneficial effects would, however, depend on species, varieties within a species, the morphological and physiological status (particularly seed vigour at the time of treatment) of the seed material being studied. As such, priming would continue to remain 'a shotgun approach' for the present. Perhaps, each material (seed sample) for a more accurate approach, would need a preliminary laboratory test for the optimum timing to stop priming prior to radicle emergence of the pregerminative metabolism phase because undue advancement would lead to desiccation intolerance with consequent adverse effects. Further, additives and their dosages, temperature, and other variables should be given due consideration before treatment.

The question is whether the farmers will be able to do that or they would have to get that done by others. In the developing countries mostly with small and marginal farmers, these issues may create genuine problems for the average farmers to take up sophisticated priming techniques on a commercial basis.

Ashraf and Foolad (2005), however, correctly observed that prior assessments must be taken up before large-scale experiments if seed priming was to be used for large-scale field planting as a commercial proposition.

While discussing the efficacy of drum priming, the reviewers cited the view of Warren and Bennet (1997) that drum priming (as used in highly mechanised western agriculture) could provide a better alternative to osmotic or matrix priming for improving crop stand establishment in two genotypes of sweet corn. The applicability of drum priming for introducing microbial inoculants such as bacteria and fungi for promoting microbial population dynamics was assessed on carrot, leek and parsnip seeds by Wright *et al.* (2003). Irrespective of seed types, drum priming resulted in significant increases in the populations of *Pseudomonas* (that would account for 10 per cent of the culturable bacterial population which would also decrease little by re-drying). However, all these issues are presently not so relevant in the poor, small and marginal farmer-dominated farming systems of the tropical and subtropical countries, at least in the near future.

Some Important Observations from the Review of Paparella *et al.* (2015)

Pre-Germinative Matabolism and Seed Priming

Major changes	
De novo synthesis of nucleic acids, proteins, ATP production, repair and antioxidant mechanisms.	Macovei *et al.*, 2010, 2011, 2014; Balestrazzi *et al.*, 2011, 2012; Chen *et al.*, 2012
Hydrotime concept analysis for priming assessment	
Hydrotime concept provides a unifying model for the description of the relationship between water potential and seed germination rate and percentage also defines the time for radicle emergence.	Bradford and Still (2004); (According to Paparella *et al.* 2015, some constituents of this time consuming procedure may be streamlined with automated seed imaging for scoring germination)
Seed priming and crop productivity under adverse environmental conditions.	
Global climatic change causing drought and other stress conditions adversely affecting crop system but consequent physiological responses are still poorly understood.	Bhargava and Sawant, 2013
Selenium (Se)-mediated priming would increase germination efficiency and drought tolerance of wheat.	Nawaz *et al.*, 2013
The positive role of Se may be due to modulation of sugar flux that might act as molecular signal for the activation of plant response under drought stress.	Hartikainen, 2005; Kishor *et al.*, 2005 (Review article on regulation of proline biosynthesis).

Hydropriming, PEG-mediated osmo-priming and SA (salicylic acid-mediated priming were able to provide water stress tolerance to rice (SA more effective).	Lin and Zhang (2012); Shekari *et al.*, 2010 [recorded the beneficial effect of SA priming under water deficit at poddling stage of cowpea (*Vigna unguiculata L.*)]

Digital Image Technology (DIT) Applications

Observations

Digital images of seeds and seedlings captured by a charged coupled device (CCD) camera would replace human visual inspection and data recording for integration with molecular and biochemical data on various aspects of priming.	Fiorani and Schurr, 2013; Leil *et al.*, 2014 (DIT based information may be converted to curves to provide spatio temporal response of priming.
Calculation of germination index, seedling vigour index, leaf area index of direct seeded rice given priming treatment for germination, growth and yield	Mahajan *et al.*, 2011. Also see Dell'Aquila *et al.*, 2000 on DIT in monitoring the imbibition process in white cabbage.

Molecular Perspectives in Seed Priming

Observations

Efficient predictive methods are needed for the early determination of correct time point to stop priming and start seed drying (particularly for seeds of varying vigour and viability status, besides species cultivars and other variables). Prolonged treatment would result in unpredicted enhancement of oxidative DNA injury irreversibly affecting seed viability.	Balestrazzi *et al.*, 2010-2015; Macovei *et al.*, 2011a, 2011b.

Systems Biology Approaches for Understanding Complexities

Observations

Proteome analysis of seed priming and germination have been very effective in identifying the priming induced physiological changes. The holistic 'Systems Biology' approach has proved more effective than the otherwise reductionistic molecular-cum-functional biology approaches.	Wang *et al.*, 2015 Duque *et al.*, 2013

The beneficial effect of PEG mediated osmopriming of rapeseed was studied using a global proteomic and transcriptomic profiling. Upto 952 genes and 75 proteins were differentially expressed during the main phases of the priming procedure and the final germination of the primed seeds.	Kubala *et al.*, 2015

Main Limitations of Current Priming Techniques

Observations

Most priming protocols developed so far include at the end of the treatment a rapid re-drying of seeds for storage.	Hamner, 2001
Reduced seed longevity is a well reported disadvantage of seed priming.	Chiu *et al.*, 2002
Lettuce seed subjected to different priming methods showed high susceptibility to deterioration during storage.	Tarquis and Bradford, 1992
The deleterious effects of priming on storability, may be ascribed to the prolonged treatments that cause the loss of desiccation tolerance.	Sliwinska and Jendrzejezak, 2002
The unfavourable conditions in which the seeds are kept immediately after priming (temperature, humidity, *etc.*) negatively affect viability.	Schwember and Bradford, 2005
There are also reports on improved viability of primed seeds.	Probert *et al.*, 1991; Butler *et al.*, 2009
Incubation of seeds under slow conditions following priming with galactosyl-sucrose resulted in enhanced longevity.	Gurusinghe and Bradford, 2001

Comments

In a commercial seed lot, the vigour levels of individual seeds would differ a lot. Even, in a high vigour harvest-fresh seed lot there may be significant differences in seed samples attributable to developmental variations while still on the mother plants that may differ in various ways. As such, the so-called pre-radicle emergence time of seeds may differ considerably. The situation would be further aggravated in variously stored commercial seed lots. Those well below the critical stage during presowing priming would show post-drying storability (longevity) attributable to the long established beneficial effect of hydration-dehydration treatment on the maintenance of seed viability. Those too advanced during hydration would become desiccation intolerant with poor storability particularly under adverse conditions.

As regards *ex situ* seed conservation of wild plants, simple short-term hydration-dehydration treatment is recommended instead of conventional seed priming.

Halogenation for the Preservation of Seed Stocks in Farmers' Houses in Tamil Nadu (Dharmalingam *et al.*, 1977 as summarized by Basu *et al.*, 2010)

Treatment of freshly harvested or stored relatively high vigour seeds with the halogens iodine, chlorine or bromine, usually in the gaseous (vapour) phase for the maintenance of seed viability and vigour was originally developed in the University of Calcutta (Basu and Rudrapal, 1979, 1980; Rudrapal and Basu, 1980, 1981). The treatment was subsequently found to significantly improve crop productivity.

In the meantime, a former research scholar of Calcutta University Dr. Dolly Rudrapal joined the laboratory of Professor S. Nakamura in Japan and fully confirmed the highly significant beneficial effect of use of halogens in controlling eggplant and radish seed deterioration (Rudrapal and Nakamura,1988).

Pal and Basu (1988) very successfully retained the vigour and viability of rice seeds through the exposure of nearly harvest-fresh (1-month-old, high vigour) seeds to low concentrations of iodine and chlorine vapours for 8-72 hours and then stored under ambient warm humid conditions for 7 months. Around 80 per cent average germinability was maintained in the halogenated seeds against 65 per cent in the untreated control. Dry dressing with iodinated French chalk (1 g iodinated French chalk containing 50 mg iodine for 1 kg. of seed) or with common bleaching powder (active ingredient calcium oxychloride that would slowly release chlorine) at the rate of 2 g per kg. of seed, also slowed down the deterioration of seeds stored under ambient conditions. Grain yield of the crop raised using iodinated French chalk (1 g per kg of seed) also showed consistent statistically significant (P=0.01) improvement.

Pal *et al.* (1996) in another experiment with several 'dry' seed invigoration treatments including bleaching powder (active ingredient calcium oxychloride) besides other 'dry' treatments significantly improved germinability, field emergence and yield of wheat and sunflower. The beneficial effects of all the aforesaid seed invigoration treatments were associated with reduced production of volatile compounds (aldehydes) in comparison with the untreated control seeds suggesting the involvement of free radical induced lipid peroxidation reactions and the counteraction of the same by the dry seed invigoration treatments including halogenation.

In view of the aforesaid developments on the use of halogenation as an effective seed invigoration treatment it is considered most pertinent to highlight the initiatives taken by Dharmalingam *et al.* (1997) in the Department of Seed Technology, Tamil Nadu Agricultural University (TNAU) Coimbatore towards widespread application of halogenation by farmers themselves treating their own seeds with the help of TNAU researchers. [As a matter of fact, Dr. C. Dharmalingam, the lead researcher has been fully conversant with the development of the halogenation treatment while he was in the University of Calcutta (on study leave from TNAU) for his doctoral degree in seed invigoration].

It was certainly a bold move by the TNAU researchers to request the farmers to take up the halogenation treatment themselves with necessary technical assistance by the University. In due course of time, those farmers would certainly help others to do the same. The direct involvement of the farmers on such a big scale has been a major achievement of TNAU and the associated research and supporting staff.

After all, it is the farmers, more specifically, the small and marginal farmers who constitute the backbone of our agricultural activities. As such, supporting them technically would pay more handsome dividends than doling out financial subsidies and loan waivers. The TNAU authorities, the researchers and supporting staff have clearly proved that as would be evident from the summary report outlined by Basu *et al.*, 2010.

Physical Priming Methods (*vide* Heydecker and Coolbear, 1977)

Recently there has been a revival of interest in physical seed treatment (Araujo *et al.*, 2016) in review of their ecofriendly approach necessitated by the adverse impact of global climatic change on the environment, agriculture and many other human activities involving widespread use of a range of undesirable chemicals.

In their exhaustive review of seed treatments for improved crop performance, Heydecker and Coolbear (1977) covered physical seed invigoration treatments under the subhead 'Energy treatments' involving heat, light, magnetic fields, electrical treatments, ultrasound (sonication), ionising radiation (x-ray, γ-ray) and 'shock' treatments, *etc.*

Effect of 'Energy' Treatments on Seeds

i) Heat	
Oil palm seed germination improved by storing them for 90 days at 39.5°C.	Wonky Appiah, 1974
Okra seed germination improved by exposure to 45° C for 10 hours.	Onwueme, 1975
Four minutes exposure to moist or dry heat (70-90°C) of Leguminous *Lespedeza* seed improved germination.	Martin *et al.*, 1975
Microwave drying increased the germination of cotton seeds.	Wesley *et al.*, 1974
ii) Light	
Irradiation of seeds with concentrated sunlight has been found to increase yield and quality of the resulting crops such as peas, sugar beet, *etc.*	Babushkin *et al.*, 1974; Gulyaev *et al.*, 1973, 1975; Kotsur *et al.*, 1972
Beneficial effects of high energy flashes of light on the yield and quality of maize, wheat, sugar beet, cotton, *etc.*	Shakhov, 1974

iii) Magnetic fields	
Magnetic presowing seed treatments proved beneficial in a number of crop species.	Pittman, 1967
Improved seedling growth of winter wheat	Pittman and Ormrod 1970
Improved yield of French bean	Pittman and Ansley 1967
Improved germination of barley	Pittman and Ormrod 1971
Electromagnetic fields upto 300 oersteds beneficial for several materials.	Zolyneak and Lupascu 1972
iv) Electrical treatments	
The beneficial effects of application of electric field treatments to various cereals and vegetables, particularly in the erstwhile USSR using various field intensities and exposure times perhaps operated on a similar biochemical basis to that of the biomagnetic treatments mentioned above (Heydecker and Coolbear 1977).	
v) Ultrasound (sonication)	
Beneficial effect of sonication on sunflower seeds	Tudor and Ioana, 1973
Ultrasound effect on the recovery of spring wheat root system	Albu 1972
Increased seedling growth of cotton by seed treatment with ultrasound for 5 to 10 minutes.	Mukhamedkhanov and Shermukhamedov, 1971
vi) Ionising radiations (x-ray, γ-ray)	
Germination percentage and rate of germination of *Passiflora* increased with low radiation levels, same in radish and *Datura innoxia.*	Torne and Desai 1975; Tanaka and Umeda 1974; Singh *et al.,* 1974
Increased growth and ratio of hermaphrodite to male flowers of melons	Bisaria *et al.,* 1975
Increased seed yield and essential oil yield of coriander.	Arinshtein *et al.,* 1973
Increased yield of carrot	Brazhnikova 1974
Radio stimulation as a physical seed treatment is ripe for commercial application. Crop seeds irradiated with x-ray and γ-rays mostly at levels between 0.1-2 Krad would give consistent, if not large increases in the yield of numerous crops. Very low doses may create conditions favourable for beneficial effects.	Clemente and Caldera, 1975 (survey-cum-review article)
vii) 'Shock' treatments (alternative)	
Several 'shock' treatments can stimulate seed germination that may include the following types (*vide* Chaussat and Le Deunff, 1975).	
Kinds of shocks	
Hot thermal	Le Deunff, 1973
Extremely cold	Pollock and Roos, 1972
Ultrasound	Come, 1975
Intense light flushes	Shakhov, 1974

Hot water soak, infrared and ultrasound may cause germination of many weeds.	Holm and Miller, 1972
Energy input and its shock may trigger germination as a concept.	Hendricks and Taylorson, 1974
X-rays and thermal shocks improve germination of *Datura innoxia* seeds.	Singh *et al.,* 1974
viii) Cathodic protection of seed viability	
Placing maize seeds on a negatively charged conductor extended their viability during artificial ageing. Such cathode protection may reduce free radical attack by providing a source of electrons. The results support hypothesis of free radical damage to cellular components and are consistent with such damage being important in deteriorative senescence of seeds.	Pammenter, Adamson and Berjak, 1974 (also see the comments on the issue)

Note: Observations on the original study of Molnar (1972/73) on cathodic protection of life span of mice *vis-à-vis* cathodic extension of maize seed viability by Pammenter, Adamson and Berjak (1974).

Anticipating the involvement of free radical induced damage on longevity Molnar (1972/73) attempted to reduce free radical attack and prolong lifespan of mice by maintaining them in a cage with an applied negative potential in a manner analogous to the technique of cathodic protection of metals against corrosion. The average lifespan of mice exposed to the negative charge was 25 per cent greater than that of mice exposed to the positive charge (*vide* Molnar, K.1972/73).

The aforesaid principle of cathodic protection was followed with some modifications by Pammenter, Adamson and Berjak (1974) for extension of longevity of maize seeds. Artificially aged seeds of maize (*Zea mays*) were provided with cathodic protection by placing them embryo-side down on aluminium foils within small enclosed chambers and applying a negative potential of 300 volts to the foil. Control seeds were maintained in the same manner without the applied charge. The entire contents of the chamber were taken from each of the samples for germination and other tests. Germination of seeds was recorded after soaking the seeds overnight and placing on moist paper toweling for 84 hours.

The leachate (soak-water) consistently showed greater electrical conductivity values for control seeds than those under cathodic protection indicating greater membrane damage in control seeds. Very significantly, chromosomal aberrations were almost three times more in control than under cathodic protection (12.6 per cent in control against 4.3 per cent in protected seeds). Thus the provision of a source of free electrons to air-dry (13.6 per cent moisture content) maize seeds not only greatly reduced the loss of seed viability (compared to extension of life span of hydrated systems such as mice, the air-dry plant seed, maize was far more responsive, after 30 days the differences in germination was over 70 per cent) but also other characters positively associated with seed vigour status. Obviously, in sufficiently hydrated systems (as in animals like mice in Molnar's study or in hydrated seeds), cellular repair processes may rectify to a certain extent free radical induced damages. But corrections and metabolic repairs would not be possible in air-dry seeds making them vulnerable to loss of viability and vigour.

Cathodic protection serves as an excellent example of a physical method of seed invigoration but an innovative easily practicable and non-expensive version of the same would be necessary.

Table-1 : Summary of magnetic fields (MFs) effects on seed and seedling performance (*vide* Araujo *et al.,* 2016).

Species	*Magnetic fields*	*Effects described*	*Reference*
Triticum aestivum	30 mT	No stimulation of seed germination, neither seedling growth. Increased antioxidant potential under soil flooding.	Balakhnina *et al.* 2015
Tagetes patula	100 mT	Improved germination, seedling vigor and starch metabolism.	Afzal *et al.* 2012
Glycine max	200 and 150 mT	Increased germination parameters and seedling biomass. Plants with higher efficiency of light harvesting and biomass accumulation.	Baby *et al.* 2011; Shine *et al.* 2011
Helianthus annuus	50 and 200 mT	Increased germination and germination rate. Increased seedling length and biomass accumulation.	Vashisth and Nagarajan, 2010
Vigna radiata	5 mT	Improved germination, seedling vigor and starch metabolism.	Reddy *et al.* 2012
Solanum Lycopersicum	100 and 170 mT	Improved biomass and growth. Increased tolerance to biotic stresses.	De Souza *et al.* 2006
Cucumis sativus	200 mT	Improved germination, seedling vigor, starch and anti-oxidative metabolism.	Bhardwaj *et al.* 2012
Zea mays	100 and 200 mT	Improved seedling growth, leaf water status, and photo-synthesis in seedlings under soil water stress.	Anand *et al.* 2012
Vicia faba	0.1 mT	Improved seedling growth.	Rajendra *et al.* 2005
Quercus suber	0.015 mT	Improved sprouting rate and seedling biomass.	Celestino *et al.* 2000

Comments on the Review Paper on Physical Priming by Araujo *et al.* (2016)

Since the comprehensive review of Heydecker and Coolbear (1977) on various aspects of different seed invigoration treatments including a range of physical ones (energy treatments such as heat, light, magnetic fields, electrical treatments, ultrasound, ionizing radiation, shock treatment, *etc.*) by far the most updated review on physical priming has been made by Araujo *et al.* (2016) that greatly emphasizes the importance of physical seed priming methods and protocols on a commercially viable scale employing magnetopriming and also radiation priming based on the principle of hormesis (adaptive dose-dependent response of cells and organisms to low and moderate stresses).

Table-2 : Summary of ultroviolet (UV) effects on seed and seedling performance (*vide* Araujo *et al.*, 2016).

Species	*UV applied*	*Effects described*	*Reference*
Lactuca sativa	UV-C	Enhanced NaCl stress tolerance	Ouhbi *et al.* 2014
Phaseolus vulgaris	UV-C	Increased germination rate. Increased accumulation concentration of bioactive molecules in seed coats.	Gualardo-Plores, *et al.* 2014
Vigna radiata	UV-C; UV-A	Increased germination rate and seedling vigour. Seedlings with reduced susceptibility to root infecting fungi.	Hamid and Jawaid 2011; Hamid and Jawaid 2011;
Arachis hypogaea	UV-C	Increased germination rate and seedling vigour. Seedlings with reduced susceptibility to root infecting fungi.	Siddiqui *et al.* 2011
Vigna mungo	UV-B	Exposure of hydrated seeds significantly reduced germination, suppressed the synthesis of photo-synthetic pigments, as well as, root and shoot development . Curling and twisting of the seedlings.	Shaukat *et al.* 2013
Carthamus tinctorius	UV-B	Enhanced germination rate but subsequent growth of the seedlings is hampered.	Farokh *et al.* 2010

Table-3 : Summary of microwaves (MWs) effects on seed and seedling performance (*vide* Araujo *et al.*, 2016).

Species	*MWs applied*	*Effects described*	*References*
Lens culinaris	2.45 GHz	No impact on seed germination or rate. Seedling length is stimulated after exposure of 30 s. Higher exposure time affects negatively all parameters.	Aladjadjiyan, 2010
Raphanus sativus	10.5 and 12.5 GHz	Reduced germination, germination rate and hypocotyl growth rate.	Scialabba and Tamburallo, 2002
Oriza sativa	2.45 GHz	Increased the germination percentage and rate, as well as the primary shoot and root length.	Talei *et al.* 2013
Triticum aestivum	2.45 GHz	Reduction of seed borne *Fusarium graminearum* infestation and seed vigour.	Reddy *et al.* 1988
Solanum tuberosum	2.45-54 GHz	Highest biomass growth in seed potato germs.	Jakubowski, 2010
Hordeum vulgare	2.45 GHz	Enhanced germination and vigour index after exposure during 20 s.	Juliana et al. 2013

While the potentialities of microwaves (MWs) have been recognized, seed priming with ultraviolet radiation has been considered to be a more difficult task. Nevertheless, as with other types of radiation treatments, the effect of MWs would depend on plant species, exposure duration, frequency, *etc.* and, as such should be carefully evaluated.

Araujo *et al.* (2016) have, in particular highlighted the role of electron spin resonance (ESR also known as electron paramagnetic resonance, EPR) for studying the impact of physical seed priming treatments and have reasserted the enormous potential of ESR/EPR to monitor seed priming treatments and identify the most suitable priming protocols for different kinds of seeds. The authors have concluded that presently the most suitable physical seed priming treatments include the use of magnetic fields (MFs), microwaves (MWs) and low doses of ionizing radiations (IRs).

Finally, an integrated and multidisciplinary approach has been rightly advocated along with speeding up basic and translational research as necessary to frame proper guidelines for large-scale seed invigoration by seed operators and general farmers.

Based on our over four-decade long involvement in seed invigoration studies with crop seeds, we are convinced that all successful seed invigoration protocols act through maintenance of seed vigour and viability.

Irrespective of basic methodological differences in approach (physical, chemical, physico-chemical, physiological, biochemical, *etc.*) there would be broad similarities in the expression of different seed invigoration treatments responsible for the alleviation of various stress situations such as ageing (natural as well as accelerated), field emergence and crop productivity under stressed as well as near normal field conditions.

Physiological and biochemical parameters such as retention of cellular membrane semipermeability (as such, reduced leakage of electrolytes and metabolites), reduced lipid peroxidation and very importantly reduced emanation of volatile products, particularly aldehydes (low volatiles emanation is associated with high seed vigour bioassay scores), increased dehydrogenase enzyme activity as well as several other important enzymatic reactions are positively correlated with seed invigoration. So far, negligible or low volatile substances emanations from seeds have been found to be a very consistent index of good seed vigour in all crop seeds studied by us. As such, seed vigour is negatively correlated with emanation of volatiles. The same can be mathematically converted to a positively correlated vigour score VERS (volatiles emanation resistance score in percentage scales) with the regressional efficiency further improved through integration of germination factor (GF) which is equal to germination per cent divided by 100, eliminating thereby the contribution of non-viable seeds, see papers on quantification of seed vigour (Basu 2004, 2007) for further details. Similarly, other vigour scores such as the jute seedling bioassay score (JSBS), dehydrogenase enzyme activity score (DEAS), lipid peroxidation resistance score (LPRS), *etc.* would serve as excellent measures of the efficiency of various seed priming and invigoration treatments of different crops.

Annexure No A-1

METHODOLOGIES FOLLOWED IN SEED INVIGORATION TREATMENTS

A. Midstorage Hydration-Dehydration and Physico-chemical Treatments for Seed Invigoration

Hydration-dehydration treatments of stored seeds of a wide range of agriculturally important crop plants which included a majority of orthodox seeds of cereals, millets, oil and fibre yielding crops, vegetables and flowers were carried out employing various methods. The details of the treatments are given hereunder:

I. Hydration-Dehydration

(i) Soaking-Drying (S-D)

Stored seeds were fully immersed in twice the volume of water for 2- 6 hours (depending on the material) followed by drying back to original weight in the sun or in a seed drying cabinet in a current of hot ($35^{\circ} \pm 1^{\circ}C$) air; for each material soaking duration was standardized as it is essential that the soaking period must be compatible with drying back. The control seeds which were not hydrated were also dried along with the treated seeds in the same fashion. A majority of the orthodox seeds (except protein-rich legumes) responded very well to this treatment which was particularly effective in the carry-over seed stocks. Both the control and soaked dried seeds were then transferred to a desiccator containing calcium chloride for about a week to stabilize the moisture content to a uniform level.

(ii) Dipping-Drying (D-D)

Seeds were dipped in twice the volume of water for 2-5 minutes only (in some cases wet seeds kept covered for 2 h) and then taken out and dried back in the usual manner. In case of legumes the dipped seeds were not kept covered but slowly dried back in a current of hot air at 35 ± 1°C. Dipping-drying is effective in relatively high-vigour seeds including certain proteinaceous materials like lentil and mungbean, although the effects are not as prominent as in other non-leguminous seeds. Drying of seeds hydrated by dipping is easier than seeds soaked in water.

(iii) Spraying-Drying

Besides the above two hydration methods, a light spray with water followed by slow drying back would also prolong seed viability. In this method, seeds were spread in single layers and then sprayed with an ordinary sprayer; the amount of spray was so adjusted to bring the moisture content of the hydrated seed to about 20 per cent on fresh weight basis. Approximately 200 ml of water were required for 1 kg of seed. Although drying of seed is easier and faster (requiring only 20-30 minutes), it is less effective than other methods of hydration. This method is, however, more effective in relatively high vigour seeds.

(iv) Moisture Equilibration-Drying (ME-D)

Seed hydration could also be accomplished by exposing the seeds to a water-saturated atmosphere for 24-72 hours (depending on the material) avoiding direct contact with water. The seeds were placed in thin layers in paper trays kept on a raised platform in a closed water-saturated chamber lined internally on all sides with moist blotters giving nearly 100 per cent relative humidity at room temperature. This method of seed hydration is slow unaccompanied by leaching of substances from the seed and eliminates the possibility of soaking injury due to rapid uptake of water by the seed. Besides being very effective in most seeds, it was successful to a certain degree even in legumes.

(v) Moist Sand Conditioning-Drying (MSC-D)

Leguminous seeds such as common pulses, peas and beans are susceptible to soaking injury. Hence for such seeds direct immersion in water is to be avoided and would be required to take up moisture in vapour form in a very slow and progressive manner, thereby alleviating soaking injury. In MSC-D air-dry sand was moistened with a measured quantity of water (40 ml of water mixed with 400 g of dry sand);seed to sand ratio, 1:3 or 1:4 for most kind of seeds. The seeds were mixed with the sand and incubated in a covered container (separate container for each duration) for 16 to 18 hours depending upon the material. The moisture content in the sand was so adjusted that the final seed moisture content did not exceed 30 per cent on wet weight basis so as to avoid the advancement of germination process to an extent not compatible with drying back. The seeds being hygroscopic absorbed almost the entire moisture content from the sand making it virtually air-dry. Seeds were separated from the sand by sieving which could be reused after washing and sun drying. If desired, the seeds could be then soaked in double the volume of water or dilute solutions of chemicals for 2 hours or as required for different kinds of seeds and then dried back to original weight in a current of hot air at 35 ± 1°C in a drying cabinet. The seeds were then moisture stabilized by keeping over fused calcium chloride for about three days to a constant moisture content of 8 per cent on wet weight basis.

II. Chemical Treatments

In the hydration-dehydration treatments employing liquid water (except moisture-equilibration-drying) a range of water soluble chemicals were used in low concentrations (10^{-5}–10^{-3}M) for getting additional benefits over hydration alone. The chemicals used for the study were salts such as sodium chloride, sodium phosphate (dibasic), sodium nitrate, potassium iodide, sodium thiosulfate, EDTA (Na-salt), *etc.*; phenols *p*-hydroxybenzoic acid, tannic acid, ferulic acid; organic acids oxalic acid, citric acid; organic nutrients: glucose, fructose, sucrose, ribose, *etc.*; radioprotective, and radiosensitising chemicals – cysteine, *p*-aminobenzoic acid, hydroquinone, α-tocopherol, ascorbic acid, and caffeine, sodium azide, 2-4 dinitrophenol *etc.*

Some Important Considerations in Hydration-Dehydration Seed Treatments

In evaluating the efficacy of the hydration-dehydration treatments, due consideration was given to the following:

(i) Duration of soaking

To determine the optimum duration of soaking a fixed quantity of seeds of different crops were soaked in twice its volume of distilled water for 0.25, 0.5, 1, 1.5, 2, 2.5, 3, 4, 6 hours at room temperature (27 ± 1°C). After each time interval, seeds were removed from water and dried back to their original weight in the usual way.

In case of legumes, care was taken to determine the optimum duration of soaking to avoid injury. These seeds were soaked for various periods ranging from 5 minutes to an hour followed by drying back.

To study the influence of seed vigour on the relative effectiveness of different hydration-dehydration treatments on storability of seeds, studies were conducted using seeds of three levels of vigour for each material, the particulars of which are given below namely high vigour (HV), medium vigour (MV) and low vigour (LV). Seeds stored immediately after harvest in sealed glass bottles in a refrigerator at –4°C served as high vigour (HV) seeds; those stored under ambient humidity and temperature conditions for 6 months and 18 months were taken as medium vigour (MV) and low vigour (LV) seeds respectively. Seeds from the aforesaid vigour stocks were given various hydration-dehydration treatments.

Germination tests before and after ageing were done following the inclined glass plate blotter method of Punjabi and Basu, 1982 (see details in the extracts of the paper, Chapter 5).

(ii) Stabilization and Determination of Moisture Content

The hydrated-dehydrated seeds were dried thoroughly and moisture content was stabilized by placing them in paper packets for a week in a desiccator containing fused calcium chloride. The moisture content was determined at regular intervals following the procedure of ISTA (1976). Drying back to the basal moisture level is absolutely necessary, as storage of seeds not properly dried after treatment would do more harm than good.

(iii) Physiological Age of Seeds at the Time of Treatment

The physiological age of the seed at the time of treatment is the most important determining factor and only seeds stored for some time responded to the hydration-dehydration treatments. The soaking-drying treatments were ineffective in maintaining vigour and viability (upon subsequent storage) in harvest-fresh seeds, but were very successful in medium-vigour seeds irrespective of the method of hydration, however in many low-vigour lots only soaking-drying was effective.

B. Use of Non-Aqueous Solvents and 'Dry' Vapour Treatment for Seed Invigoration – Special Considerations

Besides the hydration-dehydration treatments, the following 'dry' treatments (employing non-aqueous solvents, volatile chemicals, finely powdered materials, *etc.*) were attempted.

Organic Solvent Infusion

It was noted that in the hydration-dehydration treatments with or without chemicals water itself showed the major effect in minimizing loss of vigour and viability during subsequent storage; the chemicals with the exception of a few showed very little or no effect due to the large masking effect of hydration alone. To evaluate the effects of the chemicals the organic solvent infusion method was attempted employing non-aqueous solvents like acetone, dichloromethane, acetic anhydride, propylene glycol and ethanol. Acetone was most suitable compared to other solvents which showed toxic effects. It, however, needs to be mentioned here that our choice of chemicals was limited because of the insolubility of several chemicals (previously used) in acetone. The chemicals were dissolved in the appropriate solvent and the seeds were immersed in solution in glass stoppered conical flasks followed by drying back in the usual way. The soaking duration varied from one hour to three and half hours. Seeds treated with solvent alone were the respective controls (besides solvent untreated control). The variously treated seeds were subjected to different ageing conditions followed by the usual germination tests after ageing to evaluate the effect of the chemicals.

'Dry' Vapour Treatment

The idea of 'dry' vapour treatment came from the consideration to use an effective chemical which could be incorporated into the seed without using any solvent. These dry vapour treatments are best suited for pulse crops because of the relative ineffectiveness of water soaking-drying and solvent permeation treatments in leguminous seeds.

The use of halogens such as chlorine, bromine and iodine in vapour form and the details of the treatments for each halogen member are given below. Seeds (10-25 g) were loosely packed in nylon net packets and placed in a desiccator containing halogen in a 8 cm diameter petri dish kept at the bottom of the desiccator.

(i) Iodine Treatment

Iodine treatment of seeds was done at 27 ± 1°C in a glass desiccator (5.5 litre-capacity) in which 250 mg of solid iodine was kept in an open petri dish at the bottom, two days prior to the experiment. A part of the iodine volatilized and partially saturated the enclosed atmosphere. The seeds were kept in the desiccator in nylon net packets for various durations. The control seeds were kept in an identical desiccator of the same capacity which did not contain any iodine.

(ii) Chlorine Treatment

Chlorination treatments of seeds was done by exposing them to chlorine gas emanated by calcium oxychloride [chlorinated lime; bleaching powder; formula $Ca(OCl)Cl$]. The relative humidity inside the desiccator was 65 per cent and exposure to calcium oxychloride at that humidity resulted in a slow release of chlorine. Separate packets of seeds were exposed to the gas for various durations. Seeds kept in a similar desiccator which did not contain any chlorinated lime served as control. After chlorination the seeds of all treatments including the controls were placed under an electric fan for 6 hours and then stored over fused calcium chloride for moisture stabilization.

(iii) Bromine Treatment

Bromination treatment of seeds was done at room temperature (27 ± 1°C) in 5.5 litre-capacity glass desiccators as described above for chlorine. Bromine was obtained by adding 10 mg of potassium bromide crystals to 5 ml of concentrated sulphuric acid in a glass vial kept at the bottom of the desiccator and seeds were exposed to the gas for different durations. The control seeds were placed in a similar desiccator which contained only 5 ml of concentrated sulphuric acid and dried along with the treated seeds in the same fashion as mentioned for chlorination.

Factors Influencing the Effectiveness of Dry Halogen Vapour Treatments

The efficacy of halogen treatment in maintaining seed germinability depends on (i) seed age, (ii) duration of treatment.

(i) Seed Age

Physiological age of seed at the time of vapour treatment may be an important factor in determining the effectiveness of the same in minimizing the loss of seed germinability. To determine the optimum physiological age for the vapour treatments, harvest fresh seeds and seeds stored under ambient conditions for different periods were used.

(ii) Duration of Treatment

The optimum concentration and duration of exposure to the halogen vapours to be effective in maintaining germinability during storage is to be standardised. Treatments may not be effective if seeds are exposed for prolonged periods especially in case of iodine and bromine and non-physiological concentrations will show adverse effects.

Dry-Dressing Treatments

(i) Dry Dressing of Seed with Chemicals

In this method attempts were made to administer a number of chemicals into the seed using inert substances as carriers such as kaolinite (aluminium silicate dehydrate), French chalk (hydrous magnesium silicate), calcium carbonate ($CaCO_3$), *etc.* The seeds of various crops were treated with these powders containing the chemicals in stoppered glass bottles; the duration of the treatment being 24 h for all seeds. The powder seed ratio needs to be standardized for each kind of seed material. The seeds after treatment were stored under ambient storage conditions or subjected to different accelerated ageing conditions to evaluate the beneficial effects of the above treatments.

(ii) Treatment with Halogenated Inert Carriers

About 10 g each of kaolinite, French chalk or calcium carbonate were exposed separately to chlorine and iodine vapours for 24 hours. The halogenated materials were used in seed dressing for 24 h after which the seeds were dried in the usual way prior to storage under ambient or accelerated ageing conditions.

(iii) Slurry Treatment

An aqueous solution of the desired chemical at 10^{-3}M was mixed with the finely powdered carrier *viz.*, French chalk, kaolinite, calcium carbonate, *etc.*, and made into a slurry which was directly used for treating various kinds of seeds. The seeds were treated with the slurry for 2 h followed by drying back to the original weight. Distilled water was used as the control. In this method seeds were treated with chemicals using minimum amount of water.

C. Germination Tests for the Various Seed Treatments

The treated seeds were subjected to artificially controlled environments of temperature and relative humidity, as well as stored under natural ambient conditions. The accelerated ageing technique as described by Basu *et al.,* 1974; Basu, 1976; Dasgupta *et al.,* 1976 was followed and germination tests were carried out by the modified germination test standardised by Punjabi and Basu, 1982. The details are provided in the extracts of the paper in Chapter 5 (Figure 5.9). For root and shoot growth early recording has been found necessary to avoid the tendency of growth under different treatments levelling off when seedlings are left to grow in distilled water for prolonged period. The data on root and shoot growth were recorded using a glass plate onto which was pasted a graph paper (mm).

D. Development of Seed Vigour Bioassay

Wheat seeds of varying vigour levels wheat (*Triticum aestivum* L. cv. 'Sonalika') were used as the stock or test material and harvest fresh very high vigour jute seeds (*Corchorus olitorius* L. cv. JRO 632) were used as the bioassay material. Although in preliminary experiments *Corchorus olitorius, Corchorus capsularis* and various other small seeded materials were used, in subsequent studies only the former (*Corchorus olitorius*, 100 seed weight, 180 mg) was employed as it was more sensitive than the other materials tested. The *Corchorus olitorius* seeds germinate more quickly and it is possible to record seedling growth data after 48-72 hr which would coincide with the major phase of volatile gas production by the stock material. The choice of bioassay material was based on the criteria that small seeds are characterised by their uniformity and rapidity in germination and higher sensitivity to gaseous emanations of germinating stock seeds of wheat. The harvest fresh jute seeds after collection were dried to a moisture content of 7 per cent and stored in small glass vials at $-10°C$ in a refrigerator to be used as bioassay materials.

Wheat seeds of different vigour status (stock material) and high vigour jute seeds (bioassay material) were germinated simultaneously in air-tight 1.9 litre capacity (outside diameter 11 cm, height 24 cm) gas jars (Figure A-1). Twenty-five wheat seeds in four rows were placed on a moist blotter (16 cm × 8.5 cm) which was wrapped on the outer surface of an empty 130 ml-capacity (outside diameter 5 cm, height 12.5 cm) glass bottle. Before placing on the blotter, wheat seeds were sterilised with 0.05 per cent mercuric chloride solution for 30 min followed by thorough washing and soaking in distilled water for 2 hr. Approximately 100 jute seeds were placed in one row on a moist blotting paper (size 30 cm × 14 cm) in a straight line leaving a margin of 2.5 cm from the upper edge of the blotter. The blotter with the jute seeds was introduced into the jar in such a way that the inner wall of the jar was uniformly lined by the blotting paper sheet. Forty millilitres of water were added to the jar so that about 0.5 cm of the blotter with jute seeds could be covered by the standing water to ensure adequate moisture supply to the germinating jute seeds. The glass bottles containing the blotter with wheat seeds of different vigour status were carefully placed inside the jars on a 6.5 cm diameter petri-dish containing 30 ml distilled water for the germinating wheat seeds. The gas jars were then securely closed by the glass lids. There was also a blank set (control) in which only jute seeds (bioassay material) were placed.

The jars were kept in a temperature controlled room at $23 \pm 1°C$ and after 48 hr-72 hr germination percentage and seedling growth of wheat and jute were recorded. Final germination percentage and seedling growth of wheat seeds was

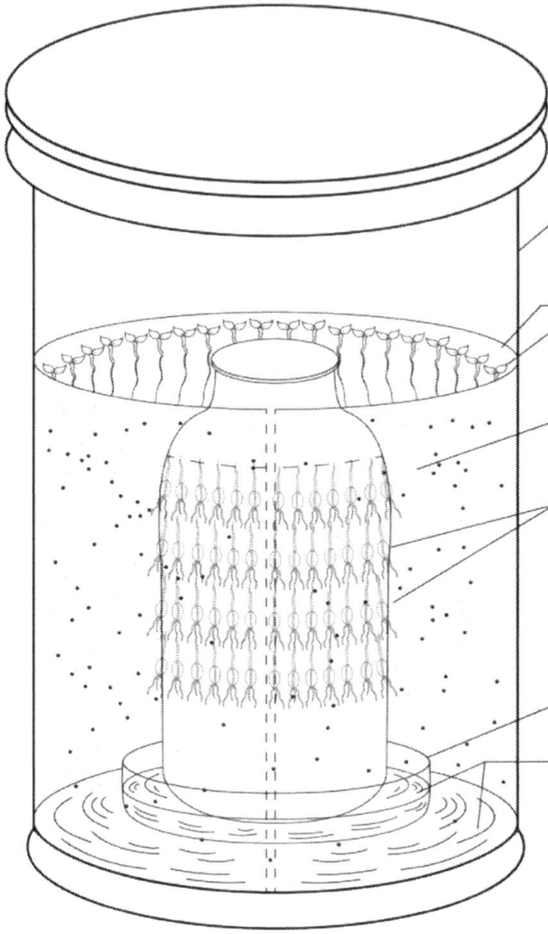

Figure A-1: The Setup for the Bioassay of Seed Vigour (*Vide* Basu, Kuchlan and Sur, 1990).

also recorded after 7 days. The experiments were repeated 5 times. Mean seedling length of wheat and jute were recorded and the data were subjected to statistical analysis. Correlation between germinability of wheat seeds (stock material) and growth of jute seeds (bioassay material) were worked out following usual statistical methodology. The jute seeds (bioassay material) are very sensitive to even very low concentration of the gaseous emanations of germinating stock seeds. However, the gas jar method, though very effective was somewhat difficult to handle as the relatively heavy gas jars occupied more space and also were not readily available in large numbers (five stock samples with 5 replications each would require at least 25 gas jars at a time).

In our subsequent experiments, the gas jars were replaced by wide-mouthed 1.3 litre capacity inexpensive glass bottles (outside diameter 11 cm, height 20 cm) with air-tight screw caps (Figure A-2, *vide* Sur and Basu, 1990). The method of placing stock seeds and the bioassay material (jute seeds) was the same as described in the earlier method. An adhesive cello tape was wrapped around the junction of the screw cap and the bottle as a further precaution against gas leak.

In this set up, the test material was wheat seeds of different vigour categories obtained by ageing or by the midstorage hydration-dehydration treatments for the maintenance of vigour and viability. The bottles with various vigour categories of wheat seeds and a blank (containing jute seeds but no wheat seeds) were kept in the temperature controlled room at 23 ± 1°C for 48 hr. The germination percentage and seedling growth of the wheat and jute seeds were recorded after 48 hr. For wheat seedling growth the total of all the seminal roots and shoot, and for jute growth the length from the tip of the shoot to the root tip was recorded. A separate setup was maintained for germination and seedling growth study of wheat in which

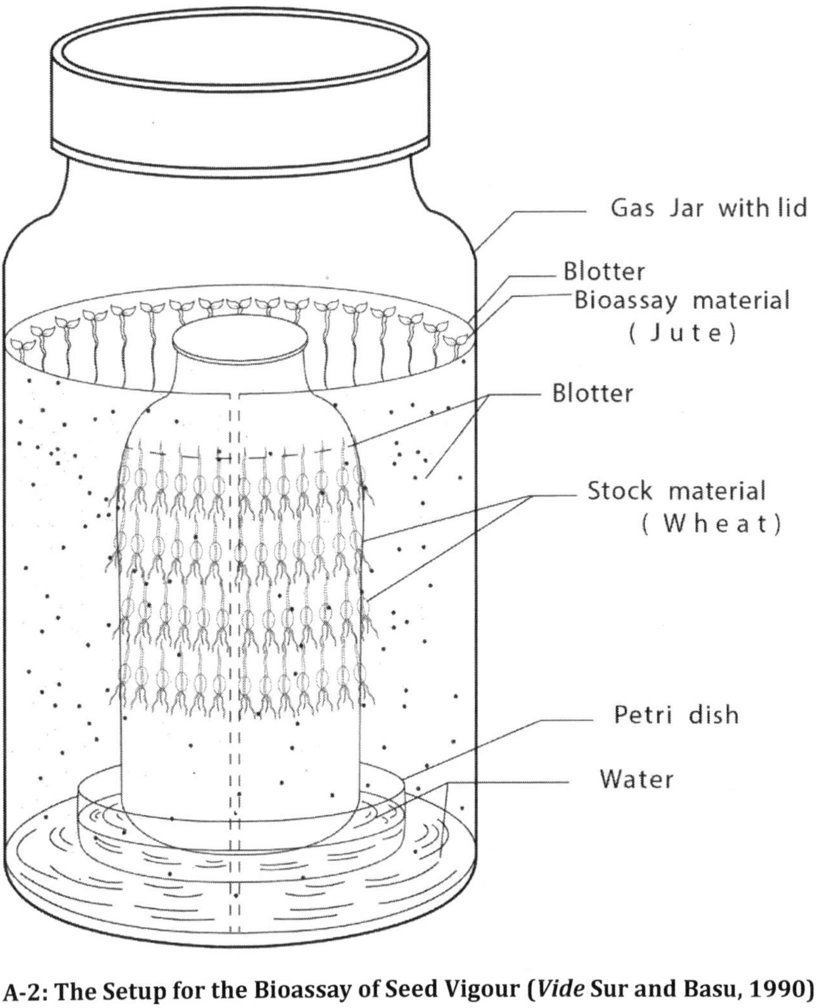

Figure A-2: The Setup for the Bioassay of Seed Vigour (*Vide* Sur and Basu, 1990).

data were recorded after seven days. Here also the bottles would require a lot of laboratory and incubator space.

In yet another study on retention of vigour and viability of stored pea seeds by midstorage hydration-dehydration seed treatments, seed vigour bioassay was done following in principle the method described above. The method was simplified by using easily available inexpensive air-tight wide-mouthed round transparent plastic containers (outside diameter 8 cm. height 6 cm. (Figure A-3). This method was equally effective. In this set up high vigour jute seeds (*Corchorus olitorius* L. cv. JRO 632) were exposed for 72 hr to the gaseous emanations of pea seeds of different treatments in air-tight containers maintained at 24 ± 1°C.

The setup consisted of placing twenty-five pea seeds on a moist blotter at the bottom of the container; twenty-five high vigour jute seeds were placed on a moistened blotter on the inner wall of the plastic containers without touching the lower moistened blotter (Figure A-3, *vide* Bhattacharya and Basu, 1990). Before placing the seeds on the blotter the pea seeds of all treatments were dry-dressed with Dithane-M-45. Data on root and shoot growth of jute seedlings were recorded after 72 hr and total seedling length was expressed as the percentages of respective blanks. The data on germinability of pea seeds were subjected to statistical analysis for evaluation of treatment effects.

The results obtained using different seeds as the stock materials variously treated or having different vigour status *viz.*, HV, HMV, LMV, *etc.* in the above three setups are discussed hereunder (I, II and III).

The jute seedling bioassay method with necessary modifications as required, for specific situations may conveniently be employed for vigour rating. An autoassay of seed vigour is another option to the jute seedling bioassay which would be

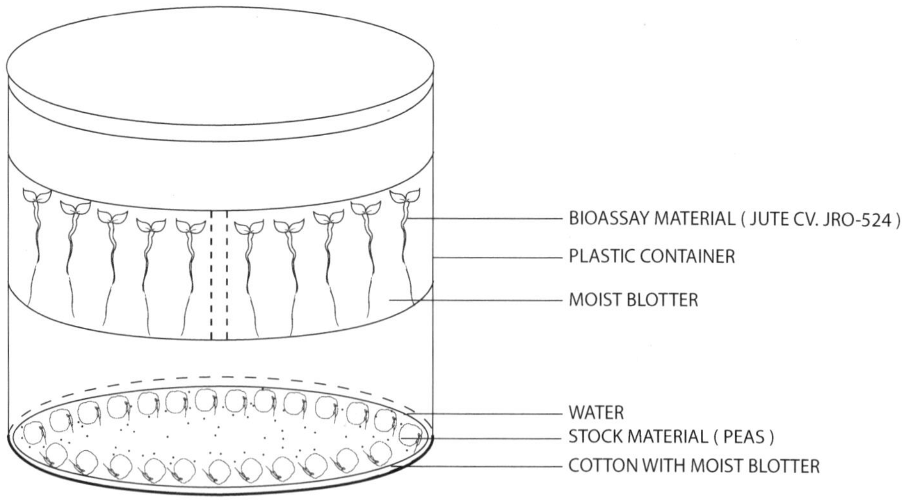

Figure A-3: The Setup for the Bioassay for Assessment of Seed Vigour (Bhattacharya and Basu, 1990)

accomplished very easily.

Autoassay of Seed Vigour

Unlike the jute seedling bioassay for the assessment of vigour of seedlots of different field crops, the autoassay (or self-expression) of seedling vigour is based on *in vitro* germinability parameters through usual germination and seedling growth tests in the laboratory. With this germination-cum-seedling evaluation test, an assessment of vigour would be possible within 4-5 days as in the jute seedling bioassay method.

Five random samples from a seedlot of a crop cultivar are placed for germination preferably on inclined glass plates using around 20-25 seeds per plate (*vide* Punjabi and Basu, 1982) using adequate number of replications of each sample (a total of 100-200 seeds for each sample, the number of seeds should not be less than one hundred). In our studies, we have used total seedling length (TSL, the sum of root and shoot lengths) as an index of vigour assessment but the same can be done with shoot length only or fresh as well as dry weight of seedlings. The total seedling length (TSL in mm) is transformed to TSL per cent by dividing the sample means with the regressionally derived TSL value at 100 per cent germination level (x variate)or any value higher than that present in the TSL data set (y variate). The TSL per cent thus obtained is the estimated seedling evaluation per cent which on multiplication by the germination factor GF (GF per cent ÷ 100 = GF) would give the germination-cum-seedling evaluation score (per cent) value. Seedling factor (SF) is obtained by dividing seedling evaluation per cent by 100 (the same way as GF is derived from germination per cent value). The germination-cum-seedling factor (GSF) is also derived in the same way from germination-cum-seedling evaluation per cent value or simply by multiplying GF and SF values. Usually, the results of vigour rating by the aforesaid autoassay method are quite close to those with the jute seedling bioassay and when high vigour (harvest-fresh) jute seeds are not readily available this method would adequately serve the purpose.

Results of Some Jute Seedling Bioassay Studies

I.

In the very first setup wheat seeds of different vigour status – high vigour (HV), high medium vigour (HMV) and low medium vigour (LMV) were used as the stock materials. Jute seeds when exposed to the gaseous emanations of the above germinating wheat seeds showed significant reduction in seedling growth in comparison to those in the blank setup (without wheat seeds). Reduction was higher in the low medium (LMV) wheat seeds as compared to the high medium vigour (HMV) wheat seeds (Table A-1).

Jute seedling growth responded in a similar manner when exposed to gaseous emanations of wheat seeds of different vigour status created by subjecting them to low humidity and high temperature conditions (dry accelerated ageing) (Table A-2).

Table A-1: Seedling Growth of Jute (Bioassay material) Seeds Exposed to Gaseous Emanations of Germinating Wheat (Stock material) Seeds of different Vigour Levels Obtained by Natural Ageing (*vide* Basu, Kuchlan and Sur, 1990; Indian Journal of Experimental biology.

Vigour Level of Wheat under Natural Ageing (Ambient conditions for different months)	*Germinability of Wheat after 48 h*						*Seedling Growth of Jute after 48 hr*			
	Germination		*Mean Root Length (mm)*	*Mean Shoot Length (mm)*	*Root/ Shoot Ratio*	*Mean Seedling Length (mm)*	*Mean Root Length (mm)*	*Mean Shoot Length (mm)*	*Root/ Shoot Ratio*	*Mean Seedling Length (mm)*
	Per cent	*Arc-sin*								
0 (HV, harvest-fresh)	97.8	81.5	116.0	11.9	9.7	127.9	17.8	5.2	3.4	23.0
6 (HMV, high-medium-vigour)	90.5	72.1	86.0	9.6	8.9	95.6	15.4	4.0	3.8	19.4
9 (LMV, low-medium-vigour)	81.8	64.8	55.0	7.3	7.5	62.3	11.6	3.2	3.6	14.8
Blank (without wheat seed)	-	-	-	-	-	-	33.6	9.6	3.5	43.2
LSD at 0.05 *P*	-	2.6	11.9	2.9	NS	11.3	4.1	1.3	NS	2.9
LSD at 0.01 *P*	-	3.7	16.9	4.1	NS	16.0	5.7	1.7	NS	6.2

(Seedling length of wheat = total length of shoot and all seminal roots per seedling).

NS = not significant.

Table A-2: Seedling Growth of Jute (Bioassay Material) Seeds Exposed to Gaseous Emanations of Germinated Wheat (Stock material) Seeds of different Vigour Levels Obtained by 'Dry Ageing' (*vide* Basu, Kuchlan and Sur, 1990)

Vigour Level of Wheat under Dry Ageing	*Germinability of Wheat after 48 hr*			*Mean Seedling Length of Jute after 48 hr (mm)*
	Germination		*Mean Seedling Length (mm)*	
	Per cent	*Arc-sin*		
High-vigour (HV)	91.3	72.8	99.8	21.8
Medium-vigour (MV)	54.0	47.3	65.0	16.8
Low-vigour (LV)	34.0	36.0	29.3	10.3
Blank (without wheat seed)	-	-	-	37.8
LSD at 0.05 *P*	-	1.8	1.8	3.2
LSD at 0.01 *P*	-	2.7	2.8	2.1

Similar results were obtained when wheat seeds which were given the effective mid-storage hydration-dehydration treatments and aged under accelerated ageing conditions for 7 days at 95 per cent RH and 40°C. Jute seeds showed maximum reduction in growth in the presence of control (untreated, aged) wheat seeds, which had poor vigour and viability. Growth of the jute seedlings was maximum in the presence of wheat seeds of soaking-drying (S-D), Dipping-drying (D-D) and moisture equilibration-drying (ME-D) treatments (Table A-3).

Table A-3: Seedling Growth of Jute (Bioassay Material) Seeds Exposed to Gaseous Emanations of Germinating Hydrated-Dehydrated Wheat (Stock material) Seeds after Accelerated Ageing at 95 per cent RH and 40°C for 7 Days (*vide* Basu, Kuchlan and Sur, 1990)

Hydration-Dehydration Treatments	*Germinability of Wheat after 48 hr*			*Mean Seedling Length of Jute after 48 hr (mm)*
	Germination		*Mean Seedling Length (mm)*	
	Per cent	*Arc-sin*		
Control	13.4	21.5	24.2	11.0
Dipping-drying (DD)	66.0	54.3	66.0	16.4
Soaking-drying (SD)	68.0	55.6	72.2	20.6
Moisture equilibration-drying (MED)	52.2	48.0	50.6	15.4
Non-aged control (without wheat seed)	98.2	82.2	111.8	24.0
Blank (without wheat seed)	-	-	-	33.2
LSD at 0.05 *P*	-	4.6	7.4	1.9
LSD at 0.01 *P*	-	6.4	10.2	2.5

Table A-4: Germinability of Wheat Seeds (Stock Material) of different Vigour Levels (High, Medium and Low), Seedling Growth of Jute (Bioassay material) and Volatile Aldehyde Production by the Stock Material (*vide* Sur and Basu, 1990; Seed Science and Technology.

Vigour Status of Wheat	*Germinability of Wheat*				*Mean Seedling Length (mm)*		*Mean Seedling Length (mm) of Jute after 48 hr*[(b)]	*Volatile Aldehyde Production after 48 hr (OD at 635 nm)*
			Germination		*After 48 hr*[(a)]	*After 7 days*[(a)]		
	Per cent	*Arc-sin*	*Per cent*	*Arc-sin*				
High-vigour (HV)	91	72.5	96	78.5	99 ± 15.5	461 ± 46.9	22 ± 1.7	0.11
					(15.7)	(10.2)	(65)	
Medium vigour (MV)	53	16.7	91	72.5	66 ± 12.6	323 ± 97.2	16 ± 2.5	0.21
					(19.0)	(30.1)	(47)	
Low-vigour (LV)	35	36.3	81	64.2	29 ± 8.7	234 ± 95.9	12 ± 2.9	0.25
					(30.0)	(41.0)	(35)	
LSD at 0.05 *P*	-	1.4	-	4.7	2.6	17.4	2.2	0.03
LSD at 0.01 *P*	-	2.1	-	6.8	3.8	25.4	3.2	0.04

Values in parentheses represent (a) CV for seedling length of wheat after 48 hr and 7 days and (b) jute seedling growth expressed as percentage of blank (taking length of jute without wheat = 100).

Table A-5: Germinability of Hydrated-Dehydrated Wheat Seeds (Stock Material) before and after Ageing, Seedling Growth of Jute (Bioassay Material) and Volatile Aldehyde Production by the Stock Material (*vide* Sur and Basu, 1990; Seed Science and Technology.

Hydration-Dehydration Treatments for Wheat	*Germinability of Wheat*				*Mean Seedling Length (mm)*		*Mean Seedling Length (mm) of Jute after 48 hours*[b]	*Volatile Aldehyde Production after 48 hours (OD at 635 nm)*
	Germination				*After 48 hours*[a]	*After 7 days*[a]		
	After 48 hours[a]		*After 7 days*[a]					
	Per cent	*Arc-sin*	*Per cent*	*Arc-sin*				
(a) Before ageing set								
Control	88	69.7	91	72.5	80 ± 11.3	304 ± 55.3	19 ± 1.9	0.21
					(14.2)	(18.2)	(56)	
Dipping-drying (D-D)	89	70.6	97	80.0	83 ± 9.8	308 ± 52.8	20 ± 2.5	0.19
					(11.8)	(17.1)	(59)	
Soaking-drying (S-D)	95	77.1	97	80.0	89 ± 9.8	327 ± 47.7	22 ± 1.7	0.15
					(11.0)	(14.6)	(65)	
Moisture equilibration-drying (ME-D)	88	69.7	93	74.7	81 ± 12.0	306 ± 55.5	20 ± 2.9	0.19
					(14.8)	18.1)	(59)	
LSD at 0.05 *P*	-	2.2	-	4.9	4.5	NS	2.2	0.04
LSD at 0.01 *P*	-	3.0	-	6.7	6.3	NS	3.0	0.06

Hydration-Dehydration Treatments for Wheat	*Germinability of Wheat*				*Mean Seedling Length (mm)*		*Mean Seedling Length (mm) of Jute after 48 hours*[b]	*Volatile Aldehyde Production after 48 hours (OD at 635 nm)*
	Germination				*After 48 hours*[a]	*After 7 days*[a]		
	After 48 hours[a]		*After 7 days*[a]					
	Per cent	*Arc-sin*	*Per cent*	*Arc-sin*				
(b) After ageing set								
Control	14	22.0	45	42.1	27 ± 17.3	192 ± 57.5	10 ± 7.8	0.37
					(64.1)	(29.9)	(29)	
Dipping-drying (D-D)	64	53.1	90	71.6	65 ± 13.0	297 ± 47.6	17 ± 1.1	0.24
					(20.0)	(16.0)	(49)	
Soaking-drying (S-D)	66	54.3	94	75.8	70 ± 13.4	309 ± 54.2	21 ± 1.9	0.23
					(19.1)	(17.5)	(60)	
Moisture equilibration-drying (ME-D)	57	49.0	86	68.0	50 ± 13.2	269 ± 51.3	15 ± 1.2	0.31
					(26.4)	(19.1)	(43)	
Non-aged control	98	81.9	98	81.9	114 ± 17.9	330 ± 61.5	23 ± 1.8	0.09
					(15.7)	(18.6)	(71)	
LSD at 0.05 *P*	-	4.0	-	5.0	6.4	20.4	1.7	0.07
LSD at 0.01 *P*	-	5.5	-	6.9	8.7	28.1	2.4	0.09

Values in parentheses represent (a) CV for seedling length of wheat after 48 hr and 7 days and (b) jute seedling growth expressed as percentage of blank (taking length of jute without wheat = 100).

Germination percentage of the bioassay material (jute seeds), however, was not significantly affected by the vigour status of the wheat seeds; it was extension growth (length of seedlings) that showed significant differences.

II.

The seed vigour of wheat (*Triticum aestivum* L. cv. 'Sonalika') of different vigour levels obtained by hydration-dehydration treatments was tested by their comparative germinability, bioassay of gaseous emanations employing high vigour jute (*Corchorus olitorius* L. cv. JRO 632) seeds and chemical assay of the volatile aldehydes in the gaseous emanations using the aldehyde trapping reagent 3-methyl-2-benzothiazolinone hydrazone (MBTH, 0.2 per cent solution).

The different hydration-dehydration treatments (S-D, D-D and ME-D) of wheat seeds showed significant beneficial effects on the maintenance of vigour and viability after accelerated ageing at 95 per cent RH, 40°C for 6 days (Table A-5). The jute seedling bioassay when conducted with the wheat seeds showed a direct correlation between growth of the jute seedling and vigour of the stock material; jute seedlings showed maximum growth in presence of germinating wheat seeds under S-D, D-D, ME-D in that order.

Table A-6: Correlation Coefficients between Various Pairs of Variates (*vide* Sur and Basu, 1990)

Independent Variates	*Dependent Variates*	
	Volatile Aldehyde Production by Wheat Seed (Stock material)	*Jute (Bioassay material) Seedling Growth*
I. Experiment with stock material of different vigour status		
Germination (per cent) of wheat	-0.972**	-0.993**
Seedling growth of wheat	-0.997**	+0.999**
Volatile aldehyde production by wheat	-	-0.993**
II. Experiment with hydrated-dehydrated stock material		
(*a*) *Before ageing set*		
Germination (per cent) of wheat	-0.876**	+0.894**
Seedling growth of wheat	-0.933**	+0.931**
Volatile aldehyde production by wheat	-	-0.941**
(*b*) *After ageing set*		
Germination (per cent) of wheat	-0.907**	+0.931**
Seedling growth of wheat	-0.937**	+0.945**
Volatile aldehyde production by wheat	-	-0.964**

**Significant at 0.01P; '*r*' values (correlation coefficient) have been calculated from data collected after 48 hours.

Also low vigour wheat seeds consistently produced larger quantities of volatile aldehydes (as trapped by MBTH solution) during the first 48 hr of germination than the high and medium vigour seeds. The results were significantly correlated with vigour of wheat seeds (Tables A-4 and A-6).

Interestingly volatile aldehyde production in hydration-dehydration treatments was lower than in control even before ageing; it was lowest in the refrigerator stored (non-aged) seed. These differences were however, not very closely related to the before ageing germinability of the variously treated wheat seeds. Ageing considerably increased the production of such aldehydes. Among the different hydration-dehydration treatments soaking drying was more effective in reducing volatile aldehyde production than the other treatments (Table A-2).

The correlation coefficient (r) values for stock seedling germinability and jute seedling growth were always positive and highly significant ($P = 0.01$, Table A-6). High negative correlations were noted between volatile aldehyde production and vigour status. Other growth inhibitory substances in gaseous emanations, besides volatile aldehydes could also be responsible for the reduction of growth of jute seedlings.

Even in the presence of MBTH, which trapped the volatile aldehydes, substantial growth reduction was still recorded. Use of mercuric perchlorate solution (0.25 M) in the bioassay system to absorb unsaturated gases like ethylene could have resulted in noticeable improvement in jute seedling growth implying thereby that ethylene was perhaps an important component of the growth inhibiting volatiles produced by the wheat seeds. Carbon dioxide produced by respiring wheat seeds, however, did not affect the growth of the jute seedlings because absorption of the gas by caustic potash pellets, did not result in any significant improvement of seedling growth of the bioassay material (Table A-7). Other unidentified volatile substances (ethanol may be one) could also have influenced growth of the bioassay material.

Table A-7: Calculated Inhibition of Jute Seedling Growth due to the different Components in the Gaseous Emanations of Medium-Vigour Wheat Seeds as Determined by Absorption of the Components by Various Gas-Trapping Reagents[(a)] (*vide* Sur and Basu, 1990)

Components of Gaseous Emanations	*Extent of Inhibition of Mean Jute Seedling Length (mm)*
Carbon dioxide	0.5 (5.7 per cent)
Ethylene	5.3 (60.9 per cent)
Volatile aldehydes	1.1 (12.6 per cent)
Unidentified volatiles	1.8 (20.7 per cent)

(a) Caustic potash pellets, mercuric perchlorate (0.25M) and MBTH (0.2 per cent) were used to absorb carbon dioxide, ethylene and volatile aldehydes respectively in the gaseous emanations of wheat seed germinated for 48 hr. The values are differences in mean jute seedling length (mm) between experimental set-ups with and without the respective trapping reagents.

Regarding the effects of moisture and temperature fluctuation on the relative growth of the bioassay material (jute) it may be pointed out that the capillary movement of moisture ensured uniform availability to the growing seedlings but temperature differences during germination influenced seedling growth of both the stock seeds (wheat) and the test (jute) materials. A temperature of 29 ± 1°C showed higher jute seedling growth than that at 23 ± 1°C. However, under both the temperatures the differences in vigour between high vigour and medium vigour stock seeds was clearly indicated by the test material (Table A-8).

Table A-8: Effect of Temperature on Growth of the Bioassay Material (*vide* Sur and Basu, 1990)

Vigour Status of Wheat	*Growth of Bioassay Material after 48 hr at 29 ± 1°C*		*Growth of Bioassay Material after 48 hr at 23 ± 1°C*	
	Total Seedling Growth of Jute (mm)	*Jute Seedling Growth as per cent of Blank*	*Total Seedling Growth of Jute (mm)*	*Jute Seedling Growth as per cent of Blank*
High-vigour (HV)	41	69	22	67
Medium vigour (MV)	24	41	15	45
Blank (without wheat)	59	100	33	100
LSD at 0.05 *P*	3.2	-	3.4	-
LSD at 0.01 *P*	4.6	-	4.9	-

III.

In another study by Bhattacharyya and Basu (1990) on retention of vigour and viability of pea seeds (*Pisum sativum* L. cv. 'Arkel' and 'Bonneville') the results were correlated with jute seedling growth bioassay following in principle the method of Basu *et al.* (1990) with minor modifications by using relatively small sized air-tight plastic containers as described earlier.

The beneficial effects of ME-D, ME-SD and dry treatments with bleaching powder and iodine were significant in both cultivars; methanol also proved to be quite effective (Tables A-9 and A-10).

In the bioassay, the volatile gases produced by the pea seeds (stock material) influenced the growth of the germinating jute seedlings (bioassay material) and such growth data expressed as the percentage of jute seedling growth in the blank (the container having no pea seeds) gave an estimate of vigour of the seeds of different treatments.

The bioassay results clearly brought out the seed invigoration effect of ME-D, ME-SD and dry treatments. Bleaching powder treatments in both 'Arkel' and 'Bonneville' exhibited vigour bioassay values as high as that shown by non-aged control seeds stored at -10°C throughout the experimental period (Tables A-9 and A-10).

Table A-9: Germinability of Pea Seed cv. *Arkel* under different Wet and Dry Treatments after Natural Ageing under Ambient Conditions for 9 months (*vide* Bhattacharya and Basu, 1990; Indian Agriculturist.

Treatment	*Germination*		*Total Seedling Length (mm)*	*Vigour Index (Germination per cent × root length in mm)*	*Vigour Bioassay (Per cent blank)*
	Per cent	*Arc-sin Values*			
Control 1 (aged)	64	53.1	124	2999	60
Soaking-drying (S-D)	46	42.7	107	1444	50
Moisture equilibration-drying (ME-D)	61	51.3	78	1758	71
Moisture equilibration-soaking-drying (ME-SD)	71	57.4	132	4682	82
Control 2 (carrier)	78	62.0	155	5328	94
Bleaching powder	83	65.6	168	5963	96
Iodine	80	63.4	160	5537	91
Methanol	83	65.6	157	5557	94
Ethanol	73	58.6	138	4480	89
Isopropanol	77	61.3	151	5430	91
Control 3 (non-aged)	90	72.5	194	8252	98
LSD at 0.05 *P*	-	4.70	6.47	-	-
LSD at 0.01 *P*	-	6.33	8.71	-	-

Treatments were given to 3-month-old seed.

SD – Soaking-drying; ME-D – Moisture equilibration-drying; ME-SD – Moisture equilibration – soaking-drying.

Table A-10: Germinability of Pea Seed cv. *Bonneville* under different Wet and Dry Treatments after Natural Ageing under Ambient Conditions for 9 Months (*vide* Bhattacharya and Basu, 1990)

Treatment	*Germination*		*Total Seedling Length (mm)*	*Vigour Index (Germination per cent × root length in mm)*	*Vigour Bioassay (Per cent blank)*
	Per cent	*Arc-sin Values*			
Control 1 (aged)	52	45.9	29	485	50
Soaking-drying (S-D)	35	36.3	21	259	55
Moisture equilibration-drying (ME-D)	84	66.0	56	1400	76
Moisture equilibration-soaking-drying (ME-SD)	81	64.3	50	1466	79
Control 2 (carrier)	91	72.7	41	2033	78
Bleaching powder	97	81.9	62	2673	95
Iodine	94	77.9	61	2211	80
Methanol	97	81.6	52	2293	81
Ethanol	82	65.1	47	1694	78
Isopropanol	96	80.6	53	2108	86
Control 3 (non-aged)	99	85.0	69	3059	97
LSD at 0.05 *P*	-	5.56	5.9	-	-
LSD at 0.01 *P*	-	7.49	7.9	-	-

Treatments were given to 3-month-old seed. Other details same as in Table A-9.

Sur and Basu (1990), clearly demonstrated a negative correlation between volatile aldehyde production and vigour status of wheat (*Triticum aestivum* L. cv. 'Sonalika') seeds. The hydration-dehydration seed invigoration treatments which showed significant beneficial effects on maintenance of vigour and viability showed a substantial reduction in post-ageing volatile aldehyde production.

Mukhopadhyay, Pal and Basu (1997) treated very high vigour French bean (*Phaseolus vulgaris* L. and soybean (*Glycine max* L.) with a number of dried powdered plant materials like rhizome of turmeric (haldi, *Curcuma longa*), leaf powder of nisinda (*Vitex negundo*) and some pharmaceuticals such as ASPRO, and a vitamin C-containing formulation Celin, common bleaching powder and *p*-aminobenzoic acid, PABA for maintaining vigour and viability after storage.

In both French bean and soybean germination percentage and mean seedling length in turmeric, nisinda, Aspro, Celin, PABA, bleaching powder and moist-sand-conditioning-drying treatment were significantly higher than the respective controls (Tables A-11 and A-12). The beneficial effects were also associated with reduced production of volatile aldehydes (Table A-12). Results of the bioassay of seed vigour in which germinating jute seeds were exposed to gaseous emanations of the germinating soybean seeds confirmed that the toxic volatile gas production was less in the dry seed invigoration treatments. Maximum growth of jute seedling (bioassay material) was obtained in Aspro treatment in both French bean and soybean (Tables A-11 and A-12). All beneficial invigoration treatments reduced the emanation of volatile gaseous products from stock French bean and soybean seeds which was also confirmed by the chemical assay using MBTH solution, as the aldehyde trapping reagent (Tables A-11 and A-12).

Table A-11: Effect of Seed Invigoration Treatments on the Germinability of French Bean Seed along with Seedling Growth of Jute (Bioassay material) and Volatile Aldehyde Production from the Stock Seed after Storage in Glass Bottle under Ambient Laboratory Conditions (av. temp 26.9 ± 1°C) for 24 Months (*vide* Mukhopadhyay, Pal and Basu, 1997; Indian Agriculturist.

Invigoration Treatments (Dry-dressing of seeds with fine powder, g/kg seed) [(a)]	*Germination* [(b)]		*Mean Seedling Length (mm)* [(b)]	*Seed Vigour Bioassay* [(c)]	*Volatile Aldehyde Production after 48 hr (OD at 635 nm)*
	Per cent	*Arc-sin Value*			
Control (untreated)	68	55.5	121	67	0.30
Turmeric (2g)	80	63.4	143	73	0.18
Nisinda (2g)	88	69.7	145	75	0.13
Aspro	89	70.6	156	79	0.09
Celin	87	68.9	138	76	0.16
P-Aminobenzoic acid	88	69.7	154	77	0.09
Bleaching powder	84	66.4	148	74	0.13
Moist-sand-conditioning (15 per cent mc 24 hr)	75	60.0	144	75	0.20
Non-aged control	100	90.0	197	91	0.04
LSD at 0.05 *P*	-	4.0	14.5	4.1	0.06

(a) Invigoration treatments with the finely powdered materials, bleaching powder and moist-sand-conditioning-drying were given to 4-month-old French bean (cv. Contender) seeds stored in sealed glass bottle under laboratory conditions. (b) Germination percentage and seedling growth were taken after 7 days at 26 ± 1°C. (c) Volatile gases produced by the fifteen French bean seeds (stock materials) influenced the growth of jute seedlings (bioassay material); total seedling length of jute (root + shoot) taken after 48 hr was expressed as percentage of blank (taking length of jute without French bean = 100).

Table A-12: Effect of Seed Invigoration Treatment of Soybean on the Germinability along with Seedling Growth of Jute (Bioassay material) and Volatile Aldehyde Production from the Stock Seed after Storage in Glass Bottles under Ambient Temperature (Avg temp 27.1 + 0.7°C) for 24 Months (*vide* Mukhopadhyay, Pal and Basu, 1997; Indian Agriculturist, 41(20, 79-85)

Invigoration Treatments (Dry-dressing of seeds with fine powder, g/kg seed) [(a)]	*Germination*[(b)]		*Mean Seedling Length (mm)*[(b)]	*Seed Vigour Bioassay*[(c)]	*Volatile Aldehyde Production after 48 hr (OD at 635 nm)*
	Per cent	*Arc-sin Value*			
Control (untreated)	40	39.2	78	74	0.30
Nisinda (2g)	69	56.2	106	83	0.18
ASPRO (0.1g)	82	64.9	144	90	0.12
CELIN (0.2g)	77	61.3	140	88	0.14
P-Aminobenzoic acid (0.1g)	68	55.5	115	86	0.17
Bleaching powder (3g)	73	58.5	124	88	0.16
Moist-sand-conditioning (10 per cent m.c., 24 hr)-drying	47	43.3	91	81	0.25
Non-aged control (refrigerated)	90	71.6	167	93	0.09
LSD at 0.05 *P*	-	6.4	16.8	5.3	0.05

(a) Invigoration treatments were given to 1-month-old soybean (cv. Soyamax) seed stored in sealed glass bottles under laboratory conditions. (b) Germination percentage and seedling growth were taken after 5 days at 26 + 1°C. (c) Volatile gases produced by the twenty soybean seeds (stock material) influenced the growth of jute seedlings (bioassay material); total seedling length of jute (root + shoot after 48 hrs. was expressed as percentage of blank.

Annexure No. A-2

Basu, R.N. 2007. A holistic approach to quantification of seed vigour through integration of morphological and physiological-biochemical parameters. *Indian Biologist*, 39(1), 1-44. Reproduced from the *Indian Biologist*, with kind permission of the Indian Association of Biological Sciences.

(Abstract in Chapter 4)

Introduction

The paper, "Approaches to quantification of seed vigour", published in the Indian Biologist (volume 36, number 2, pp 1-23, 2004)[3], in which certain methods for transformation of negatively correlated physiological-biochemical vigour parameters (*vis-à-vis* germination percentage) to respective positive versions and bringing the test results in the percentage scale of 0-100, coupled with incorporation of germination factor (GF, an entity obtained by dividing germination per cent by 100) into the test data to form respective vigour scores have been presented, constitutes the background of this rejoinder-cum-reaffirmative article. The earlier paper has generated a mixed response; many readers have appreciated the approach but a good number of them have reservations on the methodology and its usefulness in seed research and practical utility in seed testing, crop husbandry, seed trade and commerce. In the latter category of readers, who have been specifically contacted, there are some biometricians and statisticians, who in particular, are not in favour of incorporation of GF (synonymous with viability factor, VF) into vigour test results as that is nonconventional and, as such, not readily acceptable. Some of the readers have given their views in writing while others have expressed their views verbally; nevertheless, all the major issues raised so far have been duly considered item wise in the discussion in light of additional information and evidences supplied by the tables of statistically analyzed data in the text along with additional data on the efficacy of incorporation of seedling growth (including the merger of the same with germination) and its correlation with field performance and productivity. If properly worked out, such amalgamation of morphological parameters (germination, early seedling growth in the laboratory, *etc.*) with physiological and biochemical vigour parameters would give an integrated seed vigour index that would serve as a practically useful predictive tool in crop husbandry provided that growing conditions and cultural practices for seed lots varying in vigour indices are more or less similar and favourable for the particular kind of crop.

Sources of Basic Data and Procedural Aspects

As in the earlier paper, mainly published data from different sources including data from the author's group in the University of Calcutta (also some unpublished data from the laboratory of the author) have been used in the present discourse. The ten major tables in the text, constructed from the basic data of selected authors (secondary data), cover seeds of soybean, maize, wheat, sunflower, peas and cabbage that are botanically, chemically and agriculturally quite diverse in nature, belong to the following sources:

(a) Hailstones and Smith (1988)[7]: data on lipid hydroperoxide formation in seeds of soybean and cabbage in relation to germination (Table 1).

(b) Perry (1981)[12] : report to the ISTA (International Seed Testing Association) Congress in Vienna in 1980, as the chairman of the seed vigour committee (1977-80), giving vigour test data on soybean and maize (Table 2), peas (Table 3) and wheat (Table 4) along with their respective correlations between germination per cent and soil emergence.

(c) Wilson and McDonald (1986)[16]: data on electrical conductivity and volatile aldehyde emanation of soybean and their relationship with germination, seedling growth and field emergence (Table 5).

(d) Mandal, De, Saha and Basu (2000)[8]: data on the effects of soybean seed invigoration treatments on post-storage germination, seedling growth, field emergence and productivity (Table 6).

(e) Pal, Ray, Mitra and Basu (1996)[11]: data on seed invigoration treatment of sunflower (Table 7) and wheat (Table 8) and the relationship of seed vigour with field performance.

(f) Bhat (2004)[5], Sur and Bhat (unpublished): data on maize seed invigoration treatment using relatively big and medium sized seeds and the relationship between seed vigour and respective field performance (Table 9).

(g) Mukhopadhyay, Pal and Basu (1997)[10], Mukhopadhyay (2001)[9], Basu (2004)[3]: data on seed invigoration treatment of soybean and the relationship of seed vigour with field performance and productivity (Table 10).

The arrangement of the tables is based on increasing integration of factors and coverage of more and more field performance data for the different crops. The arrangement could have been cropwise but for the range of crop seeds under study, the required full set of data on laboratory germination test, seedling growth evaluation data (length or weight of root and shoot), physiological-biochemical vigour test data, field or soil emergence data, biological and agricultural yield parameter data, *etc.* are yet to be procured, processed and statistically analyzed.

The details of the methods followed by the different authors, cultivars and experimental findings may be readily available to the readers from publications in largely circulated international journals and monographs but for other papers there may be problems. As such, necessary particulars are very briefly mentioned tablewise (starting with Table 1) with special emphasis on the large number of abbreviations in the text and certain specific calculations. Explanation of abbreviations or terms, or methods of calculations once given under a table, has usually not been repeated.

Table 1 (soybean, cabbage)

G per cent: germination per cent of nondormant seed; also G for germination.

GF: germination factor (germination per cent divided by 100, range 0 – 1); GF of nondormant seed is synonymous with the viability factor, VF.

NV per cent: nonviability per cent (100 minus germination per cent), it is the percentage of nongerminable nondormant seed, percentage of dead seeds.

NVF: nonviability factor (nonviability per cent divided by 100, range 0 – 1); NVF is synonymous with nongermination factor (NGF) of nondormant seed.

LHPO: lipid hydroperoxide, also hydroperoxide, value given as absorbance or optical density (OD = 2 – log per cent transmission).

LPRS: lipid peroxide resistance score, (1 – optical density of lipid peroxide or hydroperoxide) × 100; maximum value 100; LPRS (original) is the value as directly obtained from absorbance or OD.

LPRS (revised): also LPRS (rev); LPRS (original), on division by the LPRS reference or benchmark value is expressed as a percentage value; maximum scale value is 100.

The reference value is equal to the LPRS (orig) value at 100 per cent G + the full range of LPRS (orig) values in the data including the estimated value for 100 per cent G.

Although a very well preserved material, preferably of the same cultivar, may be used as a reference, it is very often difficult to keep a stock of the same in the refrigerator or desiccators for a long time. As such, the statistical approach as given below would serve the purpose very well (see General Discussion; Reference value).

Calculation of LPRS (Original) and LPRS (Revised) Values

Example: Absorbance value of lipid hydroperoxide level of cv. Lb is of soybean = 0.026 (first cultivar) thus, LPRS (orig) = (1 – 0.026) × 100 = 97.4.

For the calculation of LPRS (rev), the reference or benchmark value has to be calculated from the data set by regression using G per cent of each cultivar and corresponding LPRS (orig)value for the same as shown below:

Germination per cent (x) :	97	95	95	94	89	82	$\bar{x}$= 92.0
LPRS (orig) (y) :	97.4	97.6	96.3	95.0	95.0	94.0	$\bar{y}$ = 95.9

Result of Correlation-Regression Analysis

Correlation coefficient (r) = + 0.832

Coefficient of determination r^2 = 0.691

Regression coefficient A = 76.01

Regression coefficient B = 0.216

Value of y (*i.e.* LPRS orig) when x = 100 is as follows:

Putting the value of x, A and B in the equation y = A+Bx, we get y = 76.0+(0.216 × 100) =97.6

LPRS (orig) range in the data set = 97.6 – 94.0 = 3.6

So, reference or benchmark value = 97.6 + 3.6 = 101.2

LPRS (revised) values are obtained by expressing the LPRS (orig) values as percentage of the reference value (*i.e.* division of each value by 101.2 and multiplying by 100); for the cv. Lb is the LPRS (rev) is thus (97.4 ÷ 101.2) × 100 = 96.2 per cent, thus effectively making the values free from differences in units and other variables. The calculation of benchmark value in the manner stated above ensures that no value would exceed 100. In practice, the value at 100 per cent G can be obtained directly using a relatively advanced scientific pocket calculator.

The calculated reference value for cabbage is 99.5, being the sum of LPRS (orig) value at 100 per cent G (89.8) and the full range value (difference of the maximum and minimum including the value estimated for 100 per cent G, *i.e.* 92.8 – 83.1 = 9.7), so the reference value is 89.8 + 9.7 = 99.5.

Math. error: mathematical error because of division of LHPO or LPRS values by 0(zero).

Cultivars of soybean and cabbage are mentioned in Table 1.

Table 2 (soybean, maize)

AA: accelerated ageing; AA per cent, germination per cent following accelerated ageing.

CT: cold test, a test that serves as an indicator of seed vigour, especially for maize.

EC: electrical conductivity of seed-steep water; usually given in µmhos, it is negatively correlated with G per cent.

ECRS: electrical conductivity resistance score, it is positively correlated with germination and is obtained by deducting from 1 the electrical conductivity expressed in millimhos and multiplying the remainder by 100, that is (1 – EC mmhos) × 100, maximum value 100. Thus for soybean seed of lot 1 with an EC µmho value of 70, the ECRS = (1 – 0.070) × 100 = 93.0; 70 µmhos on conversion to millimhos become 0.070 mmho.

ECRS (Original): it is the value directly calculated from the given data.

ECRS (revised): ECRS (orig) when expressed as the percentage of the reference or benchmark ECRS value gives ECRS (rev) calculated in the same way as for LPRS (rev); the calculation involves estimation of ECRS (orig) for 100 per cent G and adding to that the full range (difference between the highest and lowest values, taking into consideration the value estimated for 100 per cent G) value of the same to get the benchmark or reference ECRS value. The reference value for soybean data in Table 2 is 100.26, being the sum of ECRS (orig) at 100 per cent G (= 96.26) and full range value (97.0 – 93.0 =4.0). Thus, for the first lot, the calculated ECRS (rev) is equal to 93.0 (that is 93.0 ÷100 × 100 = 93.0) which is the same as the ECRS (orig) value. This has been nearly the case with the data of Wilson and McDonald (1986)[16] quoted in the earlier paper (Basu, 2004)[3] but slightly revised and cited in this paper in Table 5, or in fact, the reference values calculated for clover[6] (98.6), cotton[1] (100.3) and soybean[16] (100.6) are so close to 100 that the ECRS (orig) values have been used as such in the earlier paper. Subsequently, however, with some data from other sources, including a few from the author's laboratory, a 3-5 per cent deviation from 100 (occasionally even more) has been recorded. Therefore, in all calculations, including those in the present paper, necessary revisions have been undertaken despite the fact that the same has no effect on the value of correlation coefficients [see clarification (iv), last item in discussion].

The same principle has been followed for the other allied vigour test scores such as volatiles emanation resistance score (VERS) in Tables 5, 7, 8 and 10, besides lipid peroxidation resistance score (LPRS) as already mentioned under Table 1. However, if for a particular seed all the lots show 100 per cent G, the aforesaid regression method would invite mathematical error, the reference value in such a situation may be calculated simply by adding the range value (*i.e.* difference between the maximum and the minimum) to the highest value of the y variate.

Soil emergence per cent : also soil em per cent, soil em, the percentage of seed emerging in natural soil in pots or in the field; seedling emergence in the field is often given as field emergence per cent (FE per cent) or field emergence per sq m (FE/m^2).

Cultivars: soybean, Lancer (lots 1-5), Hardee (lots 6-10); maize, not given.

Table 3 (peas)

Seedling eval. per cent : seedling growth evaluation per cent, also seedling evaluation per cent, seedling eval per cent; seedling growth evaluation may be based on extension growth such as length of shoot, root or of total seedling, or weight (fresh/dry weight) of shoot, root or of total seedling, whichever is more convenient as well as reproducible. The growth may be expressed as a percentage of a very well preserved material of the same cultivar that serves as a reference or statistically calculated from the data set by regression the estimated seedling growth at 100 per cent G and adding to that the full range value (highest seedling growth minus lowest seedling growth, including that estimated for 100 per cent G, if that is the highest value for seedling growth). Expression of seedling growth of different lots as percentage of the estimated reference value gives the seedling evaluation per cent for each lot. Thus, calculating the reference value of seedling evaluation in the same way as for LPRS (in this case G per cent *vis-a-vis* seedling evaluation) we get the seedling evaluation value for 100 per cent G = 96.7 to which the range value 13.7 (96.7 – 83.0 = 13.7) is added making the reference value equal to 110.4. Therefore, the seedling eval per cent (rev) for lot 1 of peas is equal to (93 ÷ 110.4) × 100= 84.2 per cent. As with physiological-biochemical vigour parameters, this process brings all values in the percentage scale freeing them from differences in units, methods and their variables.

Addition of the range value raises the value of the reference seedling growth and, as such, lowers the seedling evaluation per cent which is desirable because we endeavour to express the lot values as percentages as close to that at physiological maturity (since then, it is considered to be a down hill journey for seed vigour). The same logic would hold good for other vigour parameters. An additional very important advantage of the aforesaid process is uniformity in approach, keeping thereby the morphological and physiological-biochemical scores on identical footing enabling their integration, if necessary.

SF: seedling factor (syn. seedling growth factor); it is obtained on division of seedling evaluation per cent by 100, maximum value 1; for lot 1 of peas it is equal to 0.842 (*i.e.* 84.2 ÷ 100 = 0.842).

Germination-cum-seedling evaluation per cent : seedling evaluation per cent on multiplication by GF gives the value of this term, maximum value 100 (also derivable through multiplication of G per cent and SF).

GSF: it is obtained by dividing germination-cum-seedling evaluation per cent by 100, maximum value 1; for lot 1, GSF = 0.808 (*i.e.* 80.8 ÷ 100 = 0.808).

ECRS reference value for the peas is 100.2 being the summation of ECRS (orig) for 100 per cent G (= 98.4) and the full range value of 1.8 (*i.e.* 98.4 – 96.6 = 1.8). As such, the revised ECRS value for lot 1 is equal to (98.3 ÷ 100.2) × 100 = 98 per cent.

Cultivars: not mentioned.

Table 4 (wheat)

Hiltner per cent : also Hilt per cent, Hilt; Hiltner test (brick grit test) is an indicator of seed vigour in which seed placed for germination has to emerge through a specified layer of brick grit (that offers resistance to emergence), overlying a sand layer; the percentage of emergence of seedlings is the Hiltner per cent.

Seedling evaluation per cent : seedling eval per cent has been calculated in the same way as in Table 3 firstly by estimating seedling growth in mm at 100 per cent G by regression (=91.8) and adding to that the range value (93 – 83 = 10) to get the reference seedling growth (91.8 +10 = 101.8 mm), and then the seedling eval per cent for the respective seed lots are obtained by dividing each value by 101.8 and multiplying by 100. Thus for lot 1, the seedling eval per cent value is equal to 81.5 (*i.e.* 83 ÷ 101.8 × 100 = 81.5).

SF: obtained by division of seedling eval per cent with 100; for lot 1, it is equal to 0.815.Germination-cum-seedling evaluation per cent : derived by multiplication of seedling eval per cent by GF, for lot 1 it is equal to 74.2 (*i.e.* 81.5 × 0.91 = 74.2).

GSF: derived by division of germination-cum-seedling eval per cent with 100; thus for lot 1, it is equal to 0.742 (*i.e.* 74.2 ÷ 100 = 0.742).

Cultivars: Flanders (lots 1-4), Maria Huntsman (lots 5-7), Atou (lots 8-10).

Table 5 (soybean)

VERS: volatile emanation resistance score (a vigour score based on resistance of seeds to emanation of volatile substances/aldehydes during storage or germination), it is obtained by deducting from 1, the optical density (or absorbance) of the colour complex of certain volatile substances (aldehydes in particular, as liberated by seeds) with methyl benzothiazolinone hydrazone (the colour is developed on incubation with a dilute solution of ferricchloride) and then multiplying the remainder with 100, *i.e.* (1–OD) × 100, maximum value is 100. When volatile emanation is expressed as µg/g or µg/100 seeds necessary conversions to fractions less than 1 would be necessary; in any case as discussed before the original data would become free from units, methods and other variables upon expression as a percentage of the statistically estimated reference value. For seed lot 2 (first item) in Table 5 with volatile aldehyde expressed as µg/seed, equal to 0.022 (original data of Wilson and McDonald, 1986 in µg/100 seeds, converted to per seed basis to make the same less than1), a revised formula namely, (1 – µg aldehyde per seed) × 100 gives a VERS value of 97.8(original). From the data on G per cent and VERS (orig) in Table 5 we get by regression a VERS (orig) value of 97.7 for 100 per cent G to which the full range value of 7.0 (*i.e.* the difference between the maximum value, 97.8 and the minimum value 90.8) is added giving the VERS (reference) value of 104.7. As in case of LPRS and ECRS discussed earlier, expression of the VERS (orig) values as percentage would give the revised values which for lot 2 is equal to 93.4 per cent (*i.e.* 97.8 ÷ 104.7 × 100 = 93.4).

ECRS: the ECRS (reference) value has been estimated to be equal to 100.6 (being the sum of ECRS (orig) value, 97.84 and full range, 2.74 = 100.58) as such, the revised value for lot 2 (first item) is equal to 96.7 per cent (*i.e.* 97.3 ÷ 100.6 × 100 = 96.7).

Seedling evaluation per cent : the basic data on seedling dry weight in Table 5 has been given as g per g of seed. The reference value for seedling growth is 0.4274g, being the sum of seedling weight at 100 per cent G calculated statistically, 0.2782, and the full range value 0.1487 (difference of the maximum and minimum inclusive of the value estimated for 100 per cent G). For lot 2 the seedling eval per cent value is equal to 60.8 per cent (*i.e.* 0.26 ÷ 0.4274 × 100). Incidentally, expression of seedling weight in g per g seed weight basis is a very good approach for seedling evaluation especially when different cultivars or seed lots with different seed sizes are involved.

SF: for lot 2, the value is 0.608 (*i.e.* 60.8 ÷ 100 = 0.608).

Germination-cum-seedling evaluation per cent : this is equal to 56.5 for lot 2, being the product of seedling evaluation per cent of 60.8 and GF 0.93.

GSF: for lot 2, it is equal to 56.5 ÷ 100 = 0.565.

Cultivar: All the five seed lots, arranged by the authors in order of the superiority in field emergence (FE), belong to cv. Williams.

Tables 6–10 (seed invigoration treatments)

While the data in Tables 1-5 cover stored naturally aged seeds of different species and cultivars within a species, those in Tables 6-10 on soybean, sunflower, wheat, maize and soybean respectively, originating from experiments on seed invigoration treatments for improvement of vigour, viability and productivity deal mainly with the effects of 'dry' (also some 'wet'), physiological treatments on vigour parameters and field performance of one selected cultivar of each material.

Data on post-storage germinability, physiological-biochemical parameters, and early seedling growth (about halfway of the growth phase) under controlled laboratory conditions have been recorded prior to sowing in the field; data on field emergence and yield parameters, *etc.*, have been noted 10-15 days after sowing depending on the materials, and after harvest of the crops, respectively.

Seedling evaluation in case of all the materials is based on the total seedling length (length of shoot plus length of root) at the end of germination test on moist blotters placed on inclined glass plates kept within separate polythene packets containing enough water for the required period of growth[8]. More recently, we have noted that seedling evaluation based on length of shoot only would be a more practical approach, being less cumbersome and time-saving and is almost equally effective for the purpose, specifically when seed lots do not differ in initial seed weights as in our seed invigoration studies using a single crop cultivar.

In view of the commonality in approach the methodologies involved are briefly presented hereunder in a tabulated form.

Table number, species and cv. vigour status/age of seed at the time of treatment, etc.	*Treatments (Concentration of chemicals, pharmaceuticals, botanical materials in dry finely powdered forms in g or mg per kg of seed given within brackets)*
Table 6, soybean cv. Soyamax, harvest fresh, high-vigour seed for dry pre-storage treatments; 5- month-old high-medium vigour seed for MSC-SD treatment, storage from harvest to sowing in rubber-stoppered 2.5-litre glass bottles (unsealed) in the laboratory.	Dry treatments : dry dressing in stoppered glass bottles with bleaching powder (active ingredient calcium hypochlorite, approx. 30 per cent) (2.5g), aspro, an aspirin-containing, a.i. 56.5 per cent, pharmaceutical (100 mg); red chilli fruit powder (2g); catharanthus leaf powder (2g); moist sand conditioning of naturally stored 5-month-old seed for 24 h, sand : seed 3:1, sand with 6 per cent m.c. followed by soaking of the seed in water for 2 hours and then drying back (MSC-SD) to the original weight in a drying cabinet; stored in rubber-stoppered glass bottles before and after treatment till sowing.
Table 7, sunflower cv. Morden seed after harvest stored in stoppered (wax-sealed) bottles, all treatments given to 3-month-old seed (storage under ambient laboratory conditions).	Dry dressing with nisinda (*Vitex negundo*) leaf powder (2g); aspirin (100 mg), ascorbic acid (500 mg), bleaching powder (2.5 g); water soaking for 2 h-drying back (SD); restorage in rubber-stoppered 2.5-litre glass bottles till sowing (for 5 months).
Table 8, wheat cv. Sonalika, seed after harvest stored in rubber stoppered glass bottles (wax sealed) for 3 months before treatment.	Dry dressing with turmeric (*Curcuma longa*) rhizome powder (2g), nisinda leaf powder (1g), aspirin (100 mg), aspro (200 mg), celin (ascorbic acid) (500 mg); water soaking (3h)-drying (SD); storage for 5 months in glass bottles under ambient conditions in the laboratory till sowing.
Table 9, maize cv. Vijay composite (two seed size-based seed lots, big and medium, have been separately treated), stored as above for 3 months.	Dry-dressing given separately to 3-month-old big and medium sized seeds with aspirin (100 mg), *p*-aminobenzoic acid (200 mg) and calcium carbonate (2g); maize seeds stored in the same manner as above for 5 months till sowing.
Table 10, soybean cv. Soyamax, treatment given to 1-month-old seed (very high vigour seed).	The dry-dressing treatments given to 1-month-old seed are aspro (200 mg), *p*-aminobenzoic acid (100 mg), bleaching powder (3g); the wet treatment, moisture equilibration (24 h)- drying (MED) also given to 1-month-old seed, seed storage conditions till sowing same as above.

JSBS, jute seedling bioassay score[4,15], given in Tables 7 to 10, is based on growth of jute seedlings from very high vigour seed (bioassay material, characterized by rapid and uniform growth, often just sprouted seeds are used) grown along with germinating seeds of various kinds (stock materials being tested for the assay of vigour levels) in small air-tight containers for 48–72 hours; the total length (shoot + root) of the jute seedlings is expressed as per cent of the growth of jute seedlings in the blank set (in which the same number of sprouted jute seedlings has been placed but no stock seeds in an air-tight container of the same size). If sprouted jute seeds are used, 48 h at 27°C would be enough, otherwise for nonsprouted dry seed placement, 72 h growth would give the desired seedling growth. Physiologically, the bioassay is based on greater lipid peroxidation in less vigorous stock seeds than in high vigour seeds as the former would produce more growth inhibitory gases (mostly volatile aldehydes that are presumably products of lipid peroxidation) and consequently growth of the bioassay material is reduced more in presence of low vigour stock seeds; as such, JSBS shows good positive correlation with seed vigour. The number of jute seeds used vary from 30-50 per replication and that of stock seeds from 15-50 depending on their capacity to produce volatile growth inhibitory gases. However, for a very high vigour stock material, the JSBS value should be around 90-95 per cent of the blank. As with all other physiological-biochemical vigour scores, the JSBS (original) need to be revised (JSBS rev) in the same way as shown for LPRS in Table 1; firstly, by calculating the JSBS reference value through regression equation (G per cent *vis-a-vis* JSBS orig) to estimate JSBS (orig) value at 100 per cent G and adding to that the full range value for getting the JSBS reference value and thereafter expressing all JSBS (orig) values to their respective revised versions as percentages of the calculated reference value.

Although all the reference values for morphological and physiological-biochemical vigour parameters of seeds covered in Tables 6-10, on the basis of which seedling growth evaluation per cent, SF and GSF and the revised percentage values for ECRS, VERS and JSBS have been calculated, are based on their respective original values at 100 per cent G and full range values as discussed for Tables 1-5, to save space the original total seedling lengths (TSL values, the sum of shoot and root lengths; in case of wheat and maize, total of all the seminal roots plus shoot length), and also the original values for ECRS, VERS and JSBS have not been given and only their revised versions and factors derived from them are presented in all the five tables from 6 to 10. In fact, the original values, which are so very important, are necessary only to statistically estimate the revised values and factors that form the basis of the present approach to integrate morphological and physiological-biochemical vigour parameters and correlate the same with field performance and productivity.

For the storage of the relatively small stocks of seeds of our experiments especially of the untreated control seeds of the seed invigoration experiments that would lose vigour and viability faster than the treated seeds, we routinely use rubber-stoppered glass bottles of various sizes (5L, 2.5L, 1L, 500 ml-capacity, *etc.*) depending on seed size and volume; more recently screw-capped plastic mineral water bottles, that are not commonly reused and disposed of randomly causing environmental pollution, are being successfully used for the purpose) to maintain good vigour and viability. Otherwise, in the predominantly hot and humid climate in this part of the country, short-lived seeds such as sunflower, soybean, wheat, *etc.* would deteriorate very quickly, falling below the minimum seed certification standard for germination that being the major reason for the large-scale procurement of a wide range of short-lived seeds from other parts of the country by a number of seed trading agencies for supplying to the farmers before the usual sowing time of the crops concerned.

Broad Findings, their Interpretations and Implicatons

A complex character like seed vigour that is manifested upon a range of morphological and physiological-biochemical characteristics of the seed and seedling and even far beyond would by necessity require a holistic approach for its proper understanding and possible quantification in an integrated manner. The integration of germination factor (GF) into physiological-biochemical vigour parameters reported earlier[3], has been an attempt in that direction. In this discourse, we present further confirmatory evidences on the issue along with additional information on the efficacy of integration of seedling growth (an excellent indicator of seed vigour) with germination (more specifically with GF) and physiological-biochemical vigour test parameters to further substantiate the necessity of integration of morphological and physiological-biochemical vigour parameters. All integrated values in the different tables are based on multiplication of GF, SF or GSF with the **revised** physiological-biochemical vigour scores. The findings on the issue, their possible interpretations and implications are briefly discussed tablewise starting with the basic data of Hailstones and Smith (1988)[7] on soybean and cabbage.

One of the very early studies on lipid peroxidation in relation to viability of soybean and cabbage seeds by Hailstones and Smith (1988)[7] demonstrates a high negative correlation between germination per cent and lipid hydroperoxide level with r values of -0.88* in soybean and -0.99** in cabbage (based on regression of log y on log x, excluding the cv. Cape Spitz that showed a somewhat erratic behaviour). A linear correlation including Cape Spitz(n=6) would give the value of r = -0.430 that is not statistically significant; the r value would be equal to -0.796* excluding Cape Spitz (n=5). Conversion to LPRS does not change the value of r but makes it positive (*i.e.* r = + 0.430 at n=6 and +0.796 at n=5) but on integration of LPRS and GF by multiplication, the r values become + 0.992** at n=6 and+0.998** at n=5, both significant at 0.01 P. In case of soybean, conversion to LPRS makes the correlation with germination per cent positive (r = + 0.832*) without changing its magnitude *vis-à-vis* lipid hydroperoxide level as shown in Table 1 (r = – 0.832*) indicating statistical significance at 0.05P. However, as with cabbage, incorporation of GF into LPRS greatly raises the value of the coefficient (r = + 0.995**) making it significant at 0.01P.

The logic behind the incorporation of germination factor (GF) into vigour scores (all of which have been converted to percentage values) has been to eliminate the relative contribution of non-viable seeds in the overall score based on all the seeds in

Table 1: The Relation between Germination Percentage and Lipid Hydroperoxide Levels in 6 Cultivars of Soybean and 6 Cultivars of Cabbage (Basic data from Hailstones and Smith, 1988[7] and their various conversions for statistical analysis)

(a) Soybean cultivars

Parameters	*Ibis*	*Impala A*	*Impala B*	*Pioneer*	*Harte-Beest*	*Impala C*	*Correlation Coefficients*			
							Germination Per cent/GF		*Non-viability Per cent/NVF*	
Germination per cent	97	95	95	94	89	82	-	-	-	-
GF	0.97	0.95	0.95	0.94	0.89	0.82	-	-	-	-
Nonviability per cent	3	5	5	6	11	18	-	-	-	-
NVF	0.03	0.05	0.05	0.06	0.11	0.18	-	-	-	-
Lipid hydroperoxide levels (LHPO, absorbance)	0.026	0.024	0.037	0.050	0.050	0.060	-0.832	(69.1)	+0.832	(69.1)
LHPO × GF	0.025	0.023	0.035	0.047	0.045	0.049	-0.728	(53.0)	+0.728	(53.0)
LHPO × NVF	0.0008	0.0012	0.0019	0.0030	0.0055	0.0108	-0.994	(98.7)	+0.994	(98.7)
LHPO ÷ GF	0.0268	0.0253	0.0389	0.532	0.562	0.0732	-0.897	(80.4)	+0.897	(80.4)
LHPO ÷ NVF	0.867	0.480	0.740	0.833	0.455	0.333	0.795	(63.2)	-0.795	(63.2)
LPRS (orig)	97.4	97.6	96.3	95.0	95.0	94.0	0.832	(69.1)	-0.832	(69.1)
LPRS (rev)	96.2	96.4	95.2	93.9	93.9	92.9	+0.832	(69.1)	-0.832	(69.1)
LPRS × GF	93.3	91.6	90.4	88.3	83.6	76.2	+0.995	(99.0)	-0.995	(99.0)
LPRS × NVF	2.89	4.82	4.76	5.63	10.33	16.72	-1.0	(100.0)	1.0	(100.0)
LPRS ÷ GF	99.2	101.5	100.2	99.9	105.5	113.3	-0.987	(97.5)	+0.987	(97.5)
LPRS ÷ NVF	3207	1928	1904	1565	854	516	+0.866	(75.1)	-0.866	(75.1)

(Coefficient of determination r^2, multiplied by 100, given in parenthesis)

(b) Cabbage cultivars

Parameters	*Golden Acre*	*Glory of Enkhu-zien*	*Cape Spitz*	*Savoy Perfe-ction A*	*Savoy Perfe-ction B*	*Savoy Perfe-ction C*	*Correlation Coefficients*			
							Germination Per cent/GF		*Non-viability Per cent/NVF*	
Germination per cent	100	98	82	76	73	39	-	-	-	-
GF	1.00	.98	.82	.76	.73	.39	-	-	-	-
Nonviability per cent	0	2	18	24	27	61	-	-	-	-
NVF	0	.02	.18	.24	.27	.61	-	-	-	-
Lipid hydroperoxide levels (LHPO, absorbance)	.072	.104	.169	.097	.124	.131	-0.430	(18.5)	+0.430	(18.5)
LHPO × GF	.072	.102	.139	.074	.091	.051	-0.472	(22.3)	+0.472	(22.3)
LHPO × NVF	0	.0021	.0304	.0233	.0335	.0799	-0.986	(97.3)	+0.986	(97.3)
LHPO ÷ GF	.072	.106	.206	.128	.170	.336	-0.929	(86.3)	+0.929	(86.3)
LHPO ÷ NVF	Mathematical error (first item involves division by zero as NVF = 0)									
LPRS (orig)	92.8	89.6	83.1	90.3	87.6	86.9	+0.430	(18.5)	-0.430	(18.5)
LPRS (rev)	93.3	90.1	83.5	90.8	88.0	87.3	+0.430	(18.5)	-0.430	(18.5)
LPRS × GF	93.3	88.3	68.5	69.0	64.2	34.0	+0.992	(98.4)	-0.992	(98.4)
LPRS × NVF	0	1.80	15.03	21.79	23.76	53.25	-1.0	(100)	+1.0	(100)
LPRS ÷ GF	93.3	91.9	101.8	119.5	120.5	223.8	-0.955	(91.1)	+0.955	(91.1)
LPRS ÷ NVF	Mathematical error (first item involves division by zero as NVF = 0)									

Note: (i) LHPO or LPRS are summations each of LHPO × GF and LHPO × NVF, or LPRS × GF and LPRS × NVF, respectively (similarly, ECRS = ECRS × GF + ECRS × NVF, VERS = VERS × GF + VERS × NVF, *etc.*). (ii) $r^2 \times 100$, given in parenthesis (*i.e.* coefficient of determination multiplied by 100), accounts for the percentage of explained variation (out of an estimated total variation of 100 per cent) for the regression line, y on x; the rest would be unexplained variation); the very high coverage by the explained variation and consequent reduction of the unexplained variation is attributable to appropriate GF/NVF incorporations.

the test, both germinable and nongerminable (occurring quite often in commercial seed lots), so that a more meaningful correlation between germination per cent of a seed lot and that part of the score value attributable to germinable seeds could be worked out (mathematically accepting the hypothesis that the contribution per seed to the score value is the same for germinable and nongerminable seed; physiologically that may not be so). Nevertheless, mathematically the incorporation of GF into the test parameter in the y variate would obviously improve the correlation when G per cent is the x variate which, however, cannot be helped if one wants to have the value of correlation attributable to the germinable part of the score of the test material (*i.e.* y × GF). But it would be worthwhile to check if nonviability factor or other morpho-physiological components of vigour such as seedling growth (seedling evaluation per cent, SF or its integrated version with GF, namely GSF, *etc.*) would influence the magnitude of the correlations.

The data presented in Table 1, besides demonstrating the benefits of conversion of negative correlations to positive correlations (as the initial step for unification of scores) via LPRS transformation and GF incorporation clearly show that the effects of GF and NVF incorporations are basically identical in respect of the value of the correlation coefficients; the only difference is that they are of opposite signs. Thus in soybean the correlation coefficient in G per cent and LPRS ÷ GF is negative (r = – 0.987**) but that between nonviability per cent and LPRS ÷ GF is positive (r = + 0.987**); that in G per cent and LPRS × NVF is equal to –1.0, the same with nonviability per cent (*i.e.* NV per cent and LPRS × NVF) is equal to + 1.0. This is just because GF and NVF are themselves of positive and negative connotations respectively, with correlation coefficient of –1.0. However, in practice the use of GF is better because GF = 1 (*i.e.* NVF = 0) may be recorded in many well preserved commercial seed lots but a NVF of 1.0 (*i.e.* 0 per cent germination, GF = 0) has no place in the seed industry; further, NVF =0 would invite mathematical error in calculations involving the same as a divisor as noted in two instances in case of cabbage (Table 1B). Similar calculations using the data of Wilson and McDonald (1986)[16] and several other sources with both GF and NVF (data on that not presented in this paper) have yielded the same information. It needs to be mentioned here that all vigour test score values are summations of values attributable to viable and non-viable fractions of a seed lot. Thus, LPRS (revised per cent value) is the sum of LPRS × GF and LPRS × NVF components and the methodology followed would broadly separate them out. The integration of GF, a natural seed quality factor, with other score values is simply an arithmetic step to eliminate the contribution of non-viable seeds from the overall score value of a seed lot and as a logical outcome of that process, there has been a substantial improvement in the correlations between the pairs of variates. Perhaps, in due course the likely differential contributions by the viable and non-viable seeds of a seed lot would be estimated to further improve the correlation between germination and vigour test parameters.

The report of Perry (1981)[12] as the chairman of the ISTA vigour test committee (1977-80) to the 19th ISTA Congress in Vienna is based on collaborative international

Table 2: The Correlation of Germination and Vigour Tests of 10 Seed Lots of Soybean and 6 Seed Lots of Maize with Respective Soil Emergence as Modified by the Incorporation of Germination Factor (Basic data from Perry,1981[12]: Report of vigour test committee, 1977-80, presented to the 19th ISTA Congress, 1980, Vienna)

a) (i) Soybean seed lots: basic data and conversions

Lot	*Germination Per cent*	*Accelerated Ageing Per cent*	*Electrical Conductivity umho*	*Soil Emergence Per cent*	*AA × GF*	*ECRS*	*ECRS × GF*	*Soil Emergence × GF*
1	67^{e}	48^{e}	70^{g}	9^{f}	32.2	93.0^{g}	62.3	6.0
2	80^{c}	73^{c}	41^{c}	23^{de}	58.4	95.9^{c}	76.7	18.4
3	87^{b}	81^{b}	37^{b}	37^{b}	70.5	96.3^{b}	83.8	32.2
4	96^{a}	94^{a}	30^{a}	44^{a}	90.2	97.0^{a}	93.1	42.2
5	77^{cd}	73^{c}	54^{e}	22^{e}	56.2	94.6^{e}	72.8	16.9
6	59^{f}	31^{f}	62^{g}	9^{f}	18.3	93.8^{g}	55.3	5.3
7	42^{g}	29^{f}	58^{f}	7^{g}	12.2	94.2^{f}	39.6	2.9
8	76^{d}	59^{d}	57^{f}	24^{d}	44.8	94.3^{f}	71.7	18.2
9	75^{d}	46^{e}	50^{d}	21^{e}	34.5	95.0^{d}	71.3	15.8
10	87^{b}	77^{b}	53^{d}	31^{c}	67.0	94.7^{d}	82.4	27.0

Note: Revised ECRS same as ECRS (original) because ECRS reference value is equal to 100.

a) (ii) Correlations (correlation coefficients and $r^2 \times 100$ in parentheses) between different parameters

X \ Y	*GF*	*AA*		*AA × GF*		*ECRS*		*ECRS × GF*		*Soil Emergence*		*Soil Emergence × GF*	
G per cent	-	.927	(85.9)	.943	(89.0)	.699	(48.9)	.999	(99.8)	.915	(83.7)	.909	(82.6)
AA per cent	-	-	-	-	-	.761	(57.9)	.934	(87.1)	.919	(84.4)	.918	(84.2)
AA × GF	-	-	-	-	-	.795	(63.2)	.952	(90.6)	.952	(90.6)	.956	(91.5)
ECRS	-	-	-	-	-	-	-	-	-	.859	(73.8)	.858	(73.6)
ECRS × GF	-	-	-	-	-	-	-	-	-	.929	(86.3)	.924	(85.4)

b) (i) Maize Seedlots: basic data and conversions

Lot	*Germination Per cent*	*Cold Test*	*Accelerated Ageing Per cent*	*Soil Emergence Per cent*	*Cold Test × GF*	*AA × GF*	*Soil Emergence × GF*
1	94^{b}	80^{b}	90^{b}	89^{b}	75.2	84.6	83.7
2	89^{c}	73^{c}	82^{c}	84^{c}	65.0	73.0	74.8
3	87^{d}	69^{d}	83^{c}	83^{c}	60.0	72.2	72.2
4	78^{e}	58^{e}	74^{d}	75^{d}	45.2	57.7	58.5
5	96^{a}	91^{a}	95^{a}	92^{a}	87.4	91.2	88.3
6	91^{c}	79^{b}	85^{c}	90^{ab}	71.9	77.4	81.9

b) (ii) Correlations (correlation coefficients and $r^2 \times 100$ in parentheses) between different parameters

X | Y	*GF*	*Cold Test*		*CT × GF*		*AA*		*AA × GF*		*Soil Emergence*		*Soil Emergence × GF*	
G per cent	-	.965	(93.2)	.976	(95.3)	.961	(92.3)	.983	(96.6)	.972	(94.4)	.992	(98.4)
CT	-	-	-	-	-	.969	(93.9)	.979	(95.8)	.965	(93.2)	.977	(95.5)
CT × GF	-	-	-	-	-	.976	(95.3)	.987	(97.5)	.967	(93.6)	.983	(96.6)
AA	-	-	-	-	-	-	-	-	-	.927	(86.1)	.956	(91.4)
AA × GF	-	-	-	-	-	-	-	-	-	.951	(90.5)	.978	(95.6)

Coefficient of determination r^2, multiplied by 100, in parenthesis.

Table 3: The Relation of Germination, Seedling Evaluation and Electrical Conductivity with Subsequent Soil Emergence of Peas as Modified by GF, SF and GSF Incorporation in the Calculation of Linear Correlation of different Combinations and Pairs of Variates (Basic data from Perry, 1981[12] : Report of the vigour test committee, 1977-80, presented to the 19th ISTA Congress, Vienna, 1980).

a) Basic data[12] and their various conversions

Seed Lot No.	*Germination Per cent*	*Seedling Evaluation Per cent*	*EC $uScm^{-1}g^{-1}$*	*Soil Emergence Per cent*	*GF*	*Seedling Evaluation Per cent (Revised)*	*SF*	*Germination-cum-Seedling Evaluation Per cent*	*GSF*	*ECRS*	*ECRS (Revised)*
1	96^{a}	93^{a}	17^{a}	86^{ab}	.96	84.2	.842	80.8	.808	98.3	98.1^{a}
2	94^{b}	89^{c}	28^{e}	76^{c}	.94	80.6	.806	75.8	.758	97.2	97.0^{c}
3	88^{c}	83^{d}	34^{g}	66^{d}	.88	75.2	.752	66.2	.662	96.6	96.4^{g}
4	86^{c}	83^{d}	31^{f}	76^{c}	.86	75.2	.752	64.7	.647	96.9	96.7^{f}
5	96^{a}	92^{ab}	20^{c}	88^{a}	.96	83.3	.833	80.0	.800	98.0	97.8^{c}
6	97^{a}	94^{a}	23^{d}	86^{ab}	.97	87.9	.879	85.3	.853	97.7	97.5^{d}
7	93^{b}	91^{b}	19^{b}	85^{b}	.93	82.4	.824	76.6	.766	98.1	97.9^{b}

Lot No.	*ECRS × GF (Rev.)*	*ECRS × SF*	*ECRS × GSF*	*Soil Emergence × GF*	*Soil Emergence × SF*	*Soil Emergence × GSF*
1	94.2	82.6	79.3	82.6	72.4	69.5
2	91.2	78.2	73.5	71.0	61.3	57.6
3	84.8	72.5	63.8	58.1	49.6	43.7
4	83.2	72.7	62.6	65.4	57.2	49.2
5	93.9	81.5	78.2	84.5	73.3	70.4
6	94.6	85.7	83.2	83.4	75.6	73.4
7	91.0	80.7	75.0	79.1	70.0	65.1

b) Correlations (r values) of germinability with electrical conductance and soil emergence with or without incorporation of GF, SF and GSF

X | Y	*GF*	*SF*	*GSF*	*ECRS*	*ECRS × GF*	*ECRS × SF*	*ECRS × GSF*	*Soil Emer.*	*Soil Emer. × GF*	*Soil Emer. × SF*	*Soil Emer. × GSF*
Germination per cent	-	.970	.981	.791	.997	.949	.980	.781	.892	.875	.924
Seedling evaluation per cent	-	-	.991	.887	.951	.998	.991	.885	.922	.940	.965
Germination-cum-seedling evaluation per cent	-	-	-	.827	.996	.990	.999	.935	.920	.924	.959

c) Correlations (r values) between electrical conductance and soil emergence

X | Y	*Soil Emergence*		*Soil Emergence × GF*		*Soil Emergence × SF*		*Soil Emergence × GSF*	
ECRS	.927	(85.9)	.929	(86.4)	.911	(83.0)	.843	(71.1)
ECRS × GF	.820	(67.2)	.919	(84.5)	.902	(81.3)	.943	(89.0)
ECRS × SF	.870	(75.6)	.940	(88.4)	.954	(91.1)	.976	(95.3)
ECRS × GSF	.843	(71.1)	.931	(86.7)	.934	(87.3)	.967	(93.5)

Coefficient of determination r^2, multiplied by 100, in parenthesis.

tests for germination, vigour, soil emergence, *etc.* on 10 seed lots each of wheat and soybean, 7 lots of peas and 6 lots of maize. The basic data are recalculated using the methods described in this paper and reappraised in Table 2 (soybean and maize), Table 3 (peas) and Table 4 (wheat) after integration of GF, SF and GSF, the latter two only when seedling evaluation data are available as for peas and wheat in Tables 3 and 4.

The 10 lots of soybean, five each of cv. Lancer (lots 1-5) and cv. Hardee (lots 6-10), show a wide range of differences in germinability (42-96 per cent), accelerated ageing response (29-94 per cent), electrical conductivity, (70-30 μmhos) and soil emergence (9-44 per cent). The correlation of G per cent with electrical conductivity (EC μmho) is not very high (r = – 0.699*), that with ECRS is of the same magnitude but positive (r = + 0.699*), but the r value becomes very high when ECRS is multiplied by GF (r = + 0.999**). Multiplication of ECRS by GF also improved the correlation with AA per cent as well as with AA × GF (Table 2A). That means GF incorporation into both x and y variates, even when the correlations do not directly involve G per cent as one of the variates, may substantially improve the value of correlation coefficients, especially when the nonintegrated single variates show relatively low correlation coefficients. Thus in the correlation between G per cent and soil emergence per cent in which the value of r is equal to 0.915**, multiplication of the ẏ variate by GF (*i.e.* soil emergence × GF) shows no improvement in the already very high value of r; in fact there is a slight reduction(r = 0.909**). In maize also, there has been only marginal improvements in the value of correlation coefficients as the basic nonintegrated data show very good correlations between them (Table 2, b ii). In all correlations in maize, the coefficient of determination, expressed as a percentage (*i.e.* $r^2 \times 100$), showed over 90 per cent coverage except for a single case (between AA and soil emergence, showing $r^2 \times 100 = 86.1$).

In Table 3, dealing with data on peas, besides GF two new factors, namely seedling factor (SF) and germination-cum-seedling factor (GSF), based on seedling evaluation per cent and germination-cum-seedling evaluation per cent respectively, have been taken up to assess the effects of their integration into vigour parameters. G per cent and SF show very high correlation(r = 0.970**, significant at 0.01P) and expectedly a still better correlation is seen between G per cent and GSF (r = 0.981**); the same between SF and GSF being 0.991**. It needs to be mentioned here that all correlations involving either G per cent or GF (also seedling evaluation per cent or SF, germination-cum-seedling evaluation per cent or GSF) would be identical because the respective factors are derived just by dividing the basic percentage values by 100 which would not affect the nature of correlation; as such, r values with G per cent would be exactly the same as with GF, that is also true for the other two factors, SF and GSF. Another point that may also be clarified in this connection is the designation of the factors GF, SF and GSF as morphological manifestations of vigour. That has been done for the sake of convenience only as lipid peroxidation, aldehyde emanation, electrical conductivity, *etc.* are generally considered as physiological-biochemical manifestations of seed vigour; otherwise

Table 4: The Relationship of Germination, Seedling Growth Evaluation and Electrical Conductivity of 10 Seed Lots of Wheat with their Subsequent Soil Emergence as Modified by GF, SF and GSF Incorporation for the Calculation of Bivariate Linear Correlations (Basic data from Perry, 1981[12]: Report of vigour test committee, 1977-80, presented to the 19th ISTA Congress, Vienna, 1980)

a) Basic data[12] and their various conversions

Lot	*Germination Per cent*	*Hiltner Per cent*	*Seedling Growth mm/ Seedling*	*Soil Emer. Per cent*	*GF*	*Seedling Eval. Per cent*	*SF*	*GSF*
1	91^{c}	84^{d}	83^{d}	76^{c}	0.91	81.5^{d}	.815	.742
2	97^{a}	94^{a}	92^{a}	85^{a}	0.97	90.4^{a}	.904	.877
3	96^{a}	93^{a}	93^{a}	85^{a}	0.96	91.4^{a}	.914	.877
4	92^{b}	85^{cd}	85^{cd}	79^{b}	0.92	83.5^{cd}	.835	.768
5	91^{bc}	83^{d}	83^{d}	73^{d}	0.91	81.5^{d}	.815	.742
6	91^{c}	87^{bc}	89^{b}	81^{b}	0.91	87.4^{b}	.874	.795
7	86^{d}	76^{e}	89^{b}	72^{d}	0.86	87.4^{b}	.874	.752
8	93^{b}	83^{d}	86^{c}	73^{d}	0.93	84.5^{c}	.845	.786
9	86^{d}	83^{d}	84^{cd}	77^{c}	0.86	82.5^{cd}	.825	.710
10	91^{c}	89^{b}	89^{b}	84^{a}	0.91	87.4^{b}	.874	.795

	Hilt × GF	*Hilt × SF*	*Hilt × GSF*	*Soil Emer. × GF*	*Soil Emer. × SF*	*Soil Emer. × GSF*
1	76.4	68.5	62.3	69.2	61.9	56.4
2	91.2	85.0	82.4	82.5	76.8	74.5
3	89.3	85.0	81.6	81.6	77.7	74.5
4	78.2	71.0	65.3	72.7	66.0	60.7
5	75.5	67.6	61.6	66.4	59.5	54.2
6	79.2	76.0	69.2	73.7	70.8	64.4
7	65.4	66.4	57.2	61.9	62.9	54.1
8	77.2	70.1	65.2	67.9	61.7	57.4
9	71.4	68.5	58.9	66.2	63.5	54.7
10	81.0	77.8	70.8	76.4	73.4	66.8

b) Correlations (r values) between germinability, Hiltner test and soil emergence

X | Y	*GF*	*SF*	*GSF*	*Hiltner*	*Hilt. × GF*	*Hilt × SF*	*Hilt × GSF*	*Soil Emergence*	*Soil Emergence × GF*	*Soil Emergence × SF*	*Soil Emergence × GSF*
Germination per cent	-	.518	.863	.834	.935	.789	.884	.619	.841	.637	.798
Seedling evaluation per cent	-	-	.879	.607	.615	.844	.794	.697	.706	.880	.846
Germination-cum-seedling evaluation per cent	-	-	-	.828	.887	.941	.964	.760	.888	.946	.946

c) Correlations (r values) between soil emergence and Hiltner test

X | Y	*Hiltner (Per cent)*		*Hiltner × GF*		*Hiltner × SF*		*Hiltner × GSF*	
Soil emergence per cent	.697	(48.6)	.615	(37.9)	.844	(71.3)	.870	(75.8)
Soil emergence per cent × GF	.978	(95.6)	.966	(93.3)	.966	(93.2)	.970	(94.0)
Soil emergence per cent × SF	.869	(75.5)	.822	(67.6)	.968	(93.6)	.915	(83.7)
Soil emergence per cent × GSF	.930	(86.4)	.923	(85.2)	.994	(98.9)	.982	(96.4)

Coefficient of determination r^2, multiplied by 100, in parenthesis, Hiltner test = brick grit test) (Ziegelgrus Method).

one can very pertinently argue that germination, growth, *etc.* are physiological processes outwardly manifested in sprouting, dry matter production, *etc.* In any case, as exemplified by the data in Tables 3-10 of this paper and our unpublished data on a wide range of crop seeds that include cereals, millets, pulses, oil seeds, fibre crops, summer and winter vegetables and flower seeds show the great usefulness of germination-cum-seedling evaluation per cent as a practical and easily derivable morphological parameter of seed vigour, that can serve as a vigour score in the percentage scale; GSF, the factor derived from it is also very useful.

Coming back to the data in Table 3, the correlation between G per cent and ECRS (r = 0.791*) is greatly improved when GF is integrated into the latter (r = 0.997**), even incorporation of SF (*i.e.* between G per cent and ECRS × SF) is almost equally effective (r = 0.949**). The correlations of seedling evaluation per cent and germination-cum-seedling evaluation per cent with ECRS are better than that with G per cent alone but interestingly, that between GSF and ECRS × GSF exhibits an r value of.999**, despite the fact that the basic electrical conductivity values show highly significant differences between them with the highest reading in lot 3 being double of that in the lowest in lot 1 (34 in the former against 17 in the latter). The observations in respect of soil emergence is almost similar to those with ECRS; the values of their respective correlation coefficients *vis-à-vis* G per cent, seedling evaluation per cent and germination-cum-seedling evaluation per cent are indeed very close except in case of one instance, namely GSF with soil emergence (r = 0.935** which is valuewise much better than between GSF and ECRS with r value of 0.827**). The similarity is attributable to the very high correlation between ECRS and soil emergence (r = 0.927**); the high correlation between them has also been the reason of the apparent lack of adequate improvement in r values on integration of GF, SF and GSF in most of the correlations involving ECRS and soil emergence of the seedlots of peas.

In Table 4, the data on 10 winter seed lots, four of cv. Flanders (lots 1-4), three of cv. Maria Huntsman (lots 5-7), and three of cv. Atou (lots 8-10), show germination percentages ranging from 86 to 97, seedling growth (mm) 83-93, Hiltner percentages 76 to 94, and soil emergence percentages 73 to 85. The calculations of seedling evaluation per cent, SF, germination-cum-seedling evaluation per cent, GSF are shown in the previous section with the value of reference seedling growth estimated at 101.8, in particular, based on which SF and GSF have been calculated. The r value for the correlation between G per cent and SF is 0.518* which has gone up to 0.863** in G per cent *vis-a-vis* GSF showing significance at 0.01P. G per cent and Hiltner per cent show good correlation (r = 0.834**); however, further improvement is noted with Hiltner per cent × GF (r = 0.935**) but not with Hiltner per cent × SF. Germination and soil emergence is moderately correlated but GF incorporation into the latter raised the level of significance to 0.01P.The best correlations are recorded in GSF and Hiltner per cent × GSF (r = 0.964**) and GSF and soil emergence per cent × GSF (r = 0.946**). Hiltner per cent and soil emergence per cent show a moderately good correlation with r value of 0.697* but there is no improvement in Hiltner per

Table 5: The Relationship of Germination, Seedling Dry Matter production, electrical conductivity and volatile semanation of 5 soybean seed lots with their subsequent field emergence, as modified by GF, SF and GSF incorporation in statistical calculations (basic data from Wilson and McDonald, 1986[16]).

a) Basic data[16] and their conversions

Seed Lot No.	*Germination Per cent*	*Seedling Evaluation Per cent dry wt. g/gram seed*	*EC µmho/ Seed*	*Volatile Emanation ug/ seed*	*Field Emergence Per cent*	*GF*	*Seedling Evaluation Per cent*	*SF*	*Germination-cum-Seedling Eval. per cent*	*GSF*	*FE × GF*	*FE × SF*	*FE × GSF*
1	93^{a}	$.26^{a}$	27^{a}	$.022^{a}$	84^{a}	.93	60.8	.608	56.5	.565	78.1	51.1	47.5
2	79^{b}	$.22^{b}$	28^{a}	$.022^{a}$	70^{b}	.79	51.5	.515	40.7	.407	55.3	36.1	28.5
3	92^{a}	$.23^{b}$	29^{a}	$.063^{b}$	48^{c}	.92	53.8	.535	49.5	.495	44.2	25.8	23.8
4	71^{bc}	$.16^{c}$	31^{a}	$.092^{c}$	28^{d}	.71	37.4	.374	26.6	.266	19.9	10.5	7.4
5	62^{c}	$.13^{d}$	49^{b}	$.086^{c}$	20^{d}	.62	30.4	.304	18.8	.188	12.4	6.1	3.8

Seed Lot No.	*ECRS (Original)*	*ECRS (Revised)*	*ECRS × GF*	*ECRS × SF*	*ECRS × GSF*	*VERS (Original)*	*VERS (Revised)*	*VERS × GF*	*VERS × SF*	*VERS × GSF*
1	97.3^{a}	96.7^{a}	89.9	58.8	54.6	97.8^{a}	93.4^{a}	86.9	56.8	52.8
2	97.2^{a}	96.6^{a}	76.3	49.7	39.3	97.8^{a}	93.4^{a}	73.8	48.1	38.0
3	97.1^{a}	96.5^{a}	88.8	51.6	47.8	93.7^{b}	89.5^{b}	82.3	48.2	44.3
4	96.9^{a}	96.3^{a}	68.4	36.0	25.6	90.8^{c}	86.7^{c}	61.6	33.5	23.1
5	95.1^{b}	94.5^{b}	58.6	28.7	17.8	91.4^{c}	87.3^{c}	54.1	26.5	16.4

b) Correlations (r values) of germinability with vigour parameters and field emergence

X | Y	*GF*	*SF*	*GSF*	*ECRS*	*ECRS × GF*	*ECRS × SF*	*ECRS × GSF*
Germination per cent	-	.970	.981	.797	.999	.961	.985
Seedling evaluation per cent	-	-	.991	.887	.957	.999	.991
Germination-cum-seedling evaluation per cent	-	-	-	.827	.985	.999	.999

X | Y	*VERS*	*VERS × GF*	*VERS × SF*	*VERS × GSF*	*FE*	*FE × GF*	*FE × SF*	*FE × GSF*
Germination per cent	.656	.990	.938	.973	.782	.849	.819	.855
Seedling evaluation per cent	.841	.988	.998	.994	.926	.957	.941	.950
Germination-cum-seedling evaluation per cent	.772	.999	.980	.998	.876	.927	.906	.931

c) Correlations (r values) of vigour parameters with field emergence

X | Y	*FE*	*FE × GF*	*FE × SF*	*FE × GSF*	*VERS*	*VERS × GF*	*VERS × SF*	*VERS × GSF*
ECRS	.735	.736	.719	.703	.608	.802	.829	.974
ECRS × GF	.788	.853	.919	.857	.662	.990	.942	.974
ECRS × SF	.926	.957	.941	.949	.841	.987	.998	.993
ECRS × GSF	.877	.928	.906	.931	.773	.999	.981	.998
VERS	.967	.929	.943	.904				
VERS × GF	.861	.914	.890	.915				
VERS × SF	.948	.971	.959	.963				
VERS × GSF	.901	.901	.928	.949				

cent × GF (r= 0.615*). However, SF and GSF incorporation into Hiltner per cent is more effective in improving the correlations. On the other hand, integration of soil emergence per cent with GF, SF and GSF has noticeably raised the value of the coefficients (Hiltner per cent *vis-a-vis* soil emergence per cent × GF, yielding r = 0.978**). As with peas and other materials, in wheat also, when the same factor (GF or SF or GSF) is integrated into both variates (x and y) there is a consistent improvement in the value of r. Mathematically, it may be argued that if there are no significant differences between the individual values of the x variate, and also of the y variate, we may expect r value close to 1 simply because the correlation coefficient of GF and GF, or SF and SF, or GSF and GSF would be 1. In view of the substantial statistically highly significant intravariate differences in the data presented in most of the tables such an argument can be effectively countered to accept that the improved correlations are the genuine outcome of integration of morphological and physiological-biochemical parameters of vigour.

In Table 5, the effect of incorporation of GF, SF and GSF, into two vigour parameters (*viz.* electrical conductivity of seed-steep water and volatile aldehyde emanation by seeds, after their necessary conversions to ECRS and VERS, respectively) has been assessed using the basic data of Wilson and McDonald (1986)[16] on soybean cv. Williams. The methods of calculating the seedling evaluation reference value (= 0.427g), the seedling evaluation per cent values for the different seed lots, revised ECRS and revised VERS values, SF, GSF, *etc.*, have already been discussed earlier and the processed data are presented in the table. The correlations between germination per cent and seedling evaluation per cent or SF or GSF and between SF and GSF are all very high (significant at 0.01P) but that with ECRS and VERS are respectively, moderate (significant at 0.05P) and poor (statistically not significant at 0.05P). Seedling evaluation per cent as well as germination-cum-seedling evaluation per cent are better correlated with ECRS and VERS than germination per cent. Integration of GF, SF and GSF into ECRS or VERS has resulted in vast improvements in their correlations with germination per cent, seedling evaluation per cent and germination-cum-seedling evaluation per cent and in all cases the r values are over 0.9 (range 0.938-0.999).The correlation of G per cent with field emergence per cent of May planting is moderate (r = 0.782*), that between seedling evaluation per cent and field emergence per cent (May planting) is much better (r = 0.926**). As with ECRS and VERS in which correlation with germination per cent has been substantially improved by integration of GF, SF and GSF into them, similar multiplication of field emergence per cent by GF, SF and GSF improved the correlation with germination per cent but to a lesser extent than in the former case. There is only a minor improvement of such integration when correlation is worked out with seedling evaluation per cent as the latter is already very well correlated with field emergence. The correlation between ECRS and FE just fails to reach the level of significance at 0.05P (r = 0.735^{NS}) and there is no improvement when FE is multiplied by GF, SF and GSF but GF incorporation into ECRS raises the value of r to 0.788* making that significant at 0.05 P while SF and GSF incorporations into ECRS give 0.926** and 0.877** respectively as the r values, both significant at 0.01 P. Unlike ECRS, FE is highly correlated with VERS, the r value for VERS *vis-a-vis* FE correlation is equal to 0.967**, significant at 0.01P, and as may be presumed there has been very little improvement with any of the integrations attempted (in fact only in VERS × SF *vis-a-vis* FE × SF, the r value is equal to 0.971**, in all other cases there are minor reductions). The important point arising out of the correlations shown in Table 5 between the two vigour scores, ECRS and VERS, is the very substantial increase in the value of r when both variates are integrated with the factors GF, SF or GSF; all the nine such correlation coefficients are indeed very close to 1, the maximum value of r. The three correlations namely, ECRS × GF and VERS × GF, ECRS × SF and VERS × SF, and ECRS × GSF and VERS × GSF, in particular, having the same integrating factors on both the variates, show r values of 0.990, 0.998 and 0.998, respectively that are so very improved from the basic nonintegrated ECRS and VERS correlation yielding a statistically nonsignificant (P = 0.05) correlation coefficient of 0.608. It appears that the integrating morphological factors (GF, SF and GSF) act in a complementary way to compensate or make up, at least partly, the deficiency of an ideal correlation. Theoretically the deficit is equal to 1– $\hat{r}$, where $\hat{r}$ is the estimated correlation coefficient of the non-integrated x and y variates. Obviously, in such a situation the effect of incorporation (*i.e.* the difference = 1 – $\hat{r}$) would be more pronounced when the estimated $\hat{r}$ value without incorporation is low, and less pronounced when the $\hat{r}$ in the latter is high.

An analysis of the data in Table 6 from our own experiments on soybean seed invigoration treatments for improvement of vigour, viability and field performance (Mandal *et al.,* 2000)[8], besides showing the effectiveness of seed invigoration on germinability, field emergence and productivity[8], clearly reveals the substantial benefits in most cases of incorporation of the morphological factors into ECRS as well as into FE and yield parameters especially so when ECRS as well as the field performance parameters are integrated with the same morphological factor as has already been noted and discussed above while analyzing the data of Wilson and McDonald (1986)[16]. In fact, the correlations between G per cent and SF, GSF, ECRS and FE of our data derived from seed invigoration treatments of Soyamax soybean are broadly similar to those shown by the seed lots of Williams soybean stored for different periods in Ohio and representing a diverse range of 'naturally' attained vigour levels. The correlation between G per cent and yield is moderate (r = 0.773*, significant at 0.05P) but that between G per cent and 1000-seed weight (TSW) is poor (r = 0.453^{NS}); however, GF incorporation into yield and TSW raised the r values to 0.963** and 0.998** respectively, both significant at 0.01P. Multiplication of ECRS by GF improved its correlation with yield but the effect was much more noticeable with SF incorporation (in ECRS × SF and yield, r = 0.918**,significant at 0.01P); the same incorporations are not effective for TSW, in which integration of TSW itself by the morphological factors improved the correlations; however, in all but one in nine correlations of ECRS and yield, and the same number in ECRS and TSW, in which both variates are integrated with the morphological factors, the coefficients of correlation are above 0.9 (only in ECRS × GF and yield × SF, and ECRS × GF and TSW × SF, the r values are 0.823** and 0.874**, respectively). The

Table 6: Correlation of Presowing Germinability and Electrical Conductivity (Vigour test) before Sowing with Subsequent Field Performance of Soybean Seed of different Invigoration Treatments as Influenced by Incorporation of GF, SF and GSF (Basic data from soybean seed invigoration treatment by Mandal *et al.*, 2000[9] and their various conversions for statistical analysis)

a) Basic data[9] and their various conversions

Sl. No.	*Treatment*	*Germination per cent*	*GF*	*Seedling Eval. Per cent*	*SF*	*GSF*	*ECRS*	*ECRS × GF*	*ECRS × SF*	*ECRS × GSF*	*FE/ m^2*	*Yield g/plant*	*1000-SW (TSW, g)*
1	Control	53^{f}	.53	38.7	.387	.205	89.5^{c}	47.4	34.6	18.3	21^{c}	18.3^{c}	70.1^{d}
2	Bleaching powder	75^{cd}	.75	49.9	.499	.374	91.0^{c}	68.2	45.4	34.0	36^{ab}	22.2^{ab}	71.9^{ab}
3	Aspro	79^{bc}	.79	62.3	.623	.492	92.9^{b}	73.4	57.9	45.7	36^{ab}	24.6^{a}`	72.0^{a}
4	Ibucon	70^{cde}	.70	54.4	.544	.381	90.9^{d}	63.6	49.4	34.6	34^{b}	23.7^{a}	72.1^{a}
5	Red chilli fruit powder	83^{ab}	.83	58.4	.584	.485	92.9^{b}	77.1	54.2	45.1	40^{a}	24.0^{bc}	71.2^{bc}
6	Catharanthus leaf powder	85^{a}	.85	56.7	.567	.482	92.8^{b}	78.9	52.6	44.7	33^{b}	22.3^{c}	71.0^{c}
7	Moist sand conditioning-drying (MSC-SD)	73^{cd}	.73	51.6	.516	.377	93.2^{a}	68.0	48.1	35.1	38^{a}	21.1^{ab}	71.9^{ab}

Sl. No.	*FE × GF*	*FE × SF*	*FE × GSF*	*Yield × GF*	*Yield × SF*	*Yield × GSF*	*TSW × GF*	*TSW × SF*	*TSWx GSF*
1	11.1	8.1	4.3	9.7	7.1	3.8	37.2	27.1	14.4
2	27.0	18.0	13.5	16.7	11.1	8.3	53.9	35.9	26.9
3	28.4	22.4	17.7	19.4	15.3	12.1	56.9	44.9	35.4
4	23.8	18.5	13.0	16.6	13.0	9.0	50.5	39.2	27.5
5	33.2	23.4	19.4	19.9	14.0	11.6	59.1	41.6	34.5
6	28.1	18.7	15.9	19.0	12.6	10.7	60.3	40.3	34.2
7	27.7	19.6	14.3	15.4	10.9	8.0	52.5	37.1	27.1

b) Correlations (r values) of germinability with electrical conductivity and field performance

X | Y	*GF*	*SF*	*GSF*	*ECRS*	*ECRS × GF*	*FE*	*FE × GF*	*Yield*	*Yield × GF*	*TSW*	*TSW × GF*
Germination per cent	-	.879	.965	.836	.999	.816	.941	.773	.963	.453	.998
Seedling evaluation per cent	-	-	.968	.803	.883	.784	.861	.937	.956	.607	.892
Germination-cum-seedling evaluation per cent	-	-	-	.845	.968	.795	.914	.872	.986	.496	.969

c) Correlations (r values) of different parameters as influenced by the incorporation of GF, SF and GSF

X | Y	*FE*	*FE × GF*	*FE × SF*	*FE × GSF*	*Yield*	*Yield × GF*	*Yield × SF*	*Yield × GSF*	*TSW*	*TSW × GF*	*TSW × SF*	*TSW × GSF*
ECRS	.792	.855	.845	.868	.588	.778	.707	.793	.428	.839	.796	.844
ECRS × GF	.819	.943	.890	.950	.763	.959	.823	.924	.447	.997	.874	.965
ECRS × SF	.793	.871	.939	.950	.918	954	.976	.985	.594	.901	.996	.978
ECRS × GSF	.796	.915	.925	.970	.861	.982	.926	.986	.485	.967	.958	.999
FE (Field emerg.)	-	-	-	-	.776	.831	.766	.785	.746	.839	.796	.805
FE × GF	-	-	-	-	.805	.933	.824	.890	.604	.951	.903	.954
FE × SF	-	-	-	-	.900	.937	.920	.932	.693	.903	.939	.932
FE × GSF	-	-	-	-	.876	.962	.932	.962	.570	.954	.939	.970
Yield	-	-	-	-	-	-	-	-	.697	.795	.944	.882
Yield × GF	-	-	-	-	-	-	-	-	.550	.970	.952	.988
Yield × SF	-	-	-	-	-	-	-	-	.635	.841	.988	.939
Yield × GSF	-	-	-	-	-	-	-	-	.528	.931	.980	.990

Table 7: The Relation between Presowing Germinability and Physiological Vigour Parameters before Sowing with Subsequent Field Performance of Sunflower Seed of different Invigoration Treatments as Influenced by Incorporation of GF, SF and GSF (basic data of Pal *et al.*, 1996[11])

a) Basic data[11] and their various conversions

Sl. No.	*Treatment*	*Germination Per cent*	*GF*	*Seedling Evaluation Per cent*	*SF*	*GSF*	*VERS*	*JSBS*	*FE Per cent*	*Yield/m Row Length (g)*	*1000 SW (TSW, g)*
1	Control	73[b]	.73	49.7	.497	.363	44.4[c]	54.3[b]	40[c]	21.3c	34[b]
2	Nisinda leaf powder	88[a]	.88	67.9	.679	.598	61.2[a]	66.6[a]	46[b]	26.0[a]	38[a]
3	Aspirin	84[a]	.84	58.2	.582	.489	57.1[ab]	65.4[a]	49[a]	23.7[ab]	40[a]
4	Ascorbic acid	86[a]	.86	57.3	.573	.493	54.4[b]	64.3[a]	50[a]	25.3[a]	39[a]
5	Bleaching powder	85[a]	.85	63.6	.636	.541	61.2[a]	67.3[a]	49[a]	24.3[ab]	38[a]
6	Water soaking-drying	86[a]	.86	58.8	.588	.506	55.1[a]	64.3[a]	42[c]	22.8[bc]	39[a]

b) Correlations (r values) between germinability, physiological-biochemical parameters and field performance data

X/Y	*GF*	*SF*	*GSF*	*VERS*	*VERS × GF*	*VERS × SF*	*VERS × GSF*	*JSBS*	*JSBS × GF*	*JSBS × SF*	*JSBS × GSF*
Germination per cent	-	.838	.924	.884	.946	.851	.894	.943	.983	.896	.935
Seedling eval. per cent	-	-	.983	.946	.943	.980	.990	.867	.874	.984	.974
Germination-cum-seedling eval. per cent	-	-	-	.954	.975	.979	.993	.919	.940	.988	.995

X/Y	*FE*	*FE × GF*	*FE × SF*	*FE × GSF*	*Yield*	*Yld × GF*	*Yld × SF*	*Yld × GSF*	*TSW*	*TSW × GF*	*TSW × SF*	*TSW × GSF*
Germination per cent	.617	.845	.745	.904	.827	.941	.850	.911	.849	.966	.943	.970
Seedling eval. per cent	.479	.686	.925	.912	.813	.863	.970	.963	.521	.715	.950	.934
Germination-cum-seedling eval. per cent	.531	.757	.891	.938	.849	.923	.967	.983	.645	.823	.981	.980

c) Correlations (r values) between physiological-biochemical parameters and field performance data as influenced by the incorporation of GF, SF and GSF into the two variates

X \ Y	*FE*	*FE × GF*	*FE × SF*	*FE × GSF*	*Yield*	*Yield × GF*	*Yield × SF*	*Yield × GSF*	*TSW*	*TSW × GF*	*TSW × SF*	*TSW × GSF*
VERS	.669	.830	.966	.959	.789	.866	.913	.923	.713	.833	.979	.961
VERSx GF	.651	.850	.918	.967	.829	.919	.925	.951	.771	.894	.995	.991
VERS × SF	.560	.746	.960	.939	.809	.865	.959	.955	.584	.753	.967	.948
VERS × GSF	.567	.769	.939	.950	.833	.899	.965	.972	.631	.800	.982	.971
JSBS	.732	.898	.888	.950	.784	.888	.850	.880	.833	.932	.967	.966
JSBS × GF	.685	.884	.839	.946	.820	.929	.870	.917	.825	.957	.973	.984
JSBS × SF	.589	.784	.947	.956	.824	.957	.894	.964	.650	.810	.985	.972
JSBS × GSF	.591	.801	.913	.958	.845	.924	.956	.975	.693	.853	.992	.989
Field emerg. (FE)	-	-	-	-	.741	.711	.587	.596	.690	.676	.616	.607
FE × GF	-	-	-	-	.862	.888	.767	.798	.823	.868	.821	.826
FE × SF	-	-	-	-	.736	.771	.886	.866	.574	.689	.917	.876
FE × GSF	-	-	-	-	.923	.955	.948	.957	.718	.848	.954	.948
Yield	-	-	-	-	-	-	-	-	.576	.737	.824	.835
Yield × GF	-	-	-	-	-	-	-	-	.617	.869	.914	.933
Yield × SF	-	-	-	-	-	-	-	-	.530	.727	.929	.922
Yield × GSF	-	-	-	-	-	-	-	-	.617	.803	.955	.983

correlations involving FE and yield, FE and TSW and yield and TSW that would come mainly under the purview of crop physiologists and agronomists, show statistical significance at 0.05P without any amalgamation with morphological factors, have produced on amalgamation r values, all significant at 0.01P;and as noted before with vigour parameters, and also in Tables 7-10, here also FE, yield and TSW when they (both the variates in a bivariate regression) are amalgamated with GF, SF or GSF, the linear correlations between them are substantially improved.

If only GF is taken into consideration, because that is almost always available in certified commercial seed lots (SF, hence GSF, have to be determined through separate laboratory tests), and because of the moderate to high correlation of GF with SF and GSF, germination factor would in most crop seeds serve as an effective indicator of seedling growth, we may highlight the role of GF integration on the following correlations of soybean in respect of its field performance.

Soybean Correlations (r values) (*vide* Table 6, n=7)

Variates		*Without GF Incorporation*	*With GF Incorporation into both the Variates*
X	*Y*		
ECRS	FE	0.792*	0.943**
ECRS	Yield	0.588^{NS}	0.959**
ECRS	TSW	0.428^{NS}	0.997**
FE	Yield	0.776*	0.933**
FE	TSW	0.746*	0.951**
Yield	TSW	0.697*	0.970**

The examples cited are not merely of theoretical interest but are indicative of the greatly enhanced predictive efficiencies of the regression equations derived from GF incorporated values for practical use in crop husbandry in general and seed production in particular. One may perhaps argue that GF incorporation is also there in seed lots with 100 per cent germination, the same is not evident because multiplication is by GF which has a value of 1. In lots, all with 100 per cent germination, however, SF (= GSF) based on seedling evaluation per cent has to be calculated to form similar predictive regression equations.

The data on seed invigoration treatments of sunflower by Pal *et al.* (1996)[11] in Table 7 show that the four 'dry' seed invigoration treatments as well as the 'wet' water-soaking-drying treatment have retained seed germinability much better than in the untreated control; no significant difference in G per cent is, however, noted between the different treatments. The vigour parameters also show a more or less similar trend; field emergence, yield and 1000 seed weights are also lower in the control than in the dry treatments but in case of water-soaking-drying there is no difference with control in field emergence and yield (though 1000-seed weight, TSW, is better than control and same as in the other treatments). This is because the vigour of the seed lot at treatment has been good enough for 'dry' treatments (which require high vigour seed for efficacy) but not so for soaking-drying that is so very responsive in medium and low vigour nonleguminous seed lots[13, 14]. The correlations between germination per cent and VERS and JSBS show r values of 0.884** and 0.943** respectively, both significant at 0.01P; and in either case GF incorporation improved the value and expectedly more in the former. All correlations of VERS and JSBS with seedling evaluation per cent and germination-cum-seedling evaluation per cent without or with incorporation of the morphological factors are significant at 0.01P. FE is not well correlated with G per cent but FE × GF and FE × GSF show good correlations; seedling evaluation per cent and germination-cum-seedling evaluation per cent are poorly correlated with FE but SF and GSF incorporations into FE give r values significant at 0.01P. All the three morphological parameters are well correlated with yield, and incorporations of GF, SF and GSF into yield data improved the value of the coefficients with the best value noted in germination-cum-seedling evaluation per cent and yield × GSF (r = 0.983**). TSW alone shows good correlation with G per cent but not with seedling evaluation per cent and germination-cum-seedling evaluation per cent but SF and GSF incorporations of the variates have given highly significant r values. The correlations of VERS and JSBS with FE, yield and TSW range from poor to moderate but upon integration of both x and y variates with the morphological factors there has been a consistent improvement of the correlation coefficient values. In all the bivariate linear regressions involving FE, yield and TSW, the best coefficients are recorded in those integrated with GSF in both x and y variates *viz.* FE × GSF and yield × GSF, FE × GSF and TSW × GSF and yield × GSF and TSW × GSF, giving r values of 0.975**, 0.989** and 0.983** respectively. As in case of soybean in Table 6, the important field performance related correlations of sunflower as influenced by GF incorporation in x and y variates, are listed below to show the respective improvements in r values.

Sunflower Correlations (r values) (*vide* Table 7, n=6)

Variates		*Without GF Incorporation*	*With GF Incorporation into both the Variates*
X	*Y*		
VERS	FE	0.669^{NS}	0.850**
VERS	Yield	0.789*	0.919**
VERS	TSW	0.713*	0.894**
JSBS	FE	0.732*	0.884**
JSBS	Yield	0.784*	0.989**
JSBS	TSW	0.833*	0.957**
FE	Yield	0.741*	0.888**
FE	TSW	0.690^{NS}	0.868**
Yield	TSW	0.576^{NS}	0.869**

The respective improvements with GSF integration of the two variates are

Table 8: The Relation of Germination, Seedling Evaluation and Vigour Test Scores at the Time of Sowing with Subsequent Field Performance of Wheat Seed of different Invigoration Treatments as Influenced by GF, SF and GSF Incorporation

a) Basic data[11] and their various conversions

Sl. No.	*Treatment*	*G Per cent*	*GF*	*SF*	*GSF*	*VERS*	*JSBS*	*FE/m²*	*Yield/m² (g)*	*1000 SW (TSW, g)*
1	Control	78^{c}	.78	.694	.541	18.3^{c}	42.0^{c}	98^{b}	284^{d}	29.4^{c}
2	Turmeric	88^{b}	.88	.785	.691	40.1^{ab}	53.4^{b}	115^{a}	320^{c}	33.6^{b}
3	Neem	92^{ab}	.92	.740	.681	44.8^{ab}	57.3^{b}	121^{a}	340^{abc}	33.6^{b}
4	Nisinda	90^{ab}	.90	.768	.691	36.6^{b}	59.5^{ab}	119^{a}	330^{bc}	32.8^{b}
5	Aspirin	92^{ab}	.92	.777	.715	46.0^{a}	59.5^{ab}	122^{a}	358^{ab}	33.6^{b}
6	Aspro	94^{a}	.94	.822	.773	47.8^{a}	63.4^{a}	124^{a}	360^{ab}	35.0^{ab}
7	Celin	91^{ab}	.91	.808	.735	44.2^{ab}	61.8^{a}	123^{a}	344^{abc}	34.4^{ab}
8	Bleaching powder	94^{a}	.94	.839	.789	48.3^{a}	64.1^{a}	123^{a}	348^{ab}	34.8^{ab}
9	Water soaking-drying	93^{a}	.93	.780	.725	44.8^{ab}	61.1^{a}	120^{a}	370^{a}	36.0^{a}

b) Correlations (r values) involving morphological parameters vigour scores and field performance data (without and with GF integration)

X | Y	*GF*	*SF*	*GSF*	*VERS*	*VERS × GF*	*JSBS*	*JSBS × GF*	*FE*	*FE × GF*	*Yield*	*Yield × GF*	*TSW*	*TSW × GF*
Germination per cent	-	.790	.938	.975	.982	.965	.980	.977	.994	.931	.971	.923	.977
Seedling evaluation per cent	-	-	.953	.810	.812	.862	.851	.797	.809	.689	.739	.802	.813
Germination-cum-seedling evaluation per cent	-	-	-	.936	.941	.961	.963	.927	.944	.848	.896	.904	.938

c) Correlations (r values) involving vigour scores and field performance

X | Y	*FE*	*FE × GF*	*FE × SF*	*FE × GSF*	*Yield*	*Yield × GF*	*Yield × SF*	*Yield × GSF*	*TSW*	*TSW × GF*	*TSW × SF*	*TSW × GSF*
VERS	.965	.975	.946	.956	.904	.944	.931	.950	.920	.962	.903	.943
VERSx GF	.965	.979	.948	.962	.917	.955	.941	.961	.923	.968	.907	.975
VERS × SF	.953	.966	.976	.981	.885	.928	.957	.968	.921	.958	.947	.972
VERS × GSF	.953	.970	.975	.983	.897	.939	.963	.976	.922	.963	.946	.975
JSBS	.968	.977	.971	.981	.896	.936	.954	.967	.894	.945	.917	.956
JSBS × GF	.974	.986	.969	.983	.912	.954	.960	.976	.907	.960	.921	.962
JSBS × SF	.929	.942.	.944	.988	.848	.892	.962	.974	.883	.922	.960	.974
JSBS × GSF	.941	.958	.988	.993	.874	.917	.970	.977	.897	.941	.957	.988
Field emerg. (FE)	-	-	-	-	.888	.932	.912	.934	.876	.937	.871	.919
FE × GF	-	-	-	-	.918	.960	.939	.962	.907	.964	.961	.945
FE × SF	-	-	-	-	.850	.897	.959	.964	.891	.932	.961	.976
FE × GSF	-	-	-	-	.878	.924	.968	.983	.904	.950	.956	.981
Yield	-	-	-	-	-	-	-	-	.922	.947	.842	.890
Yield × GF	-	-	-	-	-	-	-	-	.937	.973	.876	.927
Yield × SF	-	-	-	-	-	-	-	-	.943	.962	.965	.990
Yield × GSF	-	-	-	-	-	-	-	-	.956	.972	.956	.983

even more spectacular with the r values ranging from 0.948 to 0.989; but GSF may not be readily available.

Table 8 dealing with the data of Pal *et al.* (1996)[11] on wheat seed invigoration treatments, besides showing significant improvement of post-storage germination of treated wheat seeds, demonstrates the manifestation of the beneficial effects of the treatments on the two vigour test parameters VERS and JSBS as well as on field emergence, yield and 1000-seed weight. However, differences within the treatments are less conspicuous and statistical significance is attributed mainly due to the shortfall in the performance of control seeds. One point that needs mentioning is the positive response of the 3-month-old wheat seeds to the soaking-drying treatment (that is not normally beneficial for high vigour seed implying that the seed, though with 100 per cent germination at the time of treatment, has been of high-medium vigour level and, as such, less responsive to dry treatments that are much more effective in very high vigour seeds[13, 14]). Otherwise, the data on correlations between various parameters are in conformity with the previous findings. As the r values are in general quite high, the effects of integration of GF, SF and GSF are subdued. Nevertheless, in wheat the effects of GF incorporation on r values are shown below to suggest that the nature of data and initial values of r in particular would determine the quantum of improvements.

Wheat Correlations (r values) (*vide* Table 8, n=9)

Variates		*Without GF Incorporation*	*With GF Incorporation into both the Variates*
X	*Y*		
VERS	FE	0.965**	0.979**
VERS	Yield	0.904**	0.955**
VERS	TSW	0.920**	0.968**
JSBS	FE	0.968**	0.986**
JSBS	Yield	0.896**	0.954**
JSBS	TSW	0.894**	0.960**
FE	Yield	0.888**	0.960**
FE	TSW	0.876**	0.964**
Yield	TSW	0.922**	0.972**

Seed invigoration treatments given to 3-month-old big and medium sized maize seeds[5] (cv. Vijay composite) with aspirin, *p*-aminobenzoic acid and calcium carbonate using very fine powders of the chemicals at the rate of 100 mg, 200 mg and 2 g, respectively (Table 9), have shown significant improvement in post-storage germination per cent and seedling growth in the laboratory test just before sowing (5 months of storage after the aforesaid 'dry' treatments). Field emergence and yield are also better in the treatments but 1000-seedweight shows no improvement in case of medium sized seeds; only *p*-aminobenzoic acid has increased TSW in plants grown from big sized seed. Germination and seedling growt has well as G

Table 9: The Relation of Presowing Germination Percentage, Seedling Growth and Bioassay Vigour Score (JSBS)of Big and Medium Sized Maize Seeds of different Seed Invigoration Treatments with their Respective Field Performances (Basic data from Bhat 2004[5]; Sur and Bhat, *unpublished*)

a) Basic data[5] and conversions

Sl. No.	*Treatment*	*Germination Per cent*	*GF*	*SE Per cent*	*SF*	*GSF*	*JSBS*	*FE Per cent*	*Yield/m² (g)*	*1000 SW (TSW, g)*
	Big sized seed									
1	Control	80^c	.80	42.9	.429	.343	50.7^b	67^c	131^b	190^b
2	Aspirin	86^b	.86	49.1	.491	.422	58.5^b	75^b	173^a	193^b
3	*p*-Aminobenzoic acid (PABA)	90^{ab}	.90	63.6	.636	.572	74.0^a	83^a	173^a	203^a
4	Calcium carbonate	93^a	.93	59.2	.592	.551	71.7^a	77^b	169^a	191^b
	Medium sized seed									
1	Control	59^c	.59	30.7	.307	.181	46.8^c	45^b	100^b	189^a
2	Aspirin	65^b	.65	33.9	.339	.220	56.1^b	56^a	122^a	189^a
3	*p*-Aminobenzoic acid (PABA)	72^a	.72	42.2	.422	.304	67.8^a	60^a	128^a	191^a
4	Calcium carbonate	68^a	.68	43.6	.436	.296	66.3^a	58^a	124^a	190^a

Note: Significant differences ($P = 0.05$) within the means in a column, as depicted by scoring with different letters, a,b,c, *etc.* are based on separate calculations for big and medium sized seeds ($n = 4$ for each seed size); for the estimation of the correlations, however, all the eight pairs of variates have been pooled together ($n = 8$).

b) Correlations (r values) of germinability with JSBS and field performance

X | Y	*GF*	*SF*	*GSF*	*JSBS*	*JSBS × GF*	*JSBS × SF*	*JSBS × GSF*	*FE*	*FE × GF*
Germination per cent	-	.927	.964	.600	.892	.832	.897	.974	.991
Seedling evaluation per cent	-	-	.990	.800	.973	.972	.988	.943	.949
Germination-cum-seedling evaluation per cent	-	-	-	.732	.958	.939	.979	.961	.978

X | Y	*FE × SF*	*FE × GSF*	*Yield*	*Yield × GF*	*Yield × SF*	*Yield × GSF*	*TSW*	*TSW × GF*	*TSW × SF*	*TSW × GSF*
Germination per cent	.946	.966	.948	.980	.948	.958	.602	.992	.910	.952
Seedling evaluation per cent	.991	.961	.907	.930	.984	.975	.762	.952	.997	.991
Germination-cum-seedling evaluation per cent	.994	.987	.933	.963	.990	.992	.726	.980	.983	.998

c) Correlations (r values) of JSBS with field performance as modified by incorporation of GF, SF and GSF

X | Y	*FE*	*FE × GF*	*FE × SF*	*FE × GSF*	*Yield*	*Yield × GF*	*Yield × SF*	*Yield × GSF*	*TSW*	*TSW × GF*	*TSW × SF*	*TSW × GSF*
JSBS	.648	.627	.739	.627	.639	.621	.750	.703	.590	.631	.795	.735
JSBS × GF	.901	.906	.951	.906	.884	.897	.958	.942	.684	.909	.962	.954
JSBS × SF	.861	.863	.944	.885	.833	.846	.941	.921	.758	.867	.971	.945
JSBS × GSF	.907	.945	.977	.945	.883	.907	.975	.968	.750	.925	.985	.981
FE (Field emergence)	-	-	-	-	.966	.975	.964	.961	.725	.987	.939	.961
FE × GF	-	-	-	-	.966	.988	.972	.978	.693	.998	.941	.973
FE × SF	-	-	-	-	.936	.956	.993	.989	.792	.974	.992	.998
FE × GSF	-	-	-	-	.937	.970	.978	.990	.745	.986	.960	.988

per cent and field emergence are very well correlated but the correlation of G per cent with jute seedling bioassay score, JSBS, gives a statistically not significant (at 0.05P) r value of 0.60. Incorporation of GF, SF and GSF into JSBS raises the value of r to 0.892**, 0.832**and 0.897**, respectively; while the correlation is very high between G per cent and FE and G per cent and yield (in both cases, there has been little effect of incorporation of morphological factors), that between G per cent and TSW is poor ($r = 0.602^{NS}$) but incorporation of GF, SF and GSF raises r values to highly significant levels. The relationship between seedling evaluation per cent and JSBS and TSW is better than that with G per cent; nevertheless, incorporations are effective, more so when r values are moderate. JSBS alone shows low positive correlation with FE (r= 0.648*) and yield (r = 0.639*) but the integration of GF, SF and GSF into JSBS improves the relationship with FE and yield making all of them significant at 0.01P; when both x and y variates are integrated with the same morphological factor all the resultant correlations show r values well over 0.9. As FE and yield are highly correlated without integration such large improvements are not seen. The GF integration effect into x and y on correlations involving JSBS and data on FE, yield and TSW is shown below.

Maize Correlations (r values) (*vide* Table 9, n=8)

Variates		*Without GF Incorporation*	*With GF Incorporation into both the Variates*
X	*Y*		
JSBS	FE	0.648*	0.906**
JSBS	Yield	0.639*	0.897**
JSBS	TSW	0.590^{NS}	0.909**
FE	Yield	0.966**	0.988**
FE	TSW	0.725*	0.998**

In Table 10, the effects of incorporation of GF, SF and GSF into unified physiological-biochemical score (UPBS) on the correlations with germinability and field performance of soybean cv. Soya max are shown using the basic data of Mukhopadhyay *et al.* (1997)[10] and Mukhopadhyay (2001)[9], and their transformations as reported earlier[3]. However, the basic data transformed into ECRS, VERS, LPRS and JSBS have been further revised on the basis of their respective statistically estimated reference values (the value at 100 per cent G as obtained by regression plus the full range value) to convert all the aforesaid scores into the revised per cent values *vis-à-vis* the estimated reference. Besides their similar scale values,all the above four vigour test scores have the same physiological basis namely, lipid peroxidation, that adversely affects membrane functions and produces inhibitory lipid hydroperoxides and volatile aldehydes, *etc.* While ECRS (based on leaching of electrolytes and resistance to increased electrical conductivity of seed leachate from aged seeds) gives a measure of membrane integrity, VERS and LPRS are both indices of the capacity of the seed to resist lipid peroxidation, JSBS gives

Table 10: Unification of Morphological Vigour Parameters (GF, SF and GSF) with Physiological and Biochemical Vigour Scores (ECRS, VERS, LPRS and JSBS) and Correlations of such Unified Vigour Scores with Field Performance of Soybean (Basic data from Mukhopadhyay *et al.*, 1997[10], Mukhopadhyay, 2001[9], Transformed data from Basu, 2004[3])

a) Transformed data, revised ECRS, VERS, LPRS, JSBS and their unification

Seed Lot No. (Treatments)	*GF*	*SF*	*GSF*	*ECRS*	*VERS*	*LPRS*	*JSBS*	*UPBS (Means)*	*UPBS × GF*	*UPBS × SF*	*UPBS × GSF*
T_1 (Cont)	.84	.415	.348	67.2	68.7	68.3	65.7	67.5	56.7	28.0	23.5
T_2 (Aspirin)	.89	.512	.456	72.8	73.6	73.9	77.1	74.4	66.2	38.1	33.9
T_3 (PABA)	.86	.498	.428	71.6	72.6	72.0	71.8	72.0	61.9	35.9	30.8
T_4 (BIP)	.86	.495	.426	70.5	72.7	73.0	75.3	72.9	62.7	36.1	31.1
T_5 (ME-D)	.87	.475	.413	69.6	72.8	70.2	69.2	70.5	61.3	33.5	29.1
Physiological-biochemical score means	-	-	-	70.34	72.08	71.48	71.82	-	-	-	-

LSD for physiological-biochemical score means (treatments pooled): not significant at 0.05P; treatment means (scores pooled *i.e.* UPBS means): 5.41 at 0.01P, 3.26 at 0.05P.

Seed lot No. (Treatments)	*FE*	*FE × GF*	*FE × SF*	*FE × GSF*	*Yield*	*Yield × GF*	*Yield × SF*	*Yield × GSF*	*TSW*	*TSW × GF*	*TSW × SF*	*TSW × GSF*
T_1 (Cont)	64	53.8	26.6	22.3	354	297	147.0	123.0	109.0	91.0	45.2	37.9
T_2 (Aspirin)	81	72.1	41.5	36.9	500	445	256.0	228.0	124.0	110.0	63.5	56.5
T_3 (PABA)	75	64.5	37.4	32.1	486	418	242.0	208.0	120.0	103.0	59.8	51.4
T_4 (BIP)	76	65.4	37.6	32.4	524	451	259.0	223.0	120.0	103.0	59.4	51.1
T_5 (MED)	71	61.8	33.7	29.3	374	325	178.0	154.0	116.0	101.0	55.1	47.9

b) Correlations (r values) of germinability with different physiological-cum-biochemical scores singly and on unification

X | Y	*ECRS*	*ECRS × GF*	*ECRS × SF*	*ECRS × GSF*	*VERS*	*VERS × GF*	*VERS × SF*	*VERS × GSF*	*LPRS*	*LPRS × GF*
Germination per cent	.837	.944	.823	.879	.858	.959	.826	.883	.756	.907
Seedling evaluation per cent	.961	.931	.996	.986	.962	.923	.998	.989	.942	.939
Germination-cum-seedling evaluation per cent	.968	.968	.992	.997	.970	.963	.993	.999	.937	.968

X \ Y	LPRS × SF	LPRS × GSF	JSBS	JSBS × GF	JSBS × SF	JSBS × GSF	UPBS	UPBS × GF	UPBS × SF	UPBS × GSF
Germination per cent	.799	.863	.763	.880	.808	.858	.827	.928	.819	.871
Seedling evaluation per cent	.994	.987	.902	.914	.977	.969	.967	.964	.996	.987
Germination-cum-seedling evaluation per cent	.986	.995	.908	.943	.975	.981	.972	.975	.991	.997

c) Correlations (r values) of amalgamated morphological and physiological-cum-biochemical scores with field performance and productivity

X | Y	*FE*	*FE × GF*	*FE × SF*	*FE × GSF*	*Yield*	*Yield × GF*	*Yield × SF*	*Yield × GSF*	*TSW*	*TSW × GF*	*TSW × SF*	*TSW × GSF*
GF	.850	.903	.839	.875	.479	.567	.594	.646	.834	.921	.822	.870
SF	.964	.951	.985	.977	.839	.876	.912	.927	.984	.965	.996	.988
GSF	.967	.966	.983	.983	.792	.840	.875	.899	.987	.991	.992	.997
UPBS	.997	.985	.993	.991	.888	.932	.943	.962	.992	.974	.983	.983
UPBS × GF	.983	.997	.976	.989	.771	.829	.850	.883	.973	.995	.961	.980
UPBS × SF	.984	.971	.990	.996	.864	.864	.931	.947	.995	.976	.999	.995
UPBS × GSF	.990	.988	.996	.998	.827	.827	.902	.925	.995	.993	.994	.999

an estimate of seed vigour physiologically as the bioassay material (germinating high vigour jute seeds) assays the volatile inhibitory substances emanated by germinating stock seeds of different vigour levels (volatiles that are presumably products of lipid peroxidation) thereby serving as an indirect estimate of lipid peroxidation. Thus, in view of the similar scale values and physiological basis, the unification of the four scores has been done simply by taking their arithmetic mean and the value has been termed as the unified physiological-biochemical score (UPBS). It is of interest to note that the revised individual score means, namely ECRS, VERS, LPRS and JSBS (treatments pooled) are not significantly different at 0.05P which would imply that one score if carefully done with adequate number of replications may serve the purpose (*i.e.* any one score may represent UPBS at least in soybean). As has been done with single scores, UPBS has been integrated with the morphological factors, GF, SF and GSF; similar amalgamation of GF, SF and GSF has also been done with the field performance data (FE, yield, TSW) to study the influence of such incorporations on the correlations involving the vigour scores and field performance.

A perusal of the correlations of the individual physiological-biochemical vigour scores singly or after integration of GF, SF and GSF, with germination per cent, seedling evaluation per cent and germination-cum-seedling evaluation per cent would reconfirm the complementary nature of the amalgamating components of the integrated variates. Broadly, the trend of correlations with UPBS is similar to those noted with individual vigour scores; for example, all correlations with seedling evaluation per cent or germination-cum-seedling evaluation per cent with ECRS, VERS, LPRS, JSBS and UPBS, with or without incorporation of GF, SF and GSF, show r values above 0.9; which is expected because of the absence of significant differences between the scores; however, some departures are noted when correlations are worked out with G per cent where the complementary nature of the integrating components (GF in particular) of the y variate is more pronounced because of initial moderate correlations of G per cent with ECRS, VERS, LPRS, JSBS or UPBS; in the aforesaid correlations, because of relative ineffectiveness of SF incorporation, that with GSF has also been less pronounced.

Field emergence is poorly correlated with yield ($r = 0.479^{NS}$) and GF, SF and GSF integration with yield improved r values but without statistical significance at 0.05P. G per cent is, however, significantly correlated with FE and TSW with or without integration of GF, SF and GSF. The seedling evaluation per cent as well as germination-cum-seedling evaluation per cent are better correlated with the field data than G per cent. UPBS, singly or upon integration withG F, SF and GSF, shows high correlation with FE and TSW and moderately good correlation with yield. GSF incorporation into UPBS as well as into FE, yield and TSW has given correlation coefficients of over 0.9; all the correlations of UPBS with FE and TSW are very highly correlated with r values ranging from 0.971 to 0.999, but those with yield have ranged between 0.771 and 0.962.

Although GSF incorporation into physiological-biochemical vigour tests or into UPBS (especially when UPBS alone shows poor to moderate r values) would give a more holistic rating of seed vigour, the fact remains that in practice, in agronomic field trials, only the germination percentage of the seed (as given on the seed testing certificate) is taken into consideration; sowing in the field is rarely preceded by germination and seedling growth evaluation tests. As such, SF and consequently GSF, would be unavailable. However, a researcher or even a seed producer with a little input of information in the matter may easily do that in a simple germination test with 5-6 sublots of the seed to get, besides the GF (*i.e.* G per cent ÷ 100), the reference seedling evaluation per cent and SF, and subsequently GSF (seedling evaluation per cent of a sublot × GF).

General Discussion

The data on vigour parameters presented in the tables in their respective reappraised versions have been briefly enumerated table-wise but certain issues require a broader discussion because of general importance in the context of quantification of seed vigour. These include the issues relating to (i) reference or benchmark value for vigour parameters, (ii) effect of data conversion and incorporation of morphological factors, GF in particular,on the statistically calculated deviations between the observed and estimated values of variates, and (iii) integration of different morphological factors into the field performance data. Several other points raised by a number of interested readers have been taken up under a separate subheading namely 'clarifications of other issues'.

1. Reference Value for Vigour Parameters

Ever since we started our work on seed invigoration nearly three and a half decades back, we have been keeping requisite samples of very high vigour seeds of the same cultivar as a reference material for the assessment of the extent of invigoration in the treated seeds after accelerated or natural ageing. The reference material in the earlier days used to be stored in paper packets in glass desiccators over fused calcium chloride but subsequently such seeds after proper drying were stored in the refrigerator at 2-5°C in rubber-stoppered wax-sealed glass vials and bottles (5-10 ml-capacity vials for small seeds and 50 ml-capacity bottles for bold seeds, generally for one time use) to serve as checks in germination tests, physiological-biochemical tests, *etc.* However, with the increase in the volume of work due to inclusion of a good number of cultivars of a wide range of crop seeds, several problems were encountered that included shortage of freeze space, mixing up of old and new accessions and other extraneous factors. Occasionally, the reference materials gave poorer performance that would not be expected from a really high vigour seed implying that more frequent replacements of the seeds were necessary. These developments and the fact that the highest vigour of an individual seed is attained at its physiological maturity previous to its usual harvest time, would suggest that a harvest fresh composite seed lot, though considered to be a very high vigour one, would show an average vigour score below 100 in the percentage scale and, as such, theoretically not-suitable to be an ideal reference material. The more purposeful alternative, a statistical approach as stated earlier in this article, to determine the reference or benchmark values has been taken up against the aforesaid background. We are now using the regression method with the germination per cent as the x variate and any vigour parameter as the y variate to determine firstly the value of y at 100 per cent germination (100 per cent G) to which the full range value of y variate (the difference between the maximum and minimum, also considering the value at 100 per cent G, if that is the highest) is then added to get the reference value for expressing all the individual values in the y variate as the respective percentages of the estimated reference value. That has been the basis of getting the values for seedling evaluation per cent and SF, germination-cum-seedling evaluation per cent and GSF, as well as the revised physiological-biochemical vigour scores, *viz.*, ECRS, LPRS, VERS, JSBS, *etc.* enabling us to bring the values on similar footing and scale values as shown in the different tables.

The most important consideration in the statistical estimation of the reference value is the addition of full range values to the regressionally estimated value for 100 per cent G, the addition that is necessary to keep the seedling evaluation as well as physiological-biochemical vigour scores in the percentage scale of 0-100, as in case of germination; otherwise, if only the value at 100 per cent G is used as the reference, some of the lots may sometimes show scores higher than 100, exceeding the scale. As indicated above, the basic fact is that the highest vigour (100 per cent) would be attained at physiological maturity(PM) of the seed while still on the mother plant and since that time it would be a downhill journey for seed vigour because of field weathering and other uncontrollable factors while100 per cent germination may be retained not only till harvest but also for a prolonged period depending on post-harvest processing, drying and especially storage conditions (if unfavourable both vigour and viability would be adversely affected). The differences between individual seeds in vigour status within a seed lot would be attributable among other reasons, to inter-plant and even intraplant differences in PM of individual seeds. We may mention here that studies with a large number of crop seeds over a long time have given convincing data showing moderate but statistically highly significant ($P = 0.01$) differences in morphological as well as physiological-biochemical vigour parameters in seed lots of the same cultivar grown in the same plot of land and all showing 100 per cent germination in the laboratory. (Incidentally, when all the lots show 100 per cent germination and correlation cannot be worked out with G per cent as the x variate because of statistical problem, we simply take the highest value of y and add the full range to that to get the reference value). In any case, the main issue is how to get the apparently elusive reference values for vigour parameters at physiological maturity; the regression based approach to get y value at 100 per cent G and the addition of full range value to that has been an attempt to estimate the approximate value of the different parameters at PM, especially seedling evaluation per cent, a very good indicator of seed vigour. However, the same has been viewed as 'too strict' by some of our colleagues in agronomy and seed production. In fact, a perusal of our data on seedling evaluation per cent based on the reference values calculated by adding the full range value would show scores ranging from 65-80 per cent, often as low as 40-50 per cent only for, seed lots showing apparently quite good germination but what bothers the producers is the possibility of viewing such mathematical values as signs of poor quality in respect of vigour status. The same can be avoided if we make a compromise and accept addition of only half of the range value, thereby reducing substantially the magnitude of the reference value and consequently raising the vigour scores by 10-15 per cent or so. Because the range value is generally more in less vigorous seed stocks compared to high vigour seeds, the former would benefit more from that approach. Although it would be a good compromise between theory and practice one must keep in mind that sometimes as much as 20-25 per cent loss of vigour may take place in the post-PM pre-harvest stages due to adverse weather conditions, and in the tropical and subtropical countries in particular, in the subsequent post-harvest stacking in the field and threshing floor, inadequate drying (predominantly sun-drying) and relatively poor storage conditions that may lead to even greater loss of seed vigour (and often viability) than in the advanced agriculture of the western countries. Perhaps, for the latter, the addition of half range value that would also ensure vigour score values lying within the percentage scale(0-100) as mathematically, the divisor should exceed the highest value of the data set in y variate) and would be good enough for the 'good quality' point of view of the seed producer, may therefore be considered favourably (the same may however, be viewed as 'too liberal' by orthodox seed physiologists). The issue is, as yet, an open one; initially, we thought of adding the value of standard deviation, or 2 × SD, but the same would involve another minor calculation, after including the value for 100 per cent G in the y set of variates. It needs reminding that either of the two approaches will not show any difference in correlation between the x and y variates as the correlation coefficient would not be influenced by origin and scale of data. Nevertheless, the second option would be commercially more advantageous.

2. Effect of Data Conversion and Incorporation of Factors on the Deviations between Observed and Estimated Values

The discussion on the effect of conversion of negatively correlated vigour parameters to their positive versions (scores) on the sum of least squares, $\Sigma(y-\hat{y})^2$ and $\Sigma(x-\hat{x})$ and their corresponding mean deviations expressed as percentages of their respective means (*i.e.* $\overline{y}$ and $\overline{x}$) is made with the data of Wilson and McDonald (1986)[16] on electrical conductivity of soybean (*vide* Table 5), citing the same as an example, and conclusions drawn therefrom should broadly hold good for other scores. The salient points on the influence of data conversion and especially GF incorporation on the deviations are enumerated below.

(i) The value of r is identical for the correlations between germination per cent (x variate) and electrical conductivity (y variate) expressed either in μmho or mmho which is expected as r is independent of origin and

Data on Germination and EC of Soybean by Wilson and McDonald (1986)[16] (*vide* Table 5)

Correlation-regression analysis involving G% (x variate) and electrical conductance and their various transformations (y variate)

Sl.No.	*Y Variates*	$\overline{x}$	$\overline{y}$	*r*	$r^2 \times 100$	*Sample Covariance (n=5)*	$\Sigma(y-\hat{y})^2$	$\Sigma(x-\hat{x})^2$
1	EC µmho	79.4	32.8	-0.797	63.5	-979	122.97	412.5
							(15.12)	(11.44)
2	EC mmho	79.4	0.0328	-0.797	63.5	-0.979	1.229×10^{-4}	412.5
							(15.12)	(11.44)
3	ECRS (µmho based)	79.4	67.2	0.797	63.5	97.9	122.97	412.5
							(7.38)	(11.44)
4	ECRS (mmho based)	79.4	96.7	0.797	63.5	9.79	1.229	412.5
							(0.51)	(11.44)
5	ECRS (µmho based) × GF	79.4	54.1	0.973	94.6	190.51	46.33	41.04
							(5.63)	(3.61)
6	ECRS (mmho based) × GF	79.4	76.9	0.999	99.9	180.42	0.459	0.453
							(0.39)	(0.38)

Note: i. The values in parentheses below the sum of least squares represent mean deviations expressed as percentages of respective means of x and y variates (*i.e.*, $\overline{x}$, $\overline{y}$); those values have been calculated for easy understanding of the effect of data conversion and GF incorporation by general seed researchers; for the purpose, the sum of least squares has been divided by n (=5), the mean sum of squares is square-rooted (√) and expressed as percentages of respective x and 3 in the row. ii. The estimated values of x and y sets, x and y have been taken directly from the data display of *Casio fx-82 MS* calculator used for correlation-regression analysis and other calculations. iii. In the example cited above (vide Table 5), the value of ECRS (original) and ECRS (revised) are slightly different as the reference value is 100.6 (see table-wise calculations in the text); the calculations are based on ECRS (original) values and although the correlations would be exactly identical, marginal differences may be noted in deviations in the revised version. [For example in the regression equation involving G per cent and ECRS (revised) based on EC mmho, the mean deviations as per cent of $\overline{y}$ and $\overline{x}$ are respectively 0.513 per cent and 11.54 per cent against 0.51 per cent and 11.44 per cent of $\overline{y}$ and $\overline{x}$ respectively in G per cent versus ECRS (original, mmho based), as shown in the fourth row of the table above].

scale. The sum of least squares (*i.e.* squares of deviations) are different commensurate with their respective units but the mean deviations between the observed and estimated values of y, when expressed as the respective percentages of $\bar{y}$ for EC μmho and $\bar{y}$ for EC mmho are exactly the same.

(ii) The correlations of germination per cent with ECRS derived from EC μmho or EC mmho are the same as above but with opposite signs (positive instead of negative); the sum of least squares $\Sigma(y - \bar{y})^2$ for EC μmho and that for ECRS derived from EC mmho are equal, but the mean deviations expressed as percentages of respective means are different simply because the means are different. The deviations (sum of least squares as well as mean deviation per cent) are spectacularly reduced in ECRS based on EC mmho (mean deviation 0.51 per cent of $\bar{y}$ against 7.38 per cent of $\bar{y}$ for ECRS based on EC μmho).

(iii) Incorporation of GF into ECRS based on EC mmho, is more effective in improving the correlations with germination per cent than the corresponding correlation between G per cent and ECRS (μmho based) × GF and commensurate with that the sum of least squares and mean deviation per cent are much less in GF incorporated mmho based ECRS (mean deviation per cent for the y variate being 0.39 per cent in mmho based ECRS × GF against 5.63 per cent in μmho based ECRS × GF). As such, in practice, conversion to ECRS (mmho based) is recommended for getting better predictive values, especially when it is not intended to just compare the relative efficacies of different scales for ECRS conversions but the main goal is to make precise and effective comparisons between different seed lots in respect of vigour status.

(iv) None of the conversions from negative to positive correlations involving the y variate affects the sum of least squares or mean deviation per cent values of the x variate (*i.e.* germination per cent). But GF incorporation into ECRS by multiplication greatly reduces the value of the sum of least squares, $\Sigma(x - \bar{x})^2$, and mean deviations expressed as percentages of $\bar{x}$. The minimization of deviations is more emphatic when GF is incorported into mmho-based ECRS than in the corresponding μmho-based ECRS being 0.453 for $\Sigma(x - \bar{x})^2$ in the former against 41.04 in the latter (with respective mean deviations of 0.38 per cent and 3.61 per cent). Therefore for greater predictive efficiency for both x and y variates, conversion of basic data as well as GF incorporation by multiplication, are recommended. In practice, it would be simpler just to use three digits after the decimal point for necessary ECRS conversion as shown for the following arbitrary basic data : 9, 45, 81, 126, 288; just converted to 0.009, 0.045, 0.081, 0.126 and 0.288 without considering their units, *etc.* for the aforesaid improved regression equation.

(v) The conclusions for other vigour scores, such as LPRS and VERS (JSBS is originally positive) would be broadly similar but one can simply work with respective optical density(OD) values (within limits)[3]; otherwise, irrespective of the presence or absence of a decimal point, or their basic units and expressions on the basis of per 100 seeds, per seed or per g of seeds, *etc.*, the same technique as shown above with the arbitrary basic data would be applicable for all conversions from negatively correlated vigour parameters (*vis-à-vis* G per cent)to their positive versions (a conversion that does not change the quantitative relation between the two variates) with resultant improved regressions, especially following incorporation of morphological factors. Regarding the incorporation of SF and GSF also,the conclusions would be broadly similar to those made with GF incorporation.

3. Integration of different Morphological Factors into Field Performance Data

A very pertinent point that may be raised is the logic behind integration of GF into field data such as FE, yield and TSW, as the data ultimately come from only viable seeds that have established themselves in the field, may be at differing vigour levels, hence accepting the validity of SF integration into those data. The answer to that would be as follows:

(i) Germination per cent, seedling evaluation per cent and germination-cum-seedling evaluation per cent are themselves very useful and effective parameters of seed vigour and the respective factors derived from them namely, GF, SF and GSF exactly reflect their properties and have been so changed that the integration of the same with different scores would keep the values within the percentage scale. GF, the most important of the three factors, besides eliminating the contribution of non-viable seeds in a vigour score, very effectively acts as an alternative to SF when the latter is not available (the nearly universal high positive correlation of GF and SF is a testimony to that) and the fact remains that for most seed lots, SF is not available; although the same is relatively simple to work out, only a few would bother about taking the added trouble; SF, and as a result GSF, an excellent index of seed vigour, are not readily available. In the circumstances, GF, in practice becomes the only alternative to serve as a proxy and is always available from the seed testing certificates of commercial seed lots.

(ii) Theoretically, GF incorporation into field data may not be necessary when plant population has been kept at a fixed level through a proportionate increase in seed rate,implying a kind of GF adjustment in the very beginning, usually not taking into consideration tillering, branching and other compensatory mechanisms that may be operative in field trials. In that case, incorporation of SF into field data would serve the necessary purpose. One characteristic of those data would be similar correlations calculated on per unit area or per plant basis (*e.g.* r value for yield/plant nearly equal to yield/m^2).

In practice, however, achieving such a desired population level in the field may be a difficult proposition without sacrificing the benefits of optimum inter-plant spacing and foregoing the need for expensive thinning and gap-filling operations, simply by seed rate adjustment alone. Further, the reduced vigour of surviving seeds of a seed lot with a poor germinability cannot in most cases be compensated by the aforesaid seed adjustment. Nevertheless, SF incorporation into the data of a fixed population study would take care of that, but again keeping in mind the variable plant population scenario in commercial seed production, incorporation of GSF is recommended and if SF cannot be worked out (hence GSF also) the next best alternative, GF incorporation is suggested.

A critical perusal of the summarized versions of selected correlations involving field emergence and yield of soybean (Table 6), sunflower (Table 7) and maize (Table 9) and one vigour score of each, with or without integration of GF, SF and GSF, is given hereunder to substantiate the aforesaid arguments in favour of integration.

Soybean Correlations (r values) (*vide* Table 6, n = 7)

Field Emergence (FE)			*Yield of Seed*		
Variates		*r values*	*Variates*		*r values*
X	*Y*		*X*	*Y*	
ECRS	FE	0.792*	ECRS	yield	0.588^{NS}
ECRS × GF	FE	0.819**	ECRS × GF	yield	0.763*
ECRS × SF	FE	0.793*	ECRS × SF	yield	0.918**
ECRS × GSF	FE	0.796*	ECRS × GSF	yield	0.861*
ECRS × GF	FE × GF	0.943**	ECRS × GF	yield × GF	0.959**
ECRS × GF	FE × SF	0.890**	ECRS × GF	yield × SF	0.823**
ECRS × SF	FE × SF	0.939**	ECRS × SF	yield × SF	0.976**
ECRS × GSF	FE × GF	0.915**	ECRS × GSF	yield × GF	0.982**
ECRS × GSF	FE × SF	0.925**	ECRS × GSF	yield × SF	0.926**
ECRS × GSF	FE × GSF	0.970**	ECRS × GSF	yield × GSF	0.986**

Sunflower Correlations (r values) (*vide* Table 7, n = 6)

Field Emergence (FE)			*Yield of Seed*		
Variates		*r values*	*Variates*		*r values*
X	*Y*		*X*	*Y*	
VERS	FE	0.669^{NS}	VERS	yield	0.789*
VERS × GF	FE	0.651^{NS}	VERS × GF	yield	0.829*
VERS × SF	FE	0.560^{NS}	VERS × SF	yield	0.809*

VERS × GSF	FE	0.567^{NS}	VERS × GSF	yield	0.833*
VERS × GF	FE × GF	0.850**	VERS × GF	yield × GF	0.919**
VERS × GF	FE × SF	0.918**	VERS × GF	yield × SF	0.925**
VERS × SF	FE × SF	0.960**	VERS × SF	yield × SF	0.959**
VERS × GSF	FE × GF	0.769*	VERS × GSF	yield × GF	0.899**
VERS × GSF	FE × SF	0.939**	VERS × GSF	yield × SF	0.965**
VERS × GSF	FE × GSF	0.959**	VERS × GSF	yield × GSF	0.972**

Maize Correlations (r values) (*vide* Table 9, n = 8)

Field Emergence (FE)			*Yield of Seed*		
Variates		*r values*	*Variates*		*r values*
X	*Y*		*X*	*Y*	
JSBS	FE	0.648*	JSBS	yield	0.639*
JSBS × GF	FE	0.901**	JSBS × GF	yield	0.884**
JSBS × SF	FE	0.861**	JSBS × SF	yield	0.833**
JSBS × GSF	FE	0.907**	JSBS × GSF	yield	0.883**
JSBS × GF	FE × GF	0.906**	JSBS × GF	yield × GF	0.897**
JSBS × GF	FE × SF	0.951**	JSBS × GF	yield × SF	0.958**
JSBS × SF	FE × SF	0.944**	JSBS × SF	yield × SF	0.941**
JSBS × GSF	FE × GF	0.945**	JSBS × GSF	yield × GF	0.907**
JSBS × GSF	FE × SF	0.977**	JSBS × GSF	yield × SF	0.975**
JSBS × GSF	FE × GSF	0.945**	JSBS × GSF	yield × GSF	0.968**

The data on wheat (Table 8) show very high correlations (all r values irrespective of integration or no integration are significant at 0.01P); as such, the effect of the different integrations are not apparent enough to substantiate the oft noted approximate equivalence of GF, SF and GSF incorporations.

Although major emphasis has been given to GF, the role of SF that accounts for the variability in the performance of the surviving seedling population in the field also needs especial emphasis. GSF, being an integration of GF and SF would expectedly be influenced by both. As noted earlier in the section on table-wise findings and their interpretations, the incorporation of SF into the field data often shows similar effect to that of GF, sometimes even marginally better than the latter. While for field emergence of soybean, the correlation with ECRS × GF is better than with ECRS × SF, the correlation between the former (FE) and yield is lower than with ECRS × SF. In maize also, SF incorporation into JSBS is nearly as effective as GF for field emergence and yield. Similarly, SF incorporation into both x and y variates of soybean, sunflower and maize is also as good as that with GF incorporation (with both kind of incorporations, the r values are significant at 0.01P). The coefficients of determination, multiplied by 100, for the aforesaid correlations in all the three crops following SF integration would account for 88.2-95.3 per cent coverage by explained variations out of a total estimated variation of 100 per cent (explained as well as unexplained) attributable to a great extent to SF incorporation into the two variates.

While the direct effect of GF (besides its indirect influence on SF), through elimination of the contribution of non-viable seeds, is more discrete and easily understandable, SF accounts for the continuous variability in the surviving seedlings of a seed lot that would finally establish themselves in the field and perform as a crop. SF, as such, would require greater comprehension and a more prudential approach in respect of interpretation of its role in crop performance; the same would partly hold good for GSF of which SF is a constituent.

Clarifications of other Issues

(*Issues communicated personally in writing or verbally by a number of interested readers*)

Although the major points and issues raised by many readers have been dealt with while discussing the findings table-wise as well as in the general discussion, a few that are considered relevant but could not be fully accommodated there, often due to contextual incompatibility, are taken up hereunder for further elucidation.

(i) The necessity of conversion of negatively correlated vigour parameters like electrical conductivity to positive values has been questioned in view of the fact that the strength of the correlations (as expressed in r values) is more important than their signs (+ or –). The answer is multifaceted; firstly and conceptually, vigour itself has a distinct positive connotation and is positively related to germinability. Conversion of all vigour parameters to positive values in the percentage scale would be ideal as that is commensurate with germination per cent, the most important and useful measure of seed quality of a crop cultivar. Secondly, comparison of vigour test values from different vigour tests (not their 'r' values) would be more precise if they are converted to a positive percentage scale. Lastly, any attempt to unify different vigour test values, some of which are positively related to germination, some negatively related, would necessitate uniformity in scale values and signs and units; otherwise, it would be difficult to combine them and integrate with morphological factors, if necessary, keeping the basic parameters of positive and negative values as such.

(ii) The dependence on correlations and regressions for quantification of seed vigour has been questioned with the suggestion for other alternatives. We admit that in the previous paper as well as in the present paper we have depended so much upon that statistical procedure. However, the same is exactly not for quantification but for checking the validity of the quantified scores, especially when the scores are integrated with morphological factors or unified with other scores and then integrated with morphological factors. Seed researchers all over the world unanimously undertake correlation-regression analysis to check up the efficiency of any vigour test, especially calculating the correlation coefficients *vis-à-vis* germination per cent and field emergence and some even go up to check correlations with yield and yield-determining factors. The procedure is also followed at the ISTA level in which seed testing laboratories of member countries may be involved (see Tables 2, 3, 4). Till now, there is perhaps no alternative. What comes to the forefront is the integration of related factors to reach at a more meaningful correlation as noted in case of GF incorporation for eliminating the masking effect of non-viable seeds. Our dependence on correlation-regression equation to highlight the efficacy of integration of morphological factors with seed vigour scores and also with field performance data stands justified in the absence of any viable alternative.

(iii) A very important point that has been raised is the fact that usually the dependent variate (y) in a linear regression is not transformed; however, to improve correlations, the independent variate (x) may be transformed.

A fundamental issue in this connection is whether germination per cent should be accepted as the independent variate (x) and vigour test parameter or score as the dependent (y) variate. Physiologically, however, germination depends on good membrane functions and that is why lipid peroxidation which disrupts membrane functions would adversely affect germination, increased electrical conductivity that is negatively correlated with germination is also attributable to lipid peroxidation, so is the case with volatile aldehyde emanation which is regarded as a manifestation of lipid peroxidation. Most positively correlated vigour parameters can be similarly looked upon as manifestations of membrane functions and consequently attributable to lipid peroxidation. As such, it logically appears that the vigour parameters or scores should be treated physiologically as independent variates (x) and germination as the dependent variate (y), but that would be against the universally accepted convention of treating germination per cent as the x (independent) variate in all statistical analyses on the subject. This is because in correlation-regression analysis, the strength of the correlation is the major point of consideration, not the cause-effect relation between the variates; the independent variates (x) is the one which is data wise more stable and useful for the prediction of y based on the regression of y on x. In fact, there are many other examples in which the relatively less important but data-wise more dependable variate has been given the independent status irrespective of its physiological or biological role. Nevertheless, quite often it is very difficult to distinguish between the two variates; as for example, if one wants to study the correlation between the two vigour parameters ECRS and VERS (or between ECRS and JSBS), it is practically immaterial to assign any of them any particular status. Technically speaking, both of them are almost equally prone to similar sort of errors (although ECRS is more quickly and easily obtained than VERS) and it would be difficult to assume which one would be more stable and relatively error free than the other, the statistically conventional basis of treating one as the independent variate (x) and the other as the dependent variate (y). A reasonable way out would be to treat them on par as just 'related' variates and whichever comes first be looked upon as x and the other as the y variate in the linear regression analysis. Actually, in such linear bivariate regressions, the value of r (correlation coefficient) and consequently that of the coefficient of determination (r^2) would be exactly the same irrespective of whether it is based on regression of y on x, or of x on y. The estimated values of $\hat{x}$ and $\hat{y}$ from either of the equations would, however, be different depending on the value of r but are broadly of similar magnitude when r values are very high, close to 1. The matter assumes added importance in the present context because removal of dependent status of ECRS would enable integration with morphological factors like GF by multiplication (although in ECRS × GF it is a necessary arithmetic step to eliminate the contribution of non-viable seeds). Similarly, there should not be any objection to integration of morphological factors like GF, SF or GSF on both x and y variates to improve the correlations. We would, however, re-emphasize that the incorporation of all the aforesaid morphological factors are simple arithmetic conversions of data (applicable to both the variates) for a holistic approach to the relationship of seed vigour and related germination and field data and, therefore, are different from the conventional transformations such as log, power, exponential, inverse, *etc.* As shown in all our tables (Table 1-10), such simple and logical arithmetic conversions would greatly improve the predictive efficiencies of the regression equations based on several easy-to-do presowing laboratory tests.

(iv) Lastly, in response to a more recent query by our young researchers as to why GF incorporation on both variates would have so much influence on correlations (also, why similar multiplication of both variates by SF and GSF would improve 'r' values), we would like to state the following facts known to mathematicians and statisticians that may need some recapitulation by seed researchers.

In the linear correlation-regression analysis, the value of 'r' is not influenced by multiplication or division of x and y variates by the same number as that would not change the basic relationship of the variates and consequently 'r' value would be exactly the same as before. Thus the correlation of G per cent and ECRS would be identical to that of GF and ECRS, as GF is derived simply by dividing G per cent by 100; similarly, if all ECRS values are multiplied by the same number, the 'r' value will remain unchanged. For example, the correlation between G per cent and ECRS × GF with the data of Wilson and McDonald[16] gives a value of r = 0.999** (Table 5, n = 5), it would be unchanged (r = 0.797**) if ECRS values of the five seed lots are multiplied by the same GF value but what actually has happened is the fact that the ECRS values of five seed lots have been multiplied by five significantly different GF values of the five lots namely 0.93, 0.79, 0.92, 0.71 and 0.62. In case of correlations between the two vigour parameters, ECRS and VERS (Table 5,c), the incorporation of GF into both variates has increased the value of r from 0.608 in ECRS and VERS to 0.990 in ECRS × GF and VERS × GF because the integration with significantly different GF values (though for the same seed lot GF is the same for both ECRS and VERS) has changed the original relationship between x and y showing a substantial improvement. The same explanation would hold good for similar integrations involving SF and GSF. The complementarity of vigour scores with morphological factors has been partly covered in the deliberations under Table 5 highlighting the fact that the correlation coefficients improve substantially on integration when the basic correlations between the variates are low to moderate.

Mathematically, if we treat (p × GF) as the x set of variates and (q × GF) as the y set of variates, both x and y are just mutually related, neither dependent on another and p and q are two positively correlated vigour parameters, then the individual (non-integrated)correlation between the GFs of the x set and GFs of the y set (that are the same for both sets) would be exactly 1.0; a r value of the same magnitude would be obtained if the GFs are replaced by the corresponding NVFs (*i.e.* NVFs of x set and NVFs of y set, giving a r value of 1.0). Obviously, this highest possible r value between the GFs of x and y, has a great influence on the correlation between (p × GF) and (q × GF). The point of consideration is why the resultant r value instead of staying somewhere around the midpoint of 0.608 and 0.990 (as in the aforementioned example[16]), would always tend to be closer to 1.0, the theoretical maximum limit of the r value. Depending on the nature of the data, the correlation of p and q components of the two sets may be low to high. If there is no statistically significant difference within the different p values and for that matter within the different q values, we would expect the r value to be close to 1 but if there are significant differences within the p values and within the q values, the correlation between the two integrated sets of variates, x and y, would depend on how the integration of the two individual components changes the relation between the two sets of variates, an interaction, perhaps leading to greater normalization of the data that is duly reflected upon the resultant coefficient of correlation. The observed complementarity or reciprocity of the integrating components (having intra-component statistically significant differences) is due to their acting conjointly in the same direction and manner to substantially improve the correlations but that effect is mathematically limited to a value of r = 1.0, being the theoretical maximum value of the correlation coefficient. The above interpretation holds good for SF as well as GSF, in the latter case the r value in ECRS × GSF and VERS × GSF is 0.998 (see table 5). However, the question of complementarity and the fact that r values of duly integrated correlated variates tend to remain close to 1.0 would need further mathematical elucidation by statisticians and bio-metricians interested in the subject.

We conclude by reminding our young researchers that in the advanced high technology based agriculture of future, in the backdrop of all out efforts to maximize crop productivity from limited land, water and energy resources, would by necessity require use of seed not only with 100 per cent germination but also to achieve the goal of 'one seed one established plant in the field' to avoid the tedious labour-intensive job of thinning and gap-filling. Besides economizing on the cost of expensive seeds by curtailing (optimizing) seed rate and consequent saving on post-sowing cultural practices, the grower would be assured of increased crop productivity. Commercial seed producers of a highly competitive seed industry would have no choice but to go for production of super quality seeds for sowing, and in that event only the seedling factor (SF), which mathematically would be equal to GSF (100 per cent germination, hence GF = 1), will be the decisive factor for incorporation into the relevant vigour test scores and other associated data. The scenario would surely be a challenging one but a worthy one at that, for wilful acceptance and ultimately winning over!

Acknowledgments

My sincerest thanks are due to Professor T.M. Das, Editor, Indian Biologist for taking keen interest in seed vigour quantification and persistently advising me to write up this rejoinder to the earlier paper and for very critically scrutinizing the various queries and my response to the same through additional information and detailed clarifications. Professor Das particularly suggested that the procedural aspects and specifically the calculations should be given in such a manner that the researchers and even the young students of seed science and technology would understand the matter without difficulty.

I express my indebtedness to all the authors, including my co-authors, associates and students, for using their data on different materials (basic data) for recalculation and reappraisal following integration of morphological factors into physiological-biochemical vigour test scores and field data.

I am most thankful to many of my erstwhile colleagues in several relevant departments of the Calcutta University (CU) for their valued opinions and suggestions. Professor S. P. Mukherjee (Department of Statistics) has been helping me for more than a year going critically through published and unpublished data of our seed invigoration experiment sand giving most valuable suggestions. Professor

S. C. Bose (Applied Mathematics), Professor P. K. Sengupta (Pure Mathematics), Professor A. B. Banerjee (Biochemistry), Professor A.K. Banik (Chemical Engineering), Professor I. B. Chatterjee and Dr. G. Chakravarti of Guha Centre of Biotechnology, CU, have given useful comments and suggestions on the issue.

My colleagues, associates and students in the Agriculture Faculty have helped me in various ways through discussions, seeking clarifications on procedural aspects, calculations and various other aspects, especially the relevance of integration of morphological factors into vigour scores. I would specially thank Dr. A. K. Mandal, Dr. K. Sur, Dr. P. Pal, Dr. B.K. De and Dr. S. Kundagrami for their views, and particularly Dr. S. Chakravarti for his specific queries on the procedural aspects followed to free the vigour test scores from variations in units, methods, *etc.* prior to integration (which, I now feel, should have been given in greater details in the previous paper itself), Dr. A. Mukhopadhyay and Mr. S.A. Bhat for using some of their primary data on soybean and maize seed invigoration, respectively.

Most sincere thanks are due to Professor J. K. Ghosh, former Director of the Indian Statistical Institute, who in spite of his very busy schedule in India and abroad, has given a written critical appraisal of the issues raised by me in the paper on quantification of seed vigour and giving me a long patient hearing on my views on the subject. I have been greatly benefitted by his highly constructive suggestions and although his advice to take up 'principal component analysis' could not be implemented in this paper, I would remain so very grateful for his help and cooperation.

Scientists from several other universities, institutes and organizations who have given their views include Professor M. G. Som, Professor D. K. Bagchi, Professor N. Mukherji,Professor S. K. Sanyal, Professor P. K. Das, Dr. A. Bandyopadhyay all from the Agricultural University of the State, BCKV; Dr. M. Dadlani (IARI, New Delhi) for her keen interest,valued comments and on her own volition, circulating the previous paper to many experts in the field in India and abroad and, Dr. B. Punjabi (Guru Nanak College, Mumbai) and Professor K. Ramamoorthy (TNAU, Madurai), Professor S. K. Mukhopadhyay (Viswa Bharati, Santiniketan), Professor S. S. Sarkar (Higher Education Department, Govt. of W. Bengal), Dr. S. P. Shrivastava (Rockville, USA), Dr. S. Pradhan (Bethesda, USA), Dr. A. K. Lahiri (Indian Forest Service, now retired), Dr. H. N. Samaddar (State Agricultural Service, now retired) for their views. I thank Dr. R. K. Chattopadhyay (BIT, Mesra, Kolkata Extension Centre) for active help and cooperation and Professor P. N. Ray (Member, State Planning Board) for his keen interest in seed research and helping us in all possible ways. I would thank Mr. S. Ghosh, a bright young electrical engineering student for his rather unusual interest in seed vigour quantification and raising some interesting points.

Lastly, I record my sincere gratitude to Professor A. K. Sharma, eminent scientist and former Ghosh Professor of Botany, CU, Professor T. K. Bose, renowned horticulturist and former Director of Research, BCKV and Dr. S. L. Basak (Ex-Director, CRIJAF, ICAR) for their valued comments and words of inspiration, especially when doubts and confusions due to certain developments on the issue made me quite uneasy. I recall with gratitude Dr. Basak's advice to boldly go ahead with my idea of GF incorporation into vigour test scores in the previous paper in spite of its non-conventional stigma.

Although I have been helped by so many persons in various ways, I personally own full responsibility for the views expressed in this paper, some of which may still be considered unusual, non-conventional and controversial. I would keenly look forward to further critical comments and suggestions, specifically by senior as well as by junior researchers in the field. As much of the material covered in this discourse is being processed for inclusion in a detailed monograph on 'Seed Invigoration', it would indeed be of great advantage to the author, if the paper is scrutinized most critically and the comments and suggestions, if any, passed on to the author for further deliberations. Needless to say, such help from the readers would be most gratefully acknowledged in the said monograph.

REFERENCES

1. Andersen, A. M., Hart, J. R. and French, R.C. *Proc. Int. Seed Testing Association*, **29**(1), 81-96(1970).
2. Basu, R. N. *Seed Sci. Technol.*, **22**, 107-126 (1994).
3. Basu, R. N. *Indian Biologist*, **36**(2), 1-23 (2004).
4. Basu, R. N., Kuchlan, P. and Sur, K. *Indian J. Exp. Biol*, **28**, 166-170 (1990).
5. Bhat, S. A. *M. Sc. (Ag.) Thesis* (supervisor, K. Sur), University of Calcutta, India (2004).
6. Ching, T. M. and Schoolcraft, I. *Crop Sci.*, **8**, 407-409 (1968).
7. Hailstones, M. D. and Smith, M. T. *J. Plant Physiol.*, **133**, 452-456 (1988).
8. Mandal, A. K., De, B. K., Saha, R. and Basu, R. N. *Seed Sci. Technol.*, **28**, 349-355 (2000).
9. Mukhopadhyay, A. Ph.D. Thesis, (supervisor, R. N. Basu; associate supervisor, P. Pal), Universityof Calcutta, India (2001).
10. Mukhopadhyay, A., Pal, P. and Basu, R. N. *Indian Agriculturist*, **41**, 79-85 (1997).
11. Pal, P., Ray, I., Mitra, R. and Basu, R. N. *Indian Agriculturist*, **40**, 219-233 (1996).
12. Perry, D. A. *Seed Sci. Technol.* (19th ISTA Congress, Vienna, 1980), **9**, 115-126 (1981).
13. Rudrapal, D. and Basu, R. N. *Indian Agriculturist*, **46**, 1-9 (2002).
14. Rudrapal, D. and Basu, R. N. *Seed Sci. Technol.*, **32**, 205-212 (2004).
15. Sur, K. and Basu, R. N. *Seed Sci. Technol.*, **18**, 661-671 (1990).
16. Wilson, D. O. (Jr.) and McDonald, M. B. (Jr.). *Seed Sci. Technol.*, **14**, 259-268 (1986).

Accepted on August 9, 2006.

Annexure No A-3

Eco-sustainable Agriculture and Climate Change

The United Nations and World Bank sponsored mega project, "International Assessment of Agricultural Knowledge, Science and Technology for Development", popularly known as IAASTD, chaired by Professor Robert Watson (a former Chairman of IPCC and a co-sharer of Nobel Prize to IPCC) released on April 15, 2008 the summarized Synthesis Report of four-year-long international effort by over four hundred experts and nearly one thousand researchers and reviewers. The final report of 2500 pages is considered as a ground breaking one with the motto "Business as usual is not an option" has been a most timely warning against the out of conventional industrial agriculture in this era of global climate change whereby a shift to eco-sustainable agriculture would be certainly unavoidable. Agricultural policies need to be developed for reduction of greenhouse gases (GHGs) as well as adaptation to climate change. Besides being closer to carbon neutral, the policies should be such as to minimize trace gas emissions (F-gases in particular) and reduce natural capital degradation. IAASTD also highlighted the complementary nature of mitigation (reduction of GHG emission) and adaptation of agricultural systems to changed climatic situations. Although IPCC in their reports pointed out the significant contribution of conventional industrial agriculture along with forestry, deforestation in particular on greenhouse emission (together around 30.9 per cent), it was IAASTD that indicated the possibility of countering the same by ecologically sustainable natural resource management based agricultural practices.

Soon after the launch of the IAASTD Synthesis Report in 2008, the Institute of Science in Society (a well known nonprofit civil society organisation in UK) re-estimated the total GHG contribution of agriculture, forestry and all food related activities (including transport, storage, marketing, *etc.*) at around 34 per cent of total emission (Mae-Wan Ho, 2008). Data were put forth to show that organic agriculture and localized food systems, besides effectively mitigating a substantial part of the GHG emission would save a considerable portion of the energy being used in the conventional industrial agriculture and associated food related activities (Mae-Wan Ho, 2008, Basu, 2012; Basu *et al.*, 2013).

David Pimentel and co-authors in the University of Cornell, USA have shown that fossil fuel energy inputs for organic corn (maize) production were about 30 per cent lower than for the conventionally produced crop (Pimentel *et al.*, 2005), that is indeed a considerable energy saving.

Regarding the question of yield loss in organic agriculture, Badgley *et al.* (2007) in the University of Michigan, USA used a large data set (obtained from 293 publications on comparative studies on conventional agriculture and organic agriculture or near-organic production systems involving a wide range of food crops) showed an average yield ratio (overall organic versus conventional) of 0.914 for developed countries of the world, a minor yield reduction, and 1.74 for developing countries, indicating considerable overall superiority of organic farming systems *vis-a-vis* conventional agriculture. Perhaps in the developed countries transition from very intensive industrial agriculture to organic farming would need some more time. Nevertheless, the study categorically established that large-scale organic farming would not cause starvation as speculated by many, on the other hand organic agriculture could feed the world while doing away with the adverse impacts of conventional industrial agriculture on environment and ecosystem in general and providing toxin-free nutrient rich food for consumers in particular (Lu *et al.*, 2006; Mitchell *et al.*, 2007). A convincing meta-analysis by an international team of scientists, Baranski *et al.*, 2014 led by Professor Carlo Leifert of the University of New Castle, UK have shown highly significant overall superiority of organic products over their conventionally grown counterparts (Baranski *et al.*, 2014).

In the developing countries at the lower latitudes crop-livestock eco-sustainable farming systems are likely to provide viable options and as reported by UNCTAD and other UN agencies they have already shown their effectiveness through considerable increase, often doubling of the productivity (Schutter, 2011; Hoffman, 2011; Basu, 2011; Basu *et al.*, 2013).

We should point out here that certified organic agriculture by small holder farmers with the intention to enter the national and international markets would involve additional costs for certification and related issues requiring governmental help and assistance in various ways. The participatory guarantee scheme (PGS) introduced in some of the states of India has greatly reduced the certification costs and has been of much help to the poor small holder farmers. There has, however, been limited progress in most of the states.

Community Managed Sustainable Agriculture (CMSA) in Andhra Pradesh, India with its projected gradual shift to organic agriculture from its present widespread highly successful non-pesticidal management (NPM) practices by completely phasing out synthetic fertilizers (besides chemical insecticides as in NPM) and other synthetic chemical products from agriculture would hopefully be the biggest such effort in the country (see the World Bank publication, "Ecologically sound, economically viable: Community Managed Sustainable Agriculture in Andhra Pradesh, India" by Vijay Kumar *et al.*, 2009, pp. 1-29).

It would be worthwhile also to incorporate research and development findings from relevant fields of research such as SRI (System of Rice Intensification) that not only economizes the use of irrigation water and greatly reduces the amount of seed for sowing but also significantly boosts grain yield under conventional as well as organic management; SRI has successfully outperformed other commonly practiced methods of paddy cultivation under adverse climatic conditions. Data from northern India in the drought year of 2009 showed that conventional paddy yield declined by 31 per cent while with SRI yield declined by 13 per cent only (Adhikari *et al.*, 2010; also see Uphoff, 2003). The principles of SRI are now being researched and practiced in other crops also and we believe that it would be pertinent to choose eco-sustainable agricultural practices for all such crops right from the beginning. That would require, among other considerations, the removal of prohibited synthetic chemicals from all agricultural practices including seed treatment (specifically 'permissible' physiological seed invigoration treatment for improved crop performance). The same would not need much additional effort as we have already standardized herbal and naturally occurring substances which are very effective for seed invigoration, viability maintenance and productivity of most of our orthodox agricultural and horticultural seeds.

The permissible organic agriculture-compatible seed invigoration treatments include hydration-dehydration treatments with water only like short duration soaking-drying (SD), moisture equilibration-drying (ME-D) (also known as humidification-drying), moist sand conditioning-drying (MSC-D), moist sand conditioning-soaking drying (MSC-SD), and 'dry' treatments with dried powdered materials such as leaves of neem (*Azadirachta indica* L.), nisinda (*Vitex negundo* L.), oxalis (*Oxalis corniculata* L.), fruits of 'hot' chilli (*Capsicum annuum*), *etc.*, all of which are readily available. It is also necessary to point out in this connection that the hydration-dehydration treatments as well as many of the 'dry' treatments are broadly anti-insect, pest and antipathogenic in nature besides their newly established physiological role in seed invigoration to counter loss of vigour, viability and yield potential.

Annexure No A-4

Seed and Food Security of Small Holder Farmers

A very large part of the seed used for cultivation of major field crops by small and marginal farmers in India and many developing countries are farmer-owned and farmer-saved seeds from the previous harvest because the commercially available seeds are too expensive for them.

The high yielding varieties of crops like wheat and rice, which ushered in the green revolution in India were bred and multiplied by public sector institutions, were not expensive. However, being genetically not very stable (still segregating genome) could be saved by farmers on an average for 3-4 years prior to replacement by fresh breeder seed stock. With the entry of corporate sector controlled hybrid seeds of many crops, the cost of seed has become very high especially for vegetables and genetically modified crops (so far only cotton all over India, all of which are F_1 hybrids). Nevertheless, there is no relevance in saving F_1 hybrid seeds as they will segregate in the next generation (often very expensive seeds such as very high value vegetable and flower seeds are so bred that subsequent generations would be sterile). Obviously, farmer-saved crop seeds have to be pureline varieties or close to that.

Even then there is the problem of seed storage under ambient (temperature humidity uncontrolled) storage conditions in the usually hot and humid climate of many countries in the tropics and subtropics. Local (indigenous) as well as improved crop varieties may not store well if not well protected and seeds would lose vigour and viability adversely affecting field performance and productivity. Widespread use of simple seed invigoration technology would certainly be of great help to our farmers in preserving the yield potential of the seed saved by them.

This is all the more relevant in view of the threat posed by global climate change where adaptation to the changed scenario will bring into focus the need to use local germplasm and biodiversity as natural insurance against unpredictable (and erratic as well) climatological situations.

In any case, under no circumstances, the traditional and genuine rights of farmers to save all seeds grown by them should be compromised. Seed security is very closely associated with food security of the farmers as they are also the major consumers of their own edible produce and quite often only the surplus is sold to meet the family expenses. Still many of them remain half-starved or malnourished. In the Indian context seed security is closely linked to food security as well as the overall agricultural sovereignty of the country. It must be remembered that the livelihood security and survival of many millions of the country's small and marginal farmers who constitute the backbone of the rural economy is a matter of utmost importance. Any decision that may harm their genuine concerns on seed security and related issues may jeopardize the welfare of the country as a whole.

Corporate Control over Agriculture

In the discussion on thematic issues we have already mentioned that the seed is the backbone of seed-based agriculture and the green revolution itself has been such an example that brought a revolutionary change in agriculture of developing countries specially in south and south-east Asia where relatively low input requiring traditional agriculture was changed to high input requiring agriculture with the incorporation of lodging resistant dwarfing genes through conventional breeding. We shall not go into the merits and demerits of green revolution but with the plateauing of yield benefits, the revolutionary phase is practically over and the negative legacies are showing up. Even then green revolution was primarily institutionally managed and controlled (though indirectly patronizing agrochemical corporate interests) whereas the present day corporate interests are far more overbearing, monopolistic and all pervasive backed up by intellectual property rights (IPR), process and product patents on vital inputs – seeds and agrochemicals, often as composite packages. The monopolistic and profit-motivated approach has been further aggravated by simultaneous gradual recession of public sector institutions and international funding agencies which used to provide adequate funds earlier, greatly reduced their contributions, often to less than half. Over a dozen international institutions such as CYMMIT, IRRI, ICRISAT, *etc.*, in Asia, Africa, Latin America, collectively known as Consultative Group on International Agricultural Research (CGIAR) have suffered huge budget cuts from $6 billion in 1980 to $2.8 billion in 2006; donor countries like the USA cutting the aid from $2.3 billion to only $624 million. The multinational company, Monsanto alone has been estimated to have spent seven times as much fund as the 14 CGIAR institutes taken together (Stokes, 2008). The gradual withdrawal of the public sector from agricultural research and allied activities is indeed very unfortunate.

In this connection, it is worthwhile to recollect the personal views and reminiscences of the eminent plant physiologist and seed scientist Professor A. Carl Leopold in 1999 (he was 80 years old then but still worked with his own hands in his laboratory and making invaluable contributions in the field of seed science long after his formal retirement from active services in the Cornell University, Ithaca) in *Seed Science Research* where he recorded, "I am seriously worried about the research community becoming increasingly the handmaiden of the corporations". That was indeed a reminder of the concerns expressed by Martha Crouch (1990) in the journal *Plant Cell* on the inappropriateness of the academic community serving the needs of the corporations instead of serving the people. (see Annexure No A-5 for extracts from "Learning from seeds : A personal view" by Prof. A.C. Leopold, 1999).

Annexure No A-5

Extracts from Leopold, A.C. (1999) : Learning from seeds. A personal view. Seed Science Research, 9, 111-116.

In a fascinating reminiscence of his involvement in seed research, eminent plant physiologist Professor A.C. Leopold provided a glimpse of his wonderful personal experiences aptly titled "Learning from seeds". The lessons deserve all round circulation not only among researchers in seed science but also among agricultural scientists in general. The views are certainly worth sharing with as many interested people as possible. The short 6-page narrative is certainly a valuable contribution by the publishers of the journal, *Seed Science Research* that deserves critical appraisal over and over again.

The learning from seeds includes (i) the privilege of working with intelligent colleagues, (ii) the need for selection of promising subjects, (iii) progress of research thrives on good communication with other researchers, (iv) use of excessively complex technologies can be a serious trap. Simple experiments focused on novel questions may be more fruitful, lastly the eminent scientist mentioned (v) "I am seriously worried about the research community becoming increasingly the handmaiden of corporations. As research becomes increasingly expensive and the federal support system is increasingly limited in support funds, our faculties are turning increasingly toward corporate funding. In spite of the alleged goals of academic research to intellectual information to the community that supports, academic research is instead becoming increasingly the handmaiden of corporate interests". The author's foresight on the importance of research on seeds was evident around 1973 when as a consultant to a large agrochemical firm he suggested that "seeds offered the most promising area for plant research in the future ...... Seeds are crucial for all crop production and in spite of the fact that they commonly limit agricultural production they were very poorly understood".

Annexure No A-6

Gene-Edited Crops *vis-à-vis* Genetically Modified Crops

The recently developed gene-edited crops created through insertion of genetic material at precise locations (not random as in genetically modified crops leading to worries of potentially dangerous genetic disruptions or crossbreeding with non-GMO crops) have considerable promise in several States in the USA and a few other countries like UK and China so far unencumbered by regulatory controls applicable for GMOs (into which microbial vectors could randomly put new genes to merge with the host plant's DNA).

In an article in the Times of India, Delhi on January 29, 2017 entitled "Coming to a supermarket near you: Gene-edited veggies" by Kenneth Chang of the New York Times News Service.

One of the companies developing gene-edited crops, namely Cellectis hosted a dinner in a New York city restaurant in the first ever recorded public dinner in October, 2016 and served several dishes made from its gene-edited soybeans and potatoes with great fanfare. According to Dr. Andre Choulika, the chief executive of Collectis that was the first dinner on Earth with gene-edited foods but millions of people would consume the same in the 21st Century.

Other gene-edited crops in the pipeline in the USA include new versions of wheat with greater resistance to fungal diseases or with low carbohydrate but high dietary fibres; mushrooms that would not turn brown quickly; potatoes remaining fresh over long periods, soybeans that would give cooking oil without needing hydrogenation, *etc.* A Chinese laboratory has developed a fungus resistance wheat and tweaked barley plants for better yield have been developed by a laboratory in the UK.

However, the gene-edited successful crops would be most likely under intellectual properly rights and, as such, will not be freely and liberally available to small and marginal farmers in the developing countries. Many other issues on seed and food security may need to be addressed in the near future.

Annexure No A-7

Some Observations on Seed Invigoration Studies in the University of Calcutta in Relevant Reviews of the Subject at the International Level

1. Observations by Dr. Walter Heydecker and Dr. P. Coolbear

(i) Heydecker, W. and Coolbear, P. (1977). Seed treatments for improved performance – survey and attempted prognosis. *Seed Science and Technology*, **5**, 353-425.

"Brief soaks against deterioration in storage

We must now mention a completely different but most interesting discovery concerning a benefit from a wet seed treatment. This has been reported by Basu and Das Gupta (1974) and Das Gupta and Basu (1975). They found that three hours' soaking (immersion) of wheat grains after six months' conventional storage, in solutions of a range of chemicals, but even in water, largely counterbalanced the deterioration that would have occurred otherwise by the end of a year's uninterrupted storage. Deterioration in seed germination was equally well prevented by steeping the seeds in water as in salts. Similar results were obtained with jute (*Corchorus capsularis* L.) and rice seeds. Preliminary tests led to the tentative conclusion that probable auto-toxic substances accumulate in the seed during storage and that these substances can be leached out before they become actively harmful. However, seedling growth was also stimulated when seed treatments incorporating 10^{-4} to 10^{-2} M solutions of Na_2HPO_4 and NaCl were used." (387-388)

2. Observations by Professor E.H. Roberts, Dr. R.H. Ellis and co-workers

(i) Roberts, E.H. (1979). Seed deterioration and loss of viability. Advances in Research and Technology of Seeds, Part **4**, 25-42.

"................Villiers and Edgcumbe (1975) have shown that only intermittent periods of wetting are necessary in order to extend the period of viability, but the evidence for this was limited to a single species, lettuce. It is therefore interesting that a recent report suggests that soaking seeds of a number of crops seeds for short periods, followed by re-drying, extends longevity (Basu, 1977). Details of the techniques were described for rice and jute (Basu *et al.,* 1974) and for wheat (Basu and Das Gupta, 1974). It was suggested that the results may be due to leaching out of toxic metabolites or inhibiting fungal attack. But since the wetting treatment is said to be relatively ineffective if carried out either on fresh seeds, or on seeds which have lost considerable viability, the results would alternatively, and perhaps more easily be explained on Villiers' hypothesis."

(ii) Roberts, E.H. (1981). Physiology of ageing and its applications to drying and storage. *Seed Science and Technology*, **9**, 359-372.

"................In this context the recent work of Basu and his colleagues on a wide range of species is most interesting (*e.g.*, Basu, Chattopadhyay and Pal, 1974; Basu and Das Gupta, 1974; Basu and Dhar, 1979; Mitra and Basu, 1979; Rudrapal and Basu, 1979; Basu, Rudrapal and Pathak, 1980). In these cases it has been shown that, while in general there is no immediate effect of temporary hydration, loss of viability is delayed during subsequent storage. Furthermore although there is very interesting evidence of improved longevity using various chemicals the major effect is clearly due to water itself."

"................the information on oxygen corroborates the view that control of viability involves different principles at higher moisture contents, and corroborates other evidence which suggests that repair and turnover of the type suggested by Villiers, possibly begin to operate at moisture contents considerably less than those in fully hydrated seeds. If this is so, it would explain why Basu and his colleagues can obtain some improvements when treating seeds at 100 per cent RH as well as by contact with liquid water." (page 365)

(iii) Roberts, E.H. (1983). Loss of seed viability during storage. *Advances in Research and Technology of Seeds,* Part **8**, 9-34.

"For viability to be prolonged by high moisture contents it is not necessary for the seeds to be continuously hydrated; intermittent hydration will do (Villiers and Edgcumbe, 1975). Basu and his colleagues have shown that even a single short period of hydration can prolong the viability of many species (more recent publications include Basu and Dhar, 1979; Basu and Pal, 1978, 1979, 1980; Dharmalingam and Basu, 1978; Mitra and Basu, 1979; Rudrapal and Basu, 1979) and this has recently been confirmed under precisely controlled conditions in lettuce, although the period of prolongation is not great (Ibrahim, 1982). One interpretation of these results is that the temporary wetting enables some partial repair and turnover of damaged subcellular systems." (page 16, last para, page 17, first para)

"Effects of chemical treatments on longevity

Basu and his colleagues claim better results from hydration in the presence of several diverse chemicals, *e.g.*, sodium phosphate, sodium chloride, sodium thiosulphate, oxalic acid, *p*-hydroxybenzoic acid, cysteine, potassium iodide and several other substances, and suggest that most of the effective chemicals are antioxidants or could control free radical chain reactions (Basu and Dasgupta, 1978; Basu and Rudrapal, 1979). Certainly it would appear as though some substances can improve longevity quite independently of hydration for Basu *et al.* (1979) have shown beneficial effects of treating seeds with potassium iodide, *p*-hydroxybenzoic acid and tannic acid dissolved in anhydrous acetone." (page 17, para 2)

(iv) Rao, N.K., Roberts, E.H. and Ellis, R.H. (1987). The influence of pre- and post-storage hydration-dehydration treatments on chromosomal aberrations, seedling abnormalities and viability of lettuce seeds. *Annals of Botany*, **60**, 97-108.

"................Short periods of imbibition in water, which are too short to allow germination, followed by drying can also increase germination rate and extend longevity. Work on this subject, pioneered by R.N. Basu and his colleagues on a number of species, is extensive and has been reviewed elsewhere (Roberts, 1983). It has been supported by Goldsworthy, Fielding and Dover (1982) and Kuo and Chin (1986) in wheat and rice seeds respectively. Work in this laboratory on lettuce seed (Ibrahim, 1981) has shown that at 25°C the optimum period of a single hydration treatment for extending longevity was 2-4 h and typically such treatments increased longevity by 10-20 per cent. Successive short periods of hydration, however were capable of extending longevity by a factor of two to three." (page 98, column 1, para 2)

".........Instead of trying to repair damage after it has occurred, an alternative is to attempt to protect seeds from damage by treating them before storage. This has been attempted by either hydrating and then dehydrating seeds before storage (Basu, 1976, Savino, Haigh and de Leo, 1979) or by treating the seeds with various chemicals in aqueous solution and then dehydrating the seeds (Basu, 1976, Basu and Dasgupta, 1978). There is also some suggestion that certain chemical pre-treatments can improve longevity independently of hydration (Basu, Pan and Punjabi, 1979; Woodstock *et al.,* 1983). However, Barnes and Berjak (1978) have suggested that the main benefits derive from hydration and dehydration rather than the presence of the chemicals. The major technical problem of investigations of this type is that seed longevity is extremely sensitive to slight differences in seed moisture content (see particularly Ellis and Roberts, 1981; Ellis, Hong and Roberts, 1986), and is difficult to adjust the moisture content in different batches of seed after hydration treatments to precisely the same value, and this may explain some of the differences reported in the literature. Of the many diverse chemicals investigated by Basu (1976)the most effective for reducing subsequent seed deterioration were antioxidants and these chemicals that are known to control free radical reactions in radiobiological studies." (page 98, column 2, para 3).

".............Examples of these types of chemicals were chosen for the study reported here on the basis of their apparent efficacy in the earlier reports (particularly Basu, 1976; Basu and Dasgupta, 1978)." (page 99, column 1).

"............All this is compatible with the work reported by Basu and Pal, 1980 and Kuo and Chin, 1986 on rice (*Oryza sativa* L.). Both groups noted no improvement in subsequent longevity resulting from temporary hydration of freshly harvested high quality rice seeds, but a significant improvement was observed after similar temporary hydration treatment on the same seed lots when the treatments were applied after some deterioration had taken place. It may be that temporary hydration treatments cannot extend longevity in fresh seeds since in such cases insufficient damage has accumulated for repair to have a significant effect". (page 104, column 2)

3. Observations by Professor M.B. McDonald Jr. and Dr. D.O. Wilson Jr.

Wilson, D.O. Jr. and McDonald, M.B. Jr. (1986). The lipid peroxidation model of seed ageing. *Seed Science and Technology*, **14**, 269-300.

(i) "Basu's group at Calcutta University has found that treatment of mungbeans with iodine vapour for 16 h rendered them virtually immune to accelerated ageing (40°C, 95 per cent RH, 20 days) (Rudrapal and Basu, 1980). Seeds treated with I_2 retained a germination of close to 100 per cent while untreated seeds declined to approximately 75 per cent. Treatment also increased root and shoot length obtained following ageing. One explanation for these results may be attributed to the antimicrobial action of residual I_2. However, the beneficial effect was also noted after dry heat treatment (100°C, 5-10 per cent seed moisture, 45 minutes), conditions precluding microbial involvement. The authors attributed the effect to stabilisation of unsaturated fatty acids by addition of iodine to double bonds rendering the acyl chains less prone to peroxidation.

Other work involving iodination of mustard seeds (Basu and Rudrapal, 1980) demonstrated protection against damage under a variety of storage conditions. After ageing, iodinated seeds demonstrated reduced evidence of lipid peroxidation, improved membrane integrity, and higher activity of amylolytic and dehydrogenase enzymes during germination." (page 284)

(ii) "Chemicals which protect seed tissue from radiation damage also prolong the life of seeds under natural and accelerated ageing conditions. Basu and Dasgupta (1978) treated wheat seeds with dilute solutions of various reducing agents and radioprotective chemicals. In the case of most of the chemicals tested, a three hour soak in such compounds followed by drying resulted in an increase in viability when the seeds were later aged or irradiated. Water and solutions of KI, Na_2HPO_4, cysteine, *p*-aminobenzoic acid, oxalic acid and NaCl all produced promotive effects. Caffeine, a known radio-sensitiser, potentiated both X-ray and ageing damage. Working with lettuce (*Lactuca sativa* L.) seeds, Basu, Pan and Punjabi (1979) separated the effect of some of these chemicals from the hydration effect by infusing them into seeds in acetone solution." (page 285-286)

(iii) ".............Basu's group has conducted several studies on the hydration-dehydration phenomenon for preservation of a variety of agricultural seeds. They found, for example, that jute (*Corchorus* sp.) seeds could be maintained in a condition of high vigour and viability by periodic hydration in either a humid atmosphere or by soaking for as little as one hour (Basu, Chattopadhyay, Bandopadhyay and Basak, 1978). Similar findings have been reported for rice (*Oryza sativa* L.) (Basu and Pal, 1979; Basu and Pal, 1980), lettuce (Basu, Pan and Punjabi, 1979), wheat (Rudrapal and Basu, 1980), mustard (*Brassica* sp.) (Rudrapal and Basu, 1982) and soybeans (Saha and Basu, 1983). These responses occurred regardless of the method of water addition to the seed eliminating the possibility that substances were leached from the seeds. The hydration treatments also resulted in a radioprotective effect." (page 287, para 2)

(iv) "The fact that seed responses to hydration can be detected after only a few hours casts some doubt on the cellular repair hypothesis. Although the invigoration phenomenon is especially striking in low vigour seeds (Basu and Pal, 1980, Perl and Feder, 1981), it is in these seeds that protein synthesis is delayed for several hours during hydration by a 'limiting factor' in wheat (Marcus *et al.,* 1966). Recent work by Goldsworthy, Fielding and Dover (1982) has demonstrated significant invigoration responses of wheat seeds with a 15 minute soak period in the absence of oxygen. Haber and Randolf (1967) found that gamma ray induced ESR signals in lettuce seeds were reduced to 8 per cent of the original level by hydration-dehydration treatment. This shows that some free radical producing mechanism in the irradiated seeds was removed by the hydration treatment." (page 287, para 3)

(v) "..........Peroxidation initiators such as cumene hydroperoxide or paraquat (Poli and Gravela, 1982) or radiopotentiators such as caffeine (Basu and Dasgupta, 1978) might be infused into seeds prior to storage or germination. Antioxidants can be introduced into seeds and serve to modify lipid peroxidation." (page 289, para 2)

Annexure No A-8

Outlines of Future Researches on Seed Invigoration

1. Dry seed invigoration treatments with volatile chemicals that would readily pass into gas phase particularly chemicals with relatively strong smell such as naphthalene, α-naphthaol, β-naphethol, phenol, 2,4,6-trichlorophenol (TCP), methyl salicylate (oil of wintergreen), camphor, *etc.* need to be taken up. Also pharmaceuticals such as paracetamol (acetaminophen) a drug which has in recent times proved to be a very effective anti-inflammatory agent and comparable to aspirin which is widely used for the purpose. Aspirin (orthoacetyl salicylic acid) has already been found to be very effective as a seed invigorating pharmaceutical (trade name Aspro) in our studies on a wide range of crop seeds.
2. Screening of natural plant and animal products such as powdered plant and seed materials, milk mixed with inert carriers and dried cow dung cake powder, liquid neem and mustard oils and such oils mixed with carriers, oil cakes especially neem oil cake, mustard oil cake, brick powder for seed treatments (also as carrier), egg shell powder and its efficacy compared to chemically pure calcium carbonate); ayurvedic and homeopathic drugs as dry dressing treatments. It would be worthwhile to evaluate their efficacy as seed invigoration treatments especially for ecofriendly organic agriculture.
3. Short and medium-term conservation of germplasm of indigenous crop varieties many of which are fast disappearing due to the onslaught of high yielding varieties (HYVs) using easily practicable wet and dry physiological seed invigoration treatments.

4. Screening of agricultural and horticultural field crop cultivars for tolerance to edaphoclimatic stress situations as aggravated by global climate change, under simulated stress conditions in the laboratory employing wet and dry physiological invigoration treatments, including non-proprietary priming treatments, followed by pilot level field studies.
5. Development of easily practicable volatile aldehyde emanation related techniques for rapid assessment of seed vigour at farm or community level seed vigour and viability issues.
6. Comparative study on the bioassay and autoassay of seed vigour using various seedling evaluation indices such as seedling length, seedling fresh weight, seedling dry weight.
7. Development of low cost easily practicable non-proprietary (open access) presowing seed priming technology employing moist sand conditioning for strictly regulated seed hydration followed by light air-drying. MSC-D has already been very successfully used for midstorage treatments in a range of crop seeds and have already shown great promise as a seed priming treatment for a number of materials but may need critical attention for further refinements. Also, the issue of combining such priming treatments with other invigoration treatments is considered very important.
   (a) Studies on interactions involving initial seed vigour levels and durations of hydration on the efficacy of seed priming. Harvest-fresh high vigour seed (HV), seeds stored for around half of the storage period under ambient conditions giving approximately medium vigour (MV) level and those stored till the approximate sowing time as the low vigour (LV) material and five hydration durations (0, 24, 48, 72 and 96 hours) with necessary modifications depending on concerned materials.
   (b) Standardization of different combinations of prestorage, midstorage and presowing seed priming treatments
      i) Combined prestorage-cum-delayed midstorage hydration-dehydration treatments.
      ii) Combined prestorage-cum-presowing seed priming treatments.
      iii) Combined prestorage-cum-midstorage hydration-dehydration treatments-cum-presowing seed priming treatments.

Bibliography

Abdalla, F.H. and Roberts, E.H. (1969). The effect of seed storage conditions on the growth and yield of barley, broad beans and peas. *Annals of Botany*, **33**, 169-184.

Abdul-Baki and Anderson, J.C. (1970). Viability and leaching of sugars from germinating barley. *Crop Science*, **10**, 31-34.

Adhikari, P., Sen, D. and Uphoff, N. (2010). System of rice intensification as a resource conserving methodology: Contributing to food security in an era of climate change. SATSA, *Mukhapatra - Annual Technical Issue*, **14**, 26-44.

Afzal, I., Mukhtar, K., Qasim, M., Basra, S.M.A., Shahid, M., and Ilaq, Z. (2012). Magnetic stimulation of marigold seed. *International Agrophysics*, **26**, 335-339.

Afzal, S.M.J. and Kesavan, P.C.(1979). Influence of seed moisture content and post-irradiation hydration temperature on the kinetics of reactivity towards oxygen or decay of oxygen-sensitive sites. *International Journal of Radiation Biology and Related Studies in Physics, Chemistry and Medicine*, **36** (2),161-176.

Aladjadjiyan, A. (2010). Effect of microwave irradiation on seeds of lentils (*Lens Culinaris*, Med.). *Romanian Journal of Biophysics*, **20**, 213-221.

Albu, N. (1972). The effect of ultrasound on spring wheat cv. Marquis, grown for recovery of the root system. *Contributii Botanice*, 355-361, (*Field Crop Abstracts*, **26**, 4525).

Alvarado, A.D. and Bradford, K.J. (1988). Priming and storage of tomato (*Lycopersicon lycopersicum*) seeds. I. Effects of storage temperature on germination rate and viability. *Seed Science and Technology*, **16**, 601-602.

Anand, A., Nagarajan, S., Verma, A.P.S., Joshi, D.K., Pathak, P.C., and Bhardwaj, J. (2012). Pre-treatment of seeds with static magnetic field ameliorates soil water stress in seedling of maize (*Zea mays* L.). *Indian Journal of Biochemistry and Biophysics*, **49**, 63-70.

Andersen, A. M., Hart, J. R. and French, R. C. (1970). *Proceedings: International Seed Testing Association*, **29** (1), 81-96.

Araujo, S.S., Paparella, S., Dondi, D., Bentivoglio, A., Carbonera, D. and Balestrazzi, A. (2016). Physical methods for seed invigoration; advantages and challenges in seed technology, *Frontiers in Plant Science*, **7**, 646.

Argerich, C. A., Bradford, K. J., & Tarquis, A. M. (1989). The effects of priming and ageing on resistance to deterioration of tomato seeds. *Journal of Experimental Botany*, **40** (5), 593-598.

Argerich, C.A. and Bradford, K.J. (1989). The effects of priming and ageing on seed vigour in tomato. *Journal of Experimental Botany*, **40** (5), 99-607.

Arinshtein, A.I., Chemarin, N.G. and Kichanova, L.A. (1973). Determination of stimulating doses of gamma rays for irradiating coriander seeds. *TrudyVNIIEfimomaslichnykh Kul tur*, **6**, 4-9, (*Horticulture Abstracts*, **46**, 3661).

Arma, S. and Watanabe, A. (1983). Long term storage of germplasm in tea (*Camellia sinensis* (L.) O. Kuntze). *Bulletin of the National Research Institute of Tea*, **19**, 29-57.

Armstrong H. and McDonald, M.B. (1992). Effects of osmoconditioning on water uptake and electrical conductivity in soybean seeds. *Seed Science and Technology*, **20**, 391-400.

Aschermann-Koch, C., Hofmann, P. and Steiner, A.M. (1992). Presowing treatment for improving quality in cereals. I. Germination and vigour. *Seed Science and Technology*, **20**, 435-440.

Ashraf, M. and Bray, C.M. (1993). DNA synthesis in osmoprimed leek (*Allium porrum* L.) seeds and evidence for repair and replication. *Seed Science Research*, **3**, 15-23.

Austin, R.B., Longden, P.C. and Hutchinson, J. (1969). Some effects of 'hardening' carrot seeds. *Annals of Botany*, **33**, 883-895.

Babushkin, K.N., Balashov, T.N. and Pogorelova, T.S. (1974). The effect of intermittent concentrated sunlight (ICS) on peas. *Problemy Fotoenergii Rastenii Kichinev, Moldavian* S.S.R.,*Stiinca*, 233-237. (*Horticulture Abstracts*, **45**, 6568).

Baby, S.M., Narayanaswamy, G.K., and Anand, A. (2011). Superoxide radical production and performance index of Photosystem II in leaves from magnetoprimed soybean seeds. *Plant Signalling Behaviour*, **6**, 1635-1637.

Badgley, C., Moghtader, J., Quintero, E., Zaken, E., Chapell, M.J., Aviles – Vasquez, K., Samulon, A. and Perfecto, I. (2007). Organic agriculture and the global food supply. *Renewable Agriculture and Food System*, **22**, 81-108.

Balakhnina, T., Bulak, P., Nosalewiez, M., Pietruszewski, S., and Wlodarczyk, T. (2015). The influence of wheat *Triticum aestivum* L. seed pre-sowing treatment with magnetic fields on germination, seedling growth, and antioxidant potential under optimal soil watering and flooding. *Acta Physiologiae Plantarum*, **37** (3): 59.

Balestrazzi, A., Confalonieri, M., Dona, M., and Carbonera, D. (2012). Genotoxic stress, DNA repair and crop productivity. In: *Crop improvement under adverse conditions.*(Eds Tuteja N., Gill, S.S.), Springer-Verlag, Berlin, pp 153-169.

Balestrazzi, A., Confalonieri, M., Macovei, A., and Carbonera, D. (2010). Seed imbibitions in *Medicago truncatula* Gaertn; expression profiles of DNA repair genes in relation to PEG-mediated stress. *Journal of Plant Physiology*, **168**, 706-713.

Balestrazzi, A., Confalonieri, M., Macovei, A., Dona, M., and Carbonera, D. (2011a). Genotoxic stress and DNA repair in plants; emerging functions and tools for improving crop productivity. *Plant Cell Reports*, **30**, 287-295.

Balestrazzi, A., Dona, M., Macovei, A., Sabatini, M.E., Pagano, A., and Carbonera, D. (2015). DNA repair and telomere maintenance during seed imbibition; correlation of transcriptional patterns, *Telomere Telomerase,* **2**, 496.

Balestrazzi, A., Macovei, A., Tava, A., Avato, P., Raimondi, E., and Carbonera, D. (2011b). Unraveling the response of plant cells to cytotoxic saponins; role of metallothionein and nitric oxide. *Plant Signalling Behaviour*, **6**, 1- 4.

Bandopadhyay, P.K., Chatopadhyay, K., Basak, S.J. and Basu, R.N. (1977). Seed treatment for vigour, viability and productivity of jute (*Corchorus* sp). In: *Proceedings of Subject Matter Seminar on Jute Production, Processing and Marketing Technology,* sponsored by Directorate of Extension, Ministry of Agriculture and Irrigation, Govt. of India, Calcutta, 23-28.

Baranski, M., Srednicka-Tober, D., Volakakis, V., Seal, C., Sanderson, R., Stewart, G.B., Benbrook, C., Biavali, B., Markellou, E., Giotis, C., Gronadzka-Ostronska, J., Rembialkowska, E., Skwarlo-Sonita, K., Janoovska, D., Niqquilis, U., Nicot, P. and Leifert, C. (2014). Higher autioxidant and lower cadmium concentrations and lower incidence of pesticide residues in organically-grown crops: a systematic literature review and meta-analyses. *British Journal of Nutrition*, **112**(5), 794-811.

Barba-Espin, G., Diaz-Vivancos, P., Job, D., Belghazi, M., Job, C., and Hernandez, J.A. (2011). Understanding the role of H_2O_2 during pea seed germination; a combined proteomic and hormone profiling approach. *Plant Cell Environment,* **34**, 1907-1919.

Barlow, E.W.R. and Haigh, A.M. (1987). Effect of seed priming on the emergence, growth and yield of UC 82B tomatoes in the field. *Acta Horticulturae*, **200**, 153-164.

Barman, T.S. (1988). Post harvest storage of tea seed. In: *Proceedings of the International Congress on Plant Physiology*, New Delhi, **2**, 1392-1395.

Barman, T.S. and Sarma, C.M. (1987). Effect of chemical treatments on the emergence of tea seed after storage. *Seed Research*, **15**, 52-56.

Barnes, G. and Berjak, P. (1978). Effect of some antioxidants on the viability of stored seeds. In: *Proceedings of Southern African Electron Microscopy Society*, **8**, 95-96.

Barua, D.N. (1961b). The significance of seed size in cultivated tea (*Camellia sinensis* L.). *Empire Journal of Experimental Agriculture*,**29**, 143-152.

Barua, D.N. (1989), Seed vs clone. In: *Science and Practice in Tea Cultures*. TRA Publication, pp.164-166.

Basu, R.N. (1978). Physico-chemical control of seed deterioration. *Seed Research*, **4**, 15-23.

Basu, R.N. (1977). Seed treatment for vigour, viability and productivity. *Indian Farming*, **27**, 27-28.

Basu, R.N. (1982). Seed viability. *Science and Culture*, **48,** 129-134.

Basu, R.N. (1990). Seed storage for extended storability. In: *Proceedings of the International Conference on Seed Science and Technology*, 1990, New Delhi.

Basu, R.N. (1993). Seed invigoration for extended storability. *Seed Research* Special Volume, 217-230.

Basu, R.N. (1994). An appraisal of research on wet and dry physiological seed treatment and their applicability with special reference to tropical and sub-tropical countries. *Seed Science and Technology*, **22**, 107-126.

Basu, R.N. (1994). Seed invigoration studies in the University of Calcutta - Glimpses of the past and present (pp.1-12). Special Felicitation Lecture on December 7-9, 1994, published by 'Indian Society of Seed Technology (ISST), New Delhi on the occasion of the All India Seed Seminar at the Tamil Nadu Agricultural University, Coimbatore.

Basu, R.N. (1995). Seed viability. In: *Seed Quality, Basic Mechanisms and Agricultural Implications* (Ed. A.S. Basra), Chapter I, The Howorth Press Inc., NY, USA, pp.1-44.

Basu, R.N. (2004). Approaches to quantification of seed vigour. *Indian Biologist*, **36**(2), 1-23.

Basu, R.N. (2007). A holistic approach to quantification of seed vigour through integration of morphological and physiological-biochemical parameters. *Indian Biologist*, **39**(1), 1-44.

Basu, R.N. (2011). Organic agriculture for food safety and security, protection of land, environment, ecosystem and mitigation of climate change. In: *Proceedings of UGC sponsored National Seminar on Organic Agriculture and Biosafety*. Gurudas College, Kolkatta, pp.1-13.

Basu, R.N. (2012). Global climate change vis-à-vis agriculture and food security. In: *Climate change: Man and Environment* (ed. G.K. Saha), Dayal Publishing House, New Delhi, pp.138-165.

Basu, R.N. and Choudhuri, N. (1983). Preservation of onion seed. *Seeds and Farms*, **IX**(4&5). 17-19.

Basu, R.N. and Dasgupta, M. (1974). Control of seed deterioration in wheat (*Triticum aestivum* L.). *Indian Agriculturist*, **18**, 285-288.

Basu, R.N. and Dasgupta, M. (1978). Control of seed deterioration by free radical controlling agents. *Indian Journal of Experimental Biology*, **16**, 1070-1073.

Basu, R.N. and Dey, G. (1983). Soaking and drying of stored sunflower seeds for maintaining viability and vigour of seedlings and increasing the yield potential. *Indian Journal of Agricultural Sciences*, **53**(7), 563-569.

Basu, R.N. and Dhar, S. (1979). Seed treatment for maintaining vigour, viability and productivity of sugar beet (*Beta valgaris* L.). *Seed Science and Technology*, **7**(2), 225-233.

Basu, R.N. and Mandal, A.K. (1985). In search of low cost technologies for improving seed performance. In: *Transferable Technology for Rural Development*, (ed. D.K. Dasgupta), pp.33-52. Associated Publishing Company, New Delhi.

Basu, R.N. and Mandal, K. (1986). Viability of horticultural seeds and seed invigoration treatments for vegetables and flowers. In: *Propagation of Tropical and Subtropical Horticultural Crops*, (eds. T.K. Bose, S.K. Mitra and M.K. Sadhu), pp.571-596. Naya Prakash, Calcutta.

Basu, R.N. and Pal, P. (1978). Seed treatment to maintain viability, vigour and yield potential of stored rice seed. *International Rice Research Newsletter*, **3**(2), 5.

Basu, R.N. and Pal, P. (1979). Physicochemical control of seed deterioration in rice (*Oryza sativa* L.). *Indian Journal of Agricultural Science*, **49**, 1-6.

Basu, R.N. and Pal, P. (1979). Seed treatment for vigour viability and productivity of rice (*Oryza sativa*). In: *Rice in West Bengal* – a Govt. of West Bengal Publication, **I**, 163-170.

Basu, R.N. and Pal, P. (1980). Control of rice seed deterioration by hydration-dehydration pretreatment. *Seed Science and Technology*, **8**, 151-160.

Basu, R.N. and Punjabi, B. (2017). Quantification of seed vigour: Vigour Related correlation-Regresion Equations and Data Transformations. XIV National Seminar on Food Security Through Augmented Seed Supply Under Climate Uncertainties. ICAR-IARI, New Delhi, January 28-30, pp. 79-85.

Basu, R.N. and Rudrapal, A.B. (1979). Iodine treatment of seed for the maintenance of vigour and viability. *Seed Research*, **7**, 80-82.

Basu, R.N. and Rudrapal, A.B. (1980). Iodination of mustard seed for the maintenance of vigour and viability. *Indian Journal of Experimental Biology*, **18**, 492-494.

Basu, R.N. and Rudrapal, A.B. (1982). Post-harvest seed physiology and seed invigoration treatments. In: *Proceedings of International Conference on Frontiers of Research in Agriculture* (Golden Jubilee International Conference, Indian Statistical Institute, Calcutta), pp.374-397

Basu, R.N. and Sur, K. (1988). Seed germination and viability. In: *Plant Physiological Research in India* (ed. S.P. Sen), pp.117-137. Society of Plant Physiology and Biochemistry, New Delhi.

Basu, R.N., Bandopadhyay, P.K. and Basak, S.L. (1978). Seed biology with special reference to jute. Non-subject Matter Seminar on Jute and Production Technology, sponsored by Directorate of Extension, Ministry of Agriculture and Irrigation, Government of India, Barrackpore, P-II, 20-31.

Basu, R.N., Bose, T.K., Chattopadhyay, K., Dasgupta, M., Dhar, N., Kundu, C., Mitra, R., Pal, P. and Pathak, G. (1975). Seed treatment for the maintenance of vigour and viability. *Indian Agriculturist*, **19**, 91-96.

Basu, R.N., Bose, T.K., Chattopadhyay, K., Dasgupta, M., Dhar, N., Kundu, C., Mitra, R., Pal, P. and Pathak, G., 1975. Seed treatment for the maintenance of vigour and viability. *Indian Agriculturist*, **19**, 91-96.

Basu, R.N., Chattopadhyay, K. and Pal, P. (1974). Maintenance of seed viability in rice (*Oryza sativa* L.) and jute (*Corchorus capsularis* L. and *C. olitorius* L.). *Indian Agriculturist*, **18**, 75-79.

Basu, R.N., Chattopadhyay, K., Bandopadhyay, P.K. and Basak, S.L. (1978). Maintenance of vigour and viability of stored jute seeds. *Seed Research*, **6**, 1-13.

Basu, R.N., Kuchlan, P. and Sur, K. (1990). Bioassay of seed vigour. *Indian Journal of Experimental Biology*, **28**, 166-170.

Basu, R.N., Mandal, A.K. and Pal, P. (2010). Seed invigoration for seed saving by small and marginal farmers and ensuring food security. *Indian Biologist*, Special Volume, 153-167.

Basu, R.N. and Ray, Sudipta (2016). Role of organic farming to mitigate climate change and maintain food security. *Organic Farming Newsletter*, NCOF, Govt. of India,**12**(1).

Basu, R.N., Mandal, A.K. and Ray, S. (2013). Current global agricultural scenario and projected climate change impacts on food security in the late twenty first century. *Indian Biologist*, **45**(1), 1-89.

Basu, R.N., Pan, D. and Punjabi, B. (1979). Control of lettuce seed deterioration. *Indian Journal of Plant Physiology*, **22**, 247-253.

Basu, R.N., Rudrapal, A.B. and Pathak, G. (1980). Physico-chemical control of mustard seed deterioration. ISTA Congress, Vienna, Preprint, **27**, 1-8.

Basu, R.N., Rudrapal, A.B. and Pathak, G. (1985). Physico-chemical control of mustard seed deterioration. *Seed Research*, **13**(2), 39-46.

Bennett, M.A., and Waters, L., Jr. (1987). Seed hydration treatments for improved sweet corn germination and stand establishment. *Journal of the American Society of Horticultural Science*, **112**, 45-49.

Bezbarua, H.P. (1975). Development of flower, pollination and seed-set in tea in North East India. *Two leaves and a Bud*, **22**, 25-30.

Berjak, P. (1978). Viability extension and improvement of stored seeds. *South African Journal of Science*, 74(10), 365.

Berrie, A.M.M., Drennan, D.S.H. (1971). The effect of hydration-dehydration on seed germination. *New Phytologist*, **70**, 135.

Bhardwaj, J., Anand, A., and Nagarajan, S. (2012). Biochemical and biophysical changes associated with magnetopriming in germinating cucumber seeds. *Plant Physiology and Biochemistry*, **57**, 67-73.

Bhargava, S., and Sawant, K. (2013). Drought stress adaptation; metabolic adjustment and regulation of gene expression. *Plant Breeding*, **132**, 21-32.

Bhat, S.A. (2004). *M.Sc. (Agriculture) Thesis*, (Supervisor, K. Sur), University of Calcutta, India.

Bhatt, S.P. and Padayatty, J.D. (1975). Transcriptional events during early phase of germination of rice embroys. *Nature* (London), **256**, 227-228.

Bhattacharya, A.K., Rahman, F. and Basu, R.N. (1994). Preservation of tea (*Camellia sinensis* L.) seed viability. *Seed Research*, **22**(2), 108-111.

Bhattacharya, S., Chowdhury, R., & Mandal, A. K. (2015). Seed invigoration treatments for improved germinability and field performance of soybean (*Glycine max* L Merill). *Indian Journal of Agricultural Research*, **49**(1), 32-38.

Bhattacharyya, A.K. and Basu, R.N. (1990). Retention of vigour and viability of stored pea (*Pisum sativum* L.) seed. *Indian Agriculturist*, **34**, 187-193.

Bhattachyaryya, A.K. and Basu, R.N. (1992). Retention of vigour and viability of jackfruit (*Artocarpus heterophyllus* L.) seed. *Indian Agriculturist*, **36**(2), 65-74.

Bhattachyaryya, A.K., Lahiri, A.K. and Basu, R.N. (1991). Improvement of germinability of Eucalyptus species by pregermination treatments. *Indian Forester*, **117**(8), 661-663.

Bisaria, A.K., Kaushik, M.P., Sharma, J.K. and Singh, I. (1975). Effect of gamma irradiation of seeds on some morphological characters and sex expression of muskmelon (*Cucumis melo* L.). *Current Science*, **44**, 392-393.

Biswas, J., Guha, P., & Mandal, A. K. (2010). Compatibility of pesticidal formulations with dry physiological treatments for retention of vigour, viability and productivity of Bengal gram (*Cicer arietinum* L.). *Indian Agriculturist*, **54**(3/4), 179-184.

Borowski, A.M., Fritz, V.A., and Waters, L., Jr. (1991). Seed maturity influences germination and vigour of two *shrunken*-2 sweet corn hybrids. *Journal of the American Society of Horticultural Science,* **116**, 401-404.

Bradford, K.J. (1986). Manipulation of seed water relations via osmotic priming to improve germination under stress conditions. *Horticulture Science,* **21**, 1105-1112.

Bradford, K.J., and Still, D.W. (2004). Application of hydrotime analysis in seed testing. *Seed Technology,* **26**, 75-85.

Bradford, K.J., May, D.M., Hoyle, B.J., Skibinski, Z.S., Scott, S.J. and Tyler, K.B. (1988). Seed and soil treatments to improve emergence of muskmelon from cold or crusted soils. *Crop Science,* **28**, 1001-1005.

Bradford, K.J., Steiner, J.J. and Trawatha, S.E. (1990). Seed priming influence and emergence of pepper seed lots. *Crop Science,* **30**, 718-721.

Bray, C.M., Davison, P.A., Ashraf, M, and Taylor, R.M. (1989). Biochemical changes during priming of leek seeds. *Annals of Botany,* **63**, 185-193.

Brazhnikova, ZH. V. (1974). Studies on the radiobiological effects of irradiation in carrots. *Trudy Inst, Genet, Selekt., Baku,***7**, 152-156, (*Horticulture Abstracts,* **46**, 4724).

Bruggink, G.T., Ooms, J.J.J., and Van der Toorn, P. (1999). Induction of longevity in primed seeds. *Seed Science Research,* **9**, 49-53.

Buchvarov, P. and Gantcheff, T. (1984). Physiological influence of accelerated and natural ageing on free radical levels in soybean seeds. *Physiologia Plantarum,* **60**, 53-56.

Butler. L., Hay, F.R., Ellis, R.H., Smith, R.D., and Murray, T. (2009). Priming and re-drying improve the survival of mature seeds of *Digitalis purpurea* during storage. *Annals of Botany,* **103**, 1261-1270.

Callan, N. W., Mathre, D. E., and Miller, J. B. (1991). Field performance of sweet corn seed bio-primed and coated with *Pseudomonas Flourescence* AB254. *Horticulture Science,* **26**(9), 1163-1165.

Celestino, C., Picazo, M.L., and Toribio, M. (2000). Influence of chronic exposure to an electromagnetic field on germination and early growth of *Quercus suber* seeds; preliminary study. *Electromagnetic Biology and Medicine,* **19**, 115-120.

Chatterjee, B.N. and Singh, A.I. (1983). Barley production from seeds treated before sowing. *Journal of Agricultural Science,* Cambridge, **100**, 235-239.

Chaussat, R. and Le Deunff, Y. (1975). *La Germination des Semences.* Ed. Gauthier Villars.

Chen, H., Chu, P., Zhou, Y., Li, Y., Liu, J., Ding, Y., Tsang, E.W., Jiang, L., Wu, K., and Huang, S. (2012). Over expression of AtOGGI, a DNA glycosylase/AP lyase, enhances seed longevity and abiotic stress tolerance in *Arabidopsis. Journal of Experimental Botany,* **63**, 4107-4121.

Chin, H.F., Y.L. Hor and Mohd. Lassim, M.B. (1984). Identification of recalcitrant seeds. *Seed Science and Technology*, **12**, 429-436.

Ching, T.M. and Schoolcraft, I. (1968). Physiological and chemical differences in aged seeds. *Crop Science*, **8**, 407-409.

Chinoy, J.J., Gurumurti, K., Shastri, K., Abraham, P.G., Dave, I.C., Shah, P.N., Pandya, R.B. and Saxena, O.P. (1968). Effect of ascorbic acid, CCC and their interaction on germination and metabolism in peanut. *Indian Journal of Plant Physiology*, **11**, 216-225.

Chiu, K.Y., Chen, C.L., and Sung, J.M. (2002). Effect of priming temperature on storability of primed sh-2 sweet corn seed. *Crop Science*, **42**, 1996-2003.

Chiu, K.Y., Wang, C.S. and Sung, J.M. (1995). Lipid peroxidation and peroxide-scavenging enzymes associated with accelerated aging and hydration of watermelon seeds differing in ploidy. *Physiologia Plantarum*, **94**, 441-446.

Choudhuri, N. and Basu, R.N. (1990). Seed invigoration of red beet (*Beta vulgaris* L.) for improved storability and field performance. *Advances in Horticulture and Forestry*,**1**, 125-133.

Choudhuri, N. and Basu, R.N. (1988). Maintenance of seed vigour and viability of onion (*Allium cepa* L.). *Seed Science and Technology*, **16**, 51-61.

Choudhuri, N. and Basu, R.N. (1989). Effect of sugarbeet fruit inhibitors on seed germinability. *Seed Research*, **17**(1), 16-22.

Clemente, G, and Caldera, P.G. (1975). *Radiostimolazione delle sementi, Italia Agriculture*, **112**, 103-114.

Come, D. (1975). Acquisition de l'aptitude a germer. In: *Germination des Semences* (eds R. Chaussat and Y. Le Deunff) pp. 59-70. Gauthier Villars.

Coolbear, P. (1995). In: *Seed Quality, Basic Mechanisms and Agricultural Implications* (ed. A. S. Basra), Food Products Press, NY, USA, pp. 223-277.

Coolbear, P., Newell, A.J. and Bryant, J.A. (1987). An evaluation of the potential of low temperature presowing treatments of tomato seeds as a means of improving germination. *Annals of Applied Biology*, **110**, 185-194.

Coolbear, P., Slater, R.J. and Bryant, J.A. (1990). Changes in nucleic acid levels associated with improved germination performance of tomato seeds after low-temperature pre-sowing treatment. *Annals of Botany*, **65**, 187-195.

Crouch, M.L. (1990). Debating the responsibilities of plant scientists in the decade. *Plant Cell*, **2**, 275-277.

Currah, I.E., Gray, D. and Thomas, T.H. (1974). The sowing of germinating vegetable seeds using a fluid drill. *Annals of Applied Biology*, **76**, 311-318.

Dasgupta, M. (1978). Studies on the viability of wheat (*Triticum aestivum* L.) seed. Ph.D *Thesis*, University of Calcutta, India.

Dasgupta, M. and Basu, R.N. (1975). Prolonging the viability of wheat seed. 14th *All India Wheat Workers' Workshop*, BCKVV, 1-4.

Dasgupta, M., Basu, P. and Basu, R.N. (1976). Seed treatment for vigour, viability and productivity of wheat (*Triticum aestivum* L.). *Indian Agriculturist*, **20**, 265-273.

Dasgupta, M., Chattopadhyay, K., Basak, S.L. and Basu, R.N. (1977). Radioprotective action of seed invigoration treatments. *Seed Research*, **5**, 104-118.

Dawson, M. J. (1965). Effect of Seed-soaking on the Growth and Development of Crop Plants: I. Finger millet (*Eleusine Coracana* Gaertn.).*Indian Journal of Plant physiology*, **8**, 52-56

De Souza, A., Garci, D., Sueiro, L., Gilart, F., Porras, E, and Licea, L. (2006). Pre-sowing magnetic treatments of tomato seeds increase the growth and yield of plants. *Bioelectromagnetics,***27**, 247-257.

De, B. K., Mandal, A. K., and Basu, R. N. (2004). Compatibility of fungicides with pre-storage dry and mid-storage wet physiological seed treatments for storability and field performance of wheat (*Triticum aestivum* L.). *Tropical agriculture.*

De, B.K., Mandal, A.K. and Basu, R.N. (1998). Effect of dry physiological seed treatments for improved vigour viability and productivity of black gram (*Phaseolus mungo*, Roxb.). *Indian Agriculturist*, **42**(1), 13-20.

De, B.K., Mandal, A.K. and Basu, R.N. (2003). Seed invigoration treatments on different seed sizes of wheat (*Triticum aestivum* L.) for improved storability and field performance. *Seed Science and Technology*, **31**, 379-388.

Dearman, J., Brocklehurst, P.A. and Drew, R.L.K. (1987). Effects of osmotic priming and ageing on the germination and emergence of carrot and leek seed. *Annals of Applied Biology*, **111**, 717-722.

Dell' Aquila, A. (2009). New perspectives for seed germination testing through digital imaging technology. *Open Agriculture Journal*, **3**, 37-42.

Dell' Aquila, A., van Eck, J.W., and van der Heijden, G.W.A.M. (2000). The application of image analysis in monitoring the imbibition process of white cabbage (*Brassica oleracea* L.) seeds. *Seed Science Research*, **10**, 163-169.

Delouche, J.C. (1969). Planting seed qualilty. In: *Proceedings of the 1969 Beltwide Cotton Production Mechanisation Conference*, January, 9-10, New Orleans, L.A., pp. 16-18.

Demopoulos, H.B. (1973). The basis of free radical pathology. *Federation Proceedings*, **32**, 18591861.

Dey (Pathak), G. and Basu, R.N. (1985). Physicochemical control of radiation and ageing damage in mustard seed. *Indian Journal of Experimental Biology*, **23**, 167-171.

Dey, G. and Basu, R.N. (1982). Maintenance of vigour, viability and yield potential of oil seeds. In: *Pulses and Oil Seeds in West Bengal*, pp.217-222. Government of West Bengal Publication, Calcutta.

Dey, G. and Basu, R.N. (1982). Studies on the maintenance of seed viability of sunflower (*Helianthus annuus* L.). *Indian Journal of Plant Physiology*, **25**(1), 87-97.

Dey, G. and Mukerjee, R.K. (1988). Invigoration of dry seeds with physiologically active chemicals in organic solvents. *Seed Science and Technology*, **16**, 145-153.

Dharmalingam, C. and Basu, R.N. (1978). Control of seed deterioration in cotton (*Gossypium hirsutum* L.). *Current Science*, **47**, 484-487.

Dharmalingam, C. and Basu, R.N. (1987). Influence of seed size and seed coat colour on the production potential of mungbean cv. Co. 3. *Seeds and Farms*, **13**(5), 16-20.

Dharmalingam, C. and Basu, R.N. (1989). Invigoration treatment for increased production in carried-over seeds of mungbean. *Seeds and Farms*, **15**(5), 34-36.

Dharmalingam, C. and Basu, R.N. (1993). Determining optimum season for the production of quality seeds in mung bean. *Madras Agricultural Journal*, **80**(12), 684-688.

Di Girolamo, G., and Barbanti, L. (2012). Treatment conditions and biochemical processes influencing seed priming effectiveness. *Italian Journal of Agronomy*, **7**, 25.

Dias Pereira, M., Fernandes, Cunha., dos Santos Dias, D., dos Santos Dias, L., and Fontes Araujo, E. (2009). Primed carrot seeds performance under water and temperature stress. *Scientia Agricola*, **66**, 174-179.

Dobretsov, G.E., Borschevskaya, T.A., Petrov, V.A. and Vladimirov, Y.A. (1977). The increase of phospholipid bilayer rigidity after lipid peroxidation. *FEBS Letters*, **84**, 125-128.

Dobrzanska, M., Tomaszewski, M., Grzelczak, Z., Rezman, E. and Buchowicz, J. (1975). Cascade activitation of genome transcription in wheat. *Nature, London*, **244**, 507-509.

Dona, M., Balestrazzi, A., Mondoni, A., Rossi, G., Ventura, L., Buttafava, A., Macovei, A., Sabatini, M.E., Valassi, A., and Carbonera, D. (2013). DNA profiling, telomere analysis and antioxidant properties as tools for monitoring *ex situ* seed longevity. *Annals of Botany*, **111**, 987-998.

Dornbos, D.L. (Jr.) (1995). Production Environment and Seed Quality. In: *Seed Quality, Basic Mechanisms and Agricultural Implications* (ed. A. S. Basra), Food Products Press, NY, USA pp. 119-152.

Duan, L., Yang, W., Huang, C., and Liu, Q. (2011). A novel machine-vision-based facility for the automatic evaluation of yield-related traits in rice. *Plant Methods*, **7**, 44-57.

Duque, A.S., Almeida, A.M., Bernardes da Silva, A., Marques da Silva, J., Farinha, A.P., Santos, D., Fevereiro, P., and de Sousa Araujo, S. (2013). Abiotic stress responses in plants; unraveling the complexity of genes and networks to survive. In:

Abiotic stress-plant responses and application in agriculture (Eds. Leslie, C & Vahdati, L), pp. 49-101.

Durrant, M.J., Payne, P.A. and McLaren, J.S. (1983). The use of water and some inorganic salt solutions to advance sugar beet seed. I. Laboratory studies. *Annals of Applied Biology*, **103**, 507-515.

Dutt, M., and Geneve, R. (2007). Time to radical protrusion does not correlate with early seedling growth in individual seeds of *Impatiens* and *Petunia*. *Journal of the American Society of Horticulture Science*, **132**, 423-428.

Ellis, R.H. and Roberts, E.H. (1981). The quantification of ageing and survival in orthodox seeds. *Seed Science and Technology* (Netherlands) agris.fao.org.

Ellis, R. H., Hong, T. D. and Roberts, E. H. (1986). Logarithmic relationship between moisture content and longevity in sesame seeds. *Annals of Botany*, **57**(4), 499-503.

Ellis, R.H., Hong, T.D. and Roberts, E.H. (1990). Effect of moisture content and method of rehydration on the susceptibility of pea seeds to imbibition damage. *Seed Science and Technology*, **18**, 131-137.

Ells, J.E. (1963). The influence of treating tomato seed with nutrient solutions on emergence rate and seedling growth. *Proceedings of the American Society of Horticultural Science*, **83**, 684-687.

Evenari, M. (1964). Hardening treatments of seeds as a means of increasing yields under conditions of inadequate moisture. *Nature* London, **204**, 1010-1011.

Evenari, M. (1984). Seed physiology: its history from antiquity to the beginning of the 20th century. *The Botanical Review*, **50** (2), 119-142.

Farokh, P., Mahmoodzadeh, H., and Satari, T. N. (2010). Response of seed germination of safflower to UV-B radiation. *Research Journal of Environmental Sciences*, 4(1), 70-74.

Feilding, J.L. and Golsworthy, A. (1982). The evolution of volatiles in relation to ageing in wheat seed. *Seed Science and Technology*, **10**, 277-282.

Fiorani F. and Schurr U. (2013). Future scenarios for plant phenotyping. *Annual Review of Plant Biology*. **64**, 267-291.

Flood, R.G. and Sinclair, A. (1981). Fatty acid analysis of aged permeable and impermeable seeds of *Trifolium subterraneum* (subterranean clover). *Seed Science and Technology*, **9**, 475-477.

Fret, J.J., Pill, W.G. and Morneau, D.C. (1991). A comparison of priming agents for tomato and asparagus seeds. *Horticulture Science*, **26**, 1158-1159.

Fujikura, Y, and Karssen, C.M. (1992). Effects of controlled deterioration and osmopriming on protein synthesis of cauliflower seeds during early germination. *Seed Science Research*, **2**, 23-31.

Gaidau, C., Niculescu, M., Stepan, E., Epure, D.G., and Gidea, M. (2014). New mixes based on collagen extracts with bioactive properties, for treatment of seeds in sustainable agriculture. *Current Pharmaceutical Biotechnology*, **14**, 1389-2010.

Georghiou, K., Thanos, C.A. and Passam, H.C. (1987). Osmoconditioning as a means of counteracting the ageing of pepper seeds during high temperature storage. *Annals of Botany*, **6**, 279-285.

Germ, N.S. (1960). Methodology of the vigour test for wheat, rye and barley, in rolled filter paper. In: *Proceedings of International Seed Testing Association*, **16**, 209-311.

Ghassemi-Golezani, K., Hosseinzadeh-Mahootchy, A., Zehtab-Salmasi, S., and Tourchi, M. (2012). Effects of hydro-priming duration on seedling vigour and grain yield of pinto bean (*Phaseolus vulgaris* L.) cultivars. *Notulae Botanicae Horti Agrobotanici*, **38**, 109-113.

Goldsworthy, A. and Feilding, J.L. and Daver, M.B.J. (1982). Flash imbibition – a method for re-invigoration of aged wheat seed. *Seed Science and Technology*, **10**, 55-57.

Gorecki, R.J. and Harman, G.E. (1987). Effects of antioxidants on viability and vigour of ageing pea seeds. *Seed Science and Technology*, **15**, 109-117.

Grover, S., and Khan, A.S. (2014). Effect of ionizing radiation on some characteristics of seeds of wheat. *International Journal of Scientific and Technology Research*, **3**, 2277-8616.

Guajardo-Flores, D., Serna-Guerrero, D., Serna-Saldivar, S.O., and Jacobo-Velazquez, D.A. (2014). Effect of germination and UV-C radiation on the accumulation of flavonoids and saponins in black bean seed coats. *Cereal Chemistry*, **91**, 276-279.

Guha, P., Biswas, J., De, B. K. and Mandal, A. K. (2012). Post-harvest dry and wet physiological seed treatments for improved storability and field performance of okra (*Abelmoschus esculentus* L.). *Indian Journal of Agricultural Researc*h, **46**(1), 16-22.

Gulyaev, B.I., Kazakov, E.A. and Okanenko, A.S. (1973). Heliotechnical apparatus for irradiating plant seed, pollen and flowers with intermittent concentrated sunlight. *Fiziologiya IBiokhimiya Kul turnykh Rastenii*, **5**, 543-548.

Gulyaev, B.I., Kazakov, E.A. and Okanenko, A.S. (1975). Automated apparatus for irradiating biological objects with concentrated sunlight. *Fiziologiya i Biokhimiya Kulturnykh Rasentii*, **7**, 641-645. (*Horticulture Abstracts*, **46**, 8175).

Gurusinghe, S., and Bradford, K. (2001). Galactosyl-sucrose oligosaccharides and potential longevity of primed seeds. *Seed Science Research*, **11**, 121-133.

Haber, A.H. and Randolph, M.L. (1967). Gamma-ray-induced ESR signals in lettuce: evidence for seed-hydration-resistant and sensitive free radicals. *Radiation Botany*, **7**, 17-28.

Haigh, A.M. and Barlow, E.W.R. (1987). Germination and priming of tomato, carrot, onion, and sorghum seeds in a range of osmotica. *Journal of the American Society of Horticultural Science*, **112**, 202-208.

Haigh, A.M., Barlow, E.W.R., Milthorpe, F.L. and Sinclair, P.J. (1986). Field emergence of tomato, carrot and onion seeds primed in an aerated salt solution. *Journal of the American Society for Horticultural Science*, **111**, 660-665.

Hailstones, M.D. and Smith, M.T. (1988). Lipid peroxidation in relation to declining vigour in seeds of soybean (*Glycine max* L.) and cabbage (*Brassica oleracea* L.). *Journal of Plant Physiology*, **133**, 452-456.

Hailstones, M.D. and Smith, M.T. (1989). Increased lipid peroxidation during the imbibition of deteriorated seeds and the invigoration effect of ferrous-iron. *South African Journal of Science*, **85**(10), 670-671.

Hailstones, M.D. and Smith, M.T. (1989). Thermally derived volatile aldehydes in relation to viability in soybean seeds. *Seed Science and Technology*, **17**(3), pp. 649-658.

Hailstones, M.D. and Smith, M.T. (1991). Soyabean seed invigoration by ferrous sulfate: changes in lipid peroxidation, conductivity, tetrazolium reduction, DNA and protein synthesis. *Journal of Plant Physiology*, **137**, 307-311.

Halmer, P. (2004). Methods to improve seed performance in the field. In: *Handbook of seed physiology; applications to agriculture*. (eds. Arnold, R.L. and Sachez, R.A.). Haworth Press Inc, New York, pp 125-165.

Hamid, N., and Jawaid, F. (2011). Influence of seed pre-treatment by UV-A and UV-C radiation on germination and growth of mung beans. *Pakistan Journal of Chemistry*, **1**, 164-167.

Hampton, J. G. (1995). In: *Seed Vigour Testing Seminar* (ed. H. A. van de Venter), ISTA, Zurich, Switzerland, pp. 10-28.

Hampton, J. G. (1995). Production environment and seed quallity. In: *Seed Quality, Basic Mechanisms and Agricultural Implications* (ed. A. S. Basra), Food Products Press, NY, USA pp.81-118.

Hampton, J.G. and Tekrony, D. M. (1995). *Handbook of Vigour Test Methods*. ISTA, Zurich, Switzerland.

Hanson, P.R., Wainman, H.E. and Chakrabarti, B. (1987). The effects of methyl bromide on the seed of some varieties of barley. *Seed Science and Technology*, **15** (1), 155-162.

Harman, G.E. and Stasz, T.E. (1986). Influence of seed quality on soil microbes and seed rots. In: *Physiological Pathological Interactions Affecting Seed Deterioration* (ed. S.H. West), pp.11-37. Crop Science Society of America Special Publication.

Harman, G.E., Nedrow, B.L. and Nash, G. (1978). Stimulation of fungal spore germination by volatiles from aged seeds. *Canadian Journal of Botany*, **56**, 2124-2127.

Harman, G.E., Nedrow, B.L., Clark, B.E. and Mattick, L.R. (1982). Association of volatile aldehyde production during germination with poor soybean and pea seed quality. *Crop Science*, **22**, 712-716.

Harrison, J.C. (1926). The packing and storage of tea seed. *Journal of the Indian Tea Association*, 126-133.

Hartikainen, H. (2005). Biogeochemistry of selenium and its impact on food chain quality and human health. *Journal of Trace Elements in Medicine and Biology*, **18**, 309-318.

Heckly, R.T. and Dimmick, R.L. (1967). Electron paramagnetic resonance in frozen and dried biological materials. *Nature, London*, **216**, 1003-1004.

Hegarty, T.W. (1970). The possibility of increasing field establishment by seed hardening. *Horticulture Research*, **10**, 59-64.

Hegarty, T.W. (1970). The possibility of increasing field establishment by seed hardening. *Horticulture Research*, **10**, 59-64.

Helsel, D.G., Helsel, D.R. and Minor, H.C. (1986). Field studies on osmoconditioning soybeans (*Glycine max). Field Crops Research*, **14**, 291-298.

Henckel, P.A. (1946). The resistance of plants to drought and methods of increasing it. Training Institute, *Plant Physiology Academy of Sciences*, USSR, **5**, 2377.

Henckel, P.A. (1964). Physiology of plants under drought. *Annual Review of Plant Physiology*. **15**, 363-386.

Hendricks, S.B. and Taylorson, R.B. (1974). Promotion of seed germination by nitrate, nitrite, hydroxylamine and ammonium salts. *Plant Physiology*, Washington, **54**, 304-309.

Henckel, P.A. (1962). Physiology of plants under drought. U.S.S.R Academy of Sciences Moscow., 363-383.

Henckel, P.A. (1968). Methodical instructions concerning presowing tempering of plants against drought. *Ministry of Agriculture*, Moscow, (Russian).

Heydecker, W. (1960). Can we measure seedling vigour? In: *Proceedings of International Seed Testing Association*, **25**, 498-512.

Heydecker, W. (1972). Vigour. In: *Viability of Seeds* (ed. E.H. Roberts), 209-286. Chapman and Hall, London.

Heydecker, W. (1974). Germination of an idea; the priming of seeds. *Report of the University of Nottingham, School of Agriculture for* 1973/1974, pp. 50-67.

Heydecker, W. (1975). 'Seed Priming' – the treatment of the future? *Grower*, 554-555.

Heydecker, W. (1977). Stress and seed germination: an agronomic view. In: *The Physiology and Biochemistry of Seed Dormancy and Germination* (ed. A.A. Khan), pp.237-282. Elsevier/North-Holland, Amsterdam.

Heydecker, W. and Chetram, R.J. (1971). Water relations of beet root seed germination. I. Microbial factors with special reference to laboratory germination. *Annals of Botany*, **35**, 17-30.

Heydecker, W. and Coolbear, P. (1977). Seed treatments for improved performance: survey and attempted prognosis. *Seed Science and Technology*, **5**, 353-425.

Heydecker, W. and Hendy, A., (1975). Pre-treating bedding plant seed for 'instant' germination. U.K., *Commercial Grower,* **4163**, 613-615.

Heydecker, W., Higgins, J. and Turner, J.Y. (1975). Invigoration of seeds. *Seed Science and Technology,* **3**, 881-888.

Heydecker, W., Higgins, J., and Gulliver, R.L. (1973). Accelerated germination by osmotic seed treatment. *Nature,* **246**, 42-44.

Hibbard, R.P. and Miller E.V. (1928). Biochemical studies on seed viability: 1.Measures of conductance and reduction. *Plant Physiology*, **3**(3), PMC 440015.

Hiltner, L. and Ihssen, G. (1911). User das Schlechte Auflanten and dic Ans-Wintering des Getreides inforge befalls durch. *Fusarium LandWirtsch. Jb. Sayern*, **1**, 20-60, 231-78, 315-62.

Ho,M.W.(2008).Organic agriculture and localized food and energy systems for mitigating climate change. Institute of Science in Society Report.

Hoffman, U. (2011). Assuring food security in developing countries under the challenges of climate change: Key trade and development issues of a fundamental transformation of agriculture. UNCTAD Discussion Paper No.201, 15 March, 2011.

Hofmann, P., Aschermann-Koch, C. and Steiner, A.M. (1992). Presowing treatment for improving seed quality in cereals. II. Field emergence and yield. *Seed Science and Technology*, **20**, 441-446.

Holm, R.E. and Miller, M.R. (1972). Weed germination responses to chemical and physical treatments. *Weed Science*, **20**, 150-153.

Hosseinzadeh-Mahootchi, A., and Ghassemi-Golezani, K. (2013). The impact of seed priming and ageing on physiological performance of chickpea under different irrigation treatments. *Plant Breeding and Seed Science,* **67**, 13-25.

Hume, P.F. (1955). Storage of tea seed. *Tea Quarterly*, **26**, 93-95.

Ibrahim, A. E. (1981). The effect of moisture content, oxygen availability and temperature on survival of lettuce (*Lactuca sativa* L.) and onion (*Allium cepa* L.). seed. (Doctoral dissertation, University of Reading).

Isely, D. (1950). The cold test for corn. In: *Proceedings of International Seed Testing Association,* **16**, 299-311.

Isely, D. (1957). Vigour tests. In: *Proceedings of Association of Official Seed Analysts*, **47**, 176-82.

ISTA (International Seed Testing Association) (1966). International rules for seed testing. In: *Proceedings of International Seed Testing Association*, **31**, 1-152.

Juliana, C., Caprita, R., Giancarla, V., and Sorina, R. (2013). Response of barley seedlings to microwaves at 2.45 GHz. *Journal of Animal Science and Biotechnology,***46**, 185-191.

Jakubowski, T. (2010). The impact of microwave radiations at different frequencies on the weight of seed potato germs and crop of potato tubers. *Agricultural Engineering,* **6**, 57-64.

Jeng, T.L. and Sung, J.M. (1994). Hydration effect on lipid peroxidation and peroxide-scavanging enzyme activity of artificially aged peanut seed. *Seed Science and Technology*, **22**, 531-539.

Jett, L.W. and Welbaum, G.E. (1996). Changes in broccoli (*Brassica oleracea* L.) seed weight, viability, and vigour during development and following drying and priming. *Seed Science and Technology,* **24**, 127-137.

Jisha, K.C., Vijayakumari, K., and Puthur, J.T. (2013). Seed priming for abiotic stress tolerance; an overview. *Acta Physiologiae Plantarum*, **35**, 1381-1396.

Jones, L. G. and Cobb, R. D. (1963). A technique for increasing the speed of laboratory germination testing. In: *Proceedings of Association of Official Seed Analysts(AOSA),* **53**, 144-160.

Joshi, R.K., Gaur, B.K. and Notani, N.K. (1969). Recovery from gamma radiation injury in barley seeds. *Radiation Botany*, **9**, 141-145.

Joslyn, M.A. and Timmons, A. (1967). Salt-use in food processing. In: *Fundamentals of Food Processing Operations* (eds. J.L. Heid and M.A. Joslyn), 83. AVI Publishing Co., Connecticut, USA.

Kapri, B., Sengupta, A.K., De, B.K., Mandal, A.K. and Basu, R.N. (2003). Pre-storage seed invigoration treatments for improved germinability and field performance of okra (*Hibiscus esculentus*). *Indian Journal of Agricultural Sciences,* **73**(5), 276-279.

Kesavan, P.C. (1973). Effect of caffiene and cysteine applied during post-irradiation hydration on an oxygen-independent component of damage in barley seed. *Radiation Botany*, **13**, 355-359.

Kesavan, P.C. and Afzal, S.J., (1975). Dependence of chemical radioprotection on the stability of radiation-induced oxygen-sensitive sites. *International Journal of Radiation Biology and Related Studies in Physics, Chemistry and Medicine*, **28**(5), 495-499.

Kesavan, P.C., Swarma, G.J. and Afzal, S.M.J. (1978). Differential modification of oxic and anoxic radiation damage by chemicals. Simulation of the action of caffeine by certain inorganic radical scavengers. *Radiation Botany,* **75**, 18-30.

Ketring, D.L. (1991). Physiology of oil seeds: IX. Effects of water deficit on peanut seed quality. *Crop Science*, **31**, 459-463.

Khan, A.A. (1977). Preconditioning, germination and performance of seeds. In: *The Physiology and Biochemistry of Seed Dormancy and Germination* (eds. Khan, A.A.), pp. 283-318. Elsevier/North-Holland Biomedical Press, Amsterdam.

Khan, A.A. (1992). Pre-plant physiological seed conditioning. *Horticultural Review*, **4**, 131-181.

Khan, A.A., Maguire, J.D., Abawi, G.S. and Ilyas, S. (1992). Matriconditioning of vegetable seeds to improve stand establishment in early field plantings. *Journal of the American Society for Horticultural Science*, **117**, 41-47.

Khan, A.A., Miura, H., Prusinski, J. and Ilyas, S. (1990). Matriconditioning of seeds to improve emergence. In: *Proceedings of the Symposium on Stand Establishment of Horticultural Crops.*, April 4-6, 1990, pp.19-40. Minneapolis, Minnesota.

King, M.W. and Roberts, E.H. (1980). Maintenance of recalcitrant seeds in storage. In: *Recalcitrant Crop Seeds*, (eds. H.F. Chin and E.H. Roberts), pp.53-89. Tropical Press, Kuala Lumpur, Malaysia.

Kishor, P.B.K., Sangama, S., Amrutha, R.N., Laxmi, P.S., Naidu, K.R., and Rao, K.S. (2005). Regulation of proline biosynthesis degradation, uptake and transport in higher plants; its implications in plant growth and abiotic stress tolerance. *Current Science*, **88**, 424-438.

Kittock, D.L. and Law, A.G. (1968). Relationship of seedling vigour, respiration and tetrazolium chloride reduction by germinating wheat seeds. *Agronomy Journal*, 60, 286-288.

Knypl, J. S., and Khan, A. A. (1981). Osmoconditioning of soybean seeds to improve performance at suboptimal temperatures 1. *Agronomy Journal*, **73**(1), 112-116.

Knypl, J.S. (1981). Osmoconditioning of soybean (*Glycine max*) seeds to improve the field performance at early planting. *Acta Physiologia Plantarum*, **3**, 117-124.

Koostra, P.T. and Harrington, J.E. (1969). Biochemical effects of age on membranal lipids of *cucumis sativus* L. seed. In: *Proceedings of International Seed Testing Association*, **34**, 329-340.

Kotsur, N.V., Pochinok, Kh.N., Umarov, G.Ya., Okanenko, A.S., Golik, K.N. and Pogol'skaya, V.I. (1972). Increasing sugar content in sugar beet by presowing seed treatment with intensive light. *Dopovidi Akademii. Nauk. Ukraines Koi R.S.R. Seriia* B, No. **12**, 1122-1124 (*Field Crop Abstracts*, **26**, 5002).

Kranner, I., Beckett, R.P., Minibayeva, F.V., and Seal, C.E. (2010). What is stress: concepts, definitions and applications in seed science.*New Phytologist*, **188**, 655-673.

Kubala, S., Garnczarska, M., Wojtyla, L., Rucinska-Sobkowiak, R., Kubala, S., and Garnczarska, M. (2015). Deciphering priming-induced improvement of rapeseed (*Brassica napus* L.) germination through an integrated transcriptomic and proteomic approach. *Plant Science*, **231**, 94-113.

Kundagrami, S., De, B. K., and Mandal, A. K. (2008). Post-harvest seed invigoration treatments for improved storability and field performance of rice (*Oryza sativa* L.). *Crop Research*, **34**(3), 317-322.

Kundu, C. and Basu, R.N. (1981). Hydration-dehydration treatment of stored carrot seed for the maintenance of vigour, viability and productivity. *Scientia Horticulturae*, **15,** 117-125.

Kundu, C. and Basu, R.N. (1983). Control of seed deterioration and improvement of field performance of okra by hydration-dehydration treatment of stored seed. *Vegetable Science*, **10**(2), 84-93.

Kuo, W.H.J. and Chi, C. (1986). Prolonged storability of rice (*oryza sativa* L.) seeds by imbibition-hydration treatments: conditions for treatment (in Chinese). *Journal of Agricultural Association of China, New Series*, **133**, 16-29.

Lanteri, S., Belletti, P., Marzach, C., Nada, E., Quagliotti, L. and Bino, R.J. (1997). Priming-induced nuclear replication activity in pepper (*Capsicum annuum* L.) seed. Effect on germination and storability. In: *Basic and Applied Aspects of Seed Biology* (eds.R.H. Ellis, M. Black, A.J. Murdoch, and T.D. Hong), pp.451-460. Kluwer Academic Publishers, Boston.

Layek, N., De, B. K., Mishra, S. K., & Mandal, A. K. (2006). Seed invigoration treatments for improved germinability and field performance of gram (*Cicer arietinum* L.). *Legume Research-An International Journal*, **29**(4), 257-261.

Layek, N., Guha, P., De, B. K., & Mandal, A. K. (2012). Pre-storage seed invigoration treatments for the maintenance of germinability and field performance of uradbean(*Vigna Mungo* L. Hepper) *Legume Research-* **35**(3), 220-225.

Le Deunff, Y. (1973). Phytochrome et choc thermique dans la germination des semences de *Rumex crispus* L., *Comptes rendus hebdomadaires des séances de l'Académie des Sciences, Paris,* **276**, 2443-46.

Lei, L., Qin, Z., and Danfeng, H. (2014). A review of imaging techniques for plant phenotyping. *Sensors,***14**, 20078-20111.

Leopold, A. C. (1999). Learning from seeds. *Seed Science Research*, **9**(2), 111-116.

Leskovar, D.I. and Sims, W.L. (1987). Emergence and yield of tomato (*Lycopersicon esculentum* Mill) in response to presowing treatment. *Acta Horticulturae,* **200**, 145-152.

Li, X., and Zhang, L. (2012). SA and PEG-induced priming for water stress tolerance in rice seeding. *Information Technology in Agriculture,* **134**, 881-887.

Li, X., and Zhang, L. (2012). SA and PEG-induced priming for water stress tolerance in rice seeding. *InformationTechnology in Agriculture,***134**, 881-887.

Liklatchev, B.S., Zelensky, G.V., Kiashko, G.Y. and Schevchenko, Z.N. (1984). Modeling of seed ageing. *Seed Science and Technology,* **12**, 385-393.

Linderbein, W. and Bulat, H. (1955). Triebkraft Ziegelgruswert and Tetra-zolium Keimwert Saatgut-wirtzehaft, Z, 315-319.

Liu, X., Xing, D., Li, L., and Zhang, L. (2007). Rapid deterioration of seed vigour based on the level of superoxide generation during early imbibition. *Photochemical and Photobiology Sciences,* **6**, 767-774.

Liu, Y., Kermode, A., and El-Kashaby, Y.O. (2013). The role of moist-chilling and thermo-priming on the germination characteristics of white spruce (*Picea glauca*) seed. *Seed Science and Technology,* **41**, 1-15.

Lu, C. and Toepl, K., Irish, R., Fensk, R.A., Barr, D.B. and Bravo, C. (2006). Organic diets significantly lower children's dietary exposure to organophosphorus pesticides. *Environmental Health Perspectives,* **114**(2), 260-263.

Macovei, A., Balestrazzi, A., Confalonieri, M., and Carbonera, D. (2010). The *Tdpl* (tyrosyl-DNA phosphodiesterase) gene family in *Medicago truncatula* Gaertn: bioinformatic investigation and expression profiles in response to copper- and PEG-mediated stress. *Planta,* **232**, 393-407.

Macovei, A., Balestrazzi, A., Confalonieri, M., Buttafava, A., and Carbonera, D. (2011a). The *TFIIS* and *TFIIS*-like genes from *Medicago truncatula* are involved in oxidative stress response. *Gene* **470,** 20-30.

Macovei, A., Balestrazzi, A., Confalonieri, M., Fae, M., and Carbonera, D. (2011b). New insights on the barrel medic *MtOGGI* and *MtFPG* functions in relation to oxidative stress response in plant and during seed imbibition. *Plant Physiology and Biochemistry,* **49**, 1040-1050.

Macovei, A., Garg, B., Raikwar, S., Balestrazzi, A., Carbonera, D., Buttafava, A., Bremont, J.F.J., Gill, S.S,, and Tuteja, N. (2014). Synergistic exposure of rice seeds to different doses of gamma-ray and salinity stress resulted in increased antioxidant enzyme activities and gene-specific modulation of TC-NER pathway. *Biomedical Research International,* **2014**, 676-934.

Mahajan, G., Sarlach, R.S., Japinder, S., and Gill, M.S. (2011). Seed priming effects on germination, growth and yield of dry direct-seeded rice. *Journal of Crop Improvement,* **25**, 409-417.

Mandal, A,K. and Basu, R.N. (1982). Alleviation of salt stress during germination and early seedling growth of wheat by hydration-dehydration pre-treatments. *Indian Agriculturist,* **26**, 121-128.

Mandal, A.K., De, B. K., Chatterjee, B. K., & Kapri, B. (2008). Pre-storage dry seed treatments for improved vigour, viability and field performance of sesamum (*Sesamum indicum* L.). *Indian Agriculturist,* 52(3), 127-132.

Mandal, K. and Basu, R.N. (1981). Role of embryo and endosperm in rice seed deterioration. In: *Proceedings of the Indian National Science Academy,* **B47**, 109-114.

Mandal, K. and Basu, R.N. (1982). Dip-dry treatment for the maintenance of vigour and viability of stored wheat seed. *Indian Agriculturist*, **26**, 289-295.

Mandal, A.K. and Basu, R.N. (1983). Maintenance of vigour, viability and yield potential of stored wheat seed. *Indian Journal of Agricultural Sciences*, **53**(10), 905-912.

Mandal, A.K. and Basu, R.N. (1984). Preservation of wheat seed by hydration-dehydration treatment with seed protectants. *Pestology*, **8**(1), 20-23.

Mandal, A.K. and Basu, R.N. (1984). Response of different wheat cultivars to hydration-dehydration treatment for the maintenance of vigour, viability and productivity. *Seeds and Farms*, **10**(9-10), 39-42.

Mandal, A.K. and Basu, R.N. (1986). Vigour and viability of wheat seed treated with bleaching powder. *Seeds and Farms*, **12**(10), 46-47.

Mandal, A.K. and Basu, R.N.(1987a). Mid-term and presowing hydration-dehydration treatments for improved field performance of wheat. *Field Crops Research*,**15**, 259-265.

Mandal, A.K. and Basu, R.N. (1987b). Effect of sowing water-soaked and dried seeds on the yield of wheat with or without fertilizer application. *Indian Journal of Agricultural Sciences*, **57**, 517-519.

Mandal, A.K., De, B.K. and Basu, R.N. (1999). Dry-seed treatment for improved germinability and productivity of wheat (*Triticum aestivum*). *Indian Journal of Agricultural Sciences*, **69**(9), 627-630.

Mandal, A.K., De, B.K., Saha, R. and Basu, R.N. (2000). Seed invigoration treatments for improved storability, field emergence and productivity of soybean (*Glycine Max*. L. cv. Merril). *Seed Science and Technology*, 28, 349-355.

Marcus, A., Feeley, J., & Volcani, T. (1966). Protein synthesis in imbibed seeds III. Kinetics of amino acid incorporation ribosome activation, and polysome formation. *Plant physiology*, 41(7), 1167-1172.

Martin, R.E., Miller, R.L. and Cushwa, C.T. (1975). Germination response of legume seeds subjected to moist and dry heat. *Ecology*, **56**, 1441-1445.

Matthews, S. and Bradnock, W.T. (1967). The detection of seed samples of wrinkle-seeded peas (*Pisum sativum* L.) of potentially low planting value. In: *Proceedings of the International Seed Testing Association*, **32**, 553-563.

Matthews, S. and Bradnock, W.T. (1968). Relationship between seed exudation and field emergence in peas and French beans. *Horticultural Research*, **8**, 89-93.

Maude, R.B., Drew, R.L.K., Gray, D., Bujalski, W. and Nienow, A.W. (1994). The effects of storage on the germination and seedling abnormalities of leek seeds primed and dried by different methods. *Seed Science and Technology*, **22**, 299-311.

Mauromicale, G. and Cavallaro, V. (1996). Effects of seed osmopriming on germination of three herbage grasses at low temperature. *Seed Science and Technology*, **24**, 331-338.

May, L.H., Milthorpe, E.J. and Milthorpe, F.L. (1962). Presowing hardening of plants to drought. *Field Crop Abstracts*, **15**, 93-98.

Mc Donald, M.B. Jr. (1999). Seed deterioration: Physiology, repair and assessment. *Seed Science and Technology*, **27**, 177-237.

McDonald, M. B. (Jr.) (1995). In: *Seed Vigour Testing Seminar* (ed. H. A. van de Venter), ISTA, Zurich, Switzerland, pp. 88-97.

McDonald, M.B. (2000). Seed priming. In: *Seed technology and its biological basis*. eds. M. Black, J.D. Bewley). Sheffield Academic Press, Sheffield, pp 287-325.

McDonald, M.B. Jr. (1993). The history of seed vigour testing. *Journal of Seed Technology*, **17**(2), 93-100.

McDonald, M.B., Jr. (1980). Vigour test subcomittee report. *Association of Official Seed Analysts Newsletter*, **54**, 137-140.

Mexal, J., Fisher, J.T., Osteryoung, J. and Reid, C.P. (1975). Oxygen availability in polyethylene glycol solutions and its implications in plant-water relations. *Plant Physiology*, ***55***(1), pp.20-24.

Milborrow, B.V. (1963). Penetration of seeds by acetone solvents. *Nature, London*, **190**, 716-717.

Mitchell, A., Hong, Y., Koh, E., Barreet, D.M., Bryant, D.E., Denision, R.F. and Kaffka, S. (2007). Ten–year comparison of the influence of organic and conventional crop management practices on the content of flavonoids in tomatoes. *Journal of Agricultural and Food Chemistry*, **55**, 654-59.

Mitra, R. and Basu, R.N. (1979). Seed treatment for viability, vigour and productivity of tomato. *Scientia Horticulturae*, **11**, 365-369.

Mitra, R. and Basu, R.N. (1986). Maintenance of vigour, viability and yield potential of cauliflower seed. *Vegetable Science*, **13**(1), 34-40.

Mitra, R., Pal, P. and Basu, R.N. (2001). Pre-storage seed invigoration treatment for improved performance of mungbean (*Phaseolus aureus* RoxB.). *Indian Biologist,* **33**(2), 21-25.

Molnár, K. (1972). Subbiological aspects of ageing and the concept of biological cathode protection. Mechanisms of ageing and development,**1** , 319-326.

Mondoni, A., Probert, R.J., Rossi, G., Vegini, E., and Hay, F.R. (2011). Seeds of alpine plants are short lived; implications for long-term conservation. *Annals of Botany*, **107**, 171-179.

Moore, R. P. (1972). Effects of mechanical injuries on viability. In: *Viability of Seeds* (ed. E. H. Roberts), Chapman and Hall, London, pp. 94-113.

Mukhamedkhanov, O, and Shermukhamedov, K. (1971). Effect of ultrasound on development and yield of cotton cv, 108-F, *Nauchnye Trudy, Tashkentsku Selskokhozyaistvennyi Institut*. No.**22**, 45-71, (*Field Crop Abstracts,* **25**, 4165).

Mukhopadhay, A., Pal, P. and Basu, R.N. (1997). Seed invigoration treatments for French bean (*Phaseolus vulgaris* L.) and soybean (*Glycine max* L.) for improved vigour and viability after storage. *Indian Agriculturist,* **41**(2), 79-85.

Mukhopadhyay, A. (2001). *Ph.D. Thesis,* (Supervisor, R. N. Basu; Associate Supervisor, P. Pal), University of Calcutta, India.

Musgrave, M.E., Priestley, D.A. and Leopold, A.C. (1980). Methanol stress test as a test of seed vigour. *Crop Science,* **20**, 626-630.

Nadkarni, S. and Kesavan, P.C. (1975). Modification of the radiosensitivity of of barley seed by post treatment with caffeine. Kinetics of decay of caffeine-reactive oxygen-sensitive sites. *International Journal of Radiation Biology and related studies in Physics, Chemistry and Medicine,* **27**(6), 569-576.

Nawaz, F., Ashraf, M.Y., Ahmad, R., and Waraich, E.A. (2013). Selenium (Se) seed priming induced growth and biochemical changes in wheat under water deficit conditions. *Biological Trace Element Research,***151**, 284-293.

Nickerson, J.T.R. (1967). Preservatives and antioxidants. In: *Fundamentals of Food Processing Operations* (Eds. J.L. Heid and N.A. Joalyn). pp. 222, AVI Publishing Company., Connecticut, USA.

Nobbe, F. (1876) *Handbuch der Samenkunde Berlin*: Wregand, Hampel and Parey.

Oelke, E. A., Ball, R. B., Wick, C. M and Miller, M.D. (1969). Influence of grain moisture at harvest on seed yield quality and seedling vigour of rice. *Crop Science,* **9**:144-147.

Oliveira, A. B., & Gomes-Filho, E. (2016). How are germination performance and seedling establishment under abiotic stress improved by seed priming? A review. *Australian Journal of Crop Science, AJCS,* **10**(7), 1047-1051.

Onwueme, I.C. (1975). Enhancement of okra (*Hibiscus esculentus*) seedling growth and emergence by heat stress condition. *Journal of. Agricultural Science, Cambridge,* **85**, 233-239.

Osborne, D.J. (1982). Deoxyribonucleic acid integrity and repair in seed germination: the importance in viability and survival. In: *The Physiology and Biochemistry of Seed Development, Dormancy and Germination* (ed. A.A. Khan), pp.435-463, Elsevier/North-Holland, Amsterdam.

Ouhibi, C., Attia, H., Rebah, F., Msilini, N., Chebbi, M., Aarrouf, J. et al. (2014). Salt stress mitigation by seed priming with UV-C in lettuce plants, growth, antioxidant activity and phenolic compounds. *Plant Physiology and Biochemistry,* **83**, 126-133.

Ovcharov, K.E. (1977). Physiological basis of seed germination. (Translation from Russian by B.R. Sharma), pp.54-140. Amerind Publishing, New Delhi.

Owen, P. L., & Pill, W. G. (1994). Germination of osmotically primed asparagus and tomato seeds after storage up to three months. *Journal of the American Society for Horticultural Science*, **119**(3), 636-641.

Pal, P. and Basu, R.N. (1982). Control of rice seed deterioration by physico-chemical treatments. In: *Rice in West Bengal*, **3**, 159-166. Government of West Bengal Publication, Calcutta.

Pal, P. and Basu, R.N. (1988). Treatment of rice seed with iodine and chlorine for the maintenance of vigour, viability and productivity. *Indian Agriculturist*, **32**, 71-75.

Pal, P. and Basu, R.N. (1989). Storage of rice seed samples for more than a decade under ambient temperature and low humidity regimes. *Indian Agriculturist*, **33**(3), 181-183.

Pal, P. and Basu, R.N. (1989). Volatile aldehyde production in relation to seed vigour of rice (*Oryza sativa* L.). *Indian Agriculturist*, **33**, 223-225.

Pal, P. and Basu, R.N. (1993). Effect of powdered red chilli, turmeric and neem leaf on pre- and post-storage germinability of wheat seed. *Indian Agriculturist*, **37**(4), 267-271.

Pal, P. and Basu, R.N. (1994). Extension of wheat seed viability with plant leaf powders and an aspirin-containing pharmaceutical preparation. *Indian Agriculturist*, **38**(2), 145-149.

Pal, P., Mandal, A.K. and Basu, R.N. (2014). Farmer oriented seed preservation technology for conservation of crop biodiversity of major agriculture and horticultural crops of West Bengal with special emphasis on indigenous locally adopted crop variates. Report of the Research Project, Financed by Govt. of West Bengal (2012-2014).

Pal, P., Mandal, A.K. and Basu, R.N. (2017). Alleviation of the adverse effects of climate change by seed invigoration treatments. In: *Souvenir and Abstracts, Indian society for Seed Technology, XIV National Seminar*, January 28-30, 2017, IARI, New Delhi.

Pal, P., Ray, I. and Basu, R.N. (2001). Seed invigoration treatment in wheat (*Triticum aestivum* L.) as influenced by the vigour status of seed at the time of treatment. *Indian Agriculturist*, **45**(1), 1-9.

Pal, P., Ray, I., Mitra, R. and Basu, R.N. (1996). Dry physiological seed treatments for improved performance of wheat (*Triticum aestivum*) and sunflower (*Helianthus annuus*). *Indian Agriculturist*, **40**(4), 219-232.

Pammenter, N.W., Adamson, J.H. and Berjak, P. (1974). Viability of stored seed: extension by cathodic protection. *Science*, **186**, 1123-1124.

Pan, D. and Basu, R.N. (1985a). Mid-storage and presowing seed treatments for lettuce and carrot. *Scientia Horticulturae*, **25**, 11-19.

Pan, D. and Basu, R.N. (1985b). Absence of *de novo* protein synthesis in moisture equilibration-drying treatment for maintenance of lettuce (*Lactuca sativa* L.) seed viability. *Indian Journal of Experimental Biology*, **23**, 375-370.

Pan, D., Punjabi, B. and Basu, R.N. (1981). A note on the involvement of protein synthesis in hydration-dehydration treatment of lettuce seed. *Seed Research*, **9**(2), 202-205.

Panse, V. G. and Sukhatme, P. V. (1967). *Statistical Methods for Agricultural Workers.* (2nd edition) ICAR, New Delhi, India, pp. 112-121.

Paparella, S., Araújo, S.S., Rossi, G., Wijayasinghe, M., Carbonera, D. and Balestrazzi, A. (2015). Seed priming: state of the art and new perspectives. *Plant Cell Reports*, **34**(8), 1281-1293.

Parera, C.A., and Cantliffe, D.J. (1994). Pre-sowing seed priming. *Horticulture Review*, **16**, 109-141.

Passam, H.C., Lambropoulos, E. and Khah, E.M. (1997). Pepper seed longevity following production under high ambient temperature. *Seed Science and Technology*, **25**, 177-185.

Pathak, G. (1980). Seed treatment for vigour, viability and productivity of some oil-yielding crops. *Ph.D. Thesis*, Calcutta University, India.

Pathak, G. and Basu, R.N. (1980). Control of seed deterioration in sunflower (*Helianthus annuus* L.), *Current Science*, **49**, 67-69.

Pearce, R.S. and Abdul-Samad, I.M. (1980). Changes in fatty acid content of polar lipids during ageing of seeds of peanut (*Arachis hypogaea* L.). *Journal of Experimental Botany*, **31**, 1283-1290.

Penaloza, A.P.S. and Eira, M.T.S. (1993). Hydration-dehydration treatments on tomato seeds *(Lycopersicon esculentum* Mill.). *Seed Science and Technology*, **21**, 309-316.

Perl, M. and Feder, Z. (1981). Improved seedling development of pepper seeds (*capsicum annum*) by seed treatment for pregermination activities. *Seed Science and Technology*, **9**, 655-663.

Perry, D. A. (1981). *Seed Science and Technology* (19th ISTA Congress, Vienna, 1980), **9**, 115-126.

Perry, D.A. (1978). Report of the vigour test committee, 1974-1977. *Seed Science and Technology*, **6**, 159-181.

Peterson, M. L., Jones, D.B. and Rutger, J.N. (1978). Cool temperature screening of rice lines for seedling vigour of rice. *Crop Science*, **27**: 269-274.

Pihakaski, K. (1977). A simple means of germinating seeds for plant physiological studies. *Annals of Botany*. **41** : 269-270.

Pill, W.G. (1995). Low water potential and presowing germination treatments to improve seed quality. In: *Seed quality: basic mechanisms and agricultural implications*. The Haworth Press, Binghamton, NY 319-359.

Pimental, D. (2005). Impacts of organic farming on the efficiency of energy use in agriculture. The Organic Centre, Cornell University, USA, pp.1-39.

Pimentel, D., Hepperly, P., Hanson, J., Douds, D., and Seidel, R. (2005). Environmental, energetic, and economic comparisons of organic and conventional farming systems. *Bioscience*, **55**(7), 573-582.

Pittman, U.J. (1967). Biomagnetic responses in Kharkov 22 M.C. winter wheat. *Canadian Journal of Plant Science*, **47**, 389-393.

Pittman, U.J. and Anstey, T.H. (1967). Magnetic treatment and seed orientation of single-harvest snap beans, (*Phaseolus vulgaris* L.). In: *Proceedings of the American Society for Horticultural Science,* **91**, 310-214.

Pittman, U.J. and Ormrod, D.P. (1970). Physiological and chemical features of magnetically treated winter wheat seeds and resultant seedlings. *Canadian Journal of Plant Science*, **50**, 211-217.

Pittman, U.J. and Ormrod, D.P. (1971). Biomagnetic responses in germinating malting barley. *Canadian Journal of Plant Science*, **51**, 64-65.

Poli, G. and Gravela, E. (1982). Lipid peroxidation in isolated hepatocytes In: *Free radical lipid peroxidation and cancer*(eds. D.C.H.Mc Drian and T.S.Slater) Academic Press London.

Pollock, B.M. and Roos, E.E. (1972). Seed and seedling vigor, In: *Seed Biology* (ed. T.T. Kozlowski). **I**, pp, 313-387, Academic Press, New York & London.

Powell, A. A. (1988). Seed vigour and field establishment: A Review. *Advances in Research and Technology of Seeds,***11**, 29-80.

Powell, A. A. (1995). The controlled deterioration test. In: *Seed Vigour Testing Seminar* (ed. H. A. van de Venter), ISTA, Zurich, Switzerland, pp. 73-87.

Powell, A.A. and Harman, G.E. (1985). Absence of a consistent association of changes in membranal lipids with the ageing of pea seeds. *Seed Science and Technology*, **13**, 659-667.

Priestley, D.A. and Leopold, A.C. (1979). Absence of lipid oxidation in accelerated ageing of soybean seeds. *Plant Physiology*, **63**, 726-729.

Priestley, D.A. and Leopold, A.C. (1983). Lipid changes during natural ageing of soybean seeds. *Physiologia Plantarum*, **59**, 467-470.

Priestley, D.A., Werner, B.G. and Leopold, A.C. (1985a). The susceptibility of soybean seed lipids to artificially enhanced atmospheric oxidation. *Journal of Experimental Botany*, **36**, 1653-1659.

Priestley, D.A., Werner, B.G., Leopold, A.C. and McBride, M.B. (1985b). Organic free radical levels in seed and pollen; The effects of hydration and ageing. *Physiologia Plantarum*, **64**, 88-94.

Probert, R.J, Bogh, S.V., Smith, A.J., and Wechsberg, G.E. (1991). The effects of priming on seed longevity in *Ranunculus sceleratus* L. *Seed Science Research*, **1**, 243-249.

Pryor, W.A. and Lasswell, L.D. (1975). Diels-alder and 1, 4-diradical intermediates in the spontaneous polymerisation of vinyl monomers. In: *Advances in Free Radical Chemistry* (ed. G.H. Williams), **5**, 27, Elek Science, London.

Punjabi, B. and Basu, R.N. (1980). A simple method for testing germination and seedling vigour. ISTA Congress, Vienna, Preprint 77/III.

Punjabi, B. and Basu, R.N. (1982). Control of age- and irradiation induced seed deterioration in lettuce (*Lactuca sativa* L.) by hydration-dehydration treatments. In: *Proceedings of the Indian National Science Academy*, **B48**(2), 242-250.

Punjabi, B., Mandal, A.K. and Basu, R.N. (1982). Maintenance of vigour, viability and productivity of stored barley seed. *Seed Research*, **10**(1), 69-71.

Punjabi, B. and Basu, R.N. (1982). Maintenance of lettuce seed viability and counteraction of radiation damage by moisture equilibration-drying treatment, *Indian Biologist*, **14**, 9-12.

Punjabi, B., and Basu, R.N. (1982). Testing germination and seedling growth by an inclined glass plate blotter method. *Indian Journal of Plant Physiology*, **25**, 289-298.

Punjabi, B. and Basu, R.N. (1989). Relation between dormancy and viability of dry stored lettuce (*Lactuca sativa* L. cv. Suttons A1) seed. *Indian Journal of Experimental Biology*, **27**, 739-741.

Radhakrishnan, R., Khan, A.L., and Lee, I-J. (2013). Endophytic fungal pretreatments of seeds alleviates salinity stress effects in soybean plants. *Journal of Microbiology*, **51**, 850-857.

Rahman, F. (1987). Studies on cold storage of tea seed. *Journal of Plantation Crops*, **15**(2), 143.

Rajasekhar, B.G., Rama Rao, S., Krishnasastry, K.S. and Mallana, K.N. (1970). Effect of seed hardening on yield of ragi (*Eleusine coracana* L.) at different levels of fertility. *Mysore Journal of Agricultural Science*, **4**, 271-277.

Rajendra, P., Nayak, H.S., Sashidhar, R.B., Subramanyam, C., Devendranath, D., Gunasekaran, B. (2005). Effects of power frequency electromagnetic fields on growth of germinating *Vicia faba* L., the broad bean. *Electromagnetic Biology and Medicine*,**24**, 39-54.

Rajjou, L., and Debeaujon, I. (2008). Seed longevity; survival and maintenance of high germination ability of dry seeds. *Comptes Rendus Biologies*, **331**, 796-805.

Rajjou, L., Duval, M., Gallardo, K., Catusse, J., Bally, J., Job, C., and Job, D. (2012). Seed germination and vigour. *Annual Review of Plant Biology*, **63**, 507-533.

Ramamoorthy, K. and Basu, R.N. (1984). Control of seed deterioration in groundnut (*Arachis hypogaea* L.) by hydration-dehydration treatments. *Plant Physiology and Biochemistry*, **11**, 148-152.

Ramamoorthy, K. and Basu, R.N. (1996). Seed invigoration for improved vigour, viability and productivity of peanut. *East African Agricultural and Forestry Journal*, **61**(3), 261-272.

Rao, N.K., Roberts, E.H. and Ellis, R.H. (1987). The influence of pre- and post-storage hydration treatments on chromosomal aberrations, seedling abnormalities and viability of lettuce seeds. *Annals of Botany*, **60**, 97-108.

Rashid, A., Alam, B. and Chakraborty, H. (1973). Storage of tea seed. *Tea Journal of Bangladesh*, **9**, 12-15.

Ray, S.K. (1982). Maintenance of vigour, viability and yield potential of stored wheat seed. *Seed Research*, **10**, 139-142.

Ray, Sudipta, Pal, P., Mandal, A.K. and Basu, R.N. (2018). Influence of physiological seed invigoration treatment for improved germinability and extended storability of several local cultivars of rice. *Indian Biologist*. **50**(2): 1-5.

Reddy, K.V., Reshma, S.R., Jareena, S., and Nagaraju, M. (2012). Exposure of greengram seeds (*Vigna radiata* var. radiata) to static magnetic fields; effects on germination and alfa-amylase activity. *Research Journal of Seed Science*, **5**, 106-114..

Reddy, M.V.B., Raghavan, G.S.V., Kushalappa, A.C., and Paulitz, T.C. (1998). Effect of microwave treatment on quality of wheat seeds with *Fusarium graminearum*. *Journal of Agricultural Engineering Research*, **71**, 113-117.

Roberts, E.H. (1981). Physiology of ageing and its application to drying and storage. *Seed Science and Technology*, **9**, 359-372.

Roberts, E.H. (1983). Loss of seed viability during storage. *Advances in Research and Technology of Seeds*, **8**, 9-34.

Rudrapal, A.B. and Basu, R.N. (1979). Carbohydrate utilization in wheat seed in relation to viability maintenance by soaking-drying treatments. *Seed Research*, **7**, 5-10.

Rudrapal, A.B. and Basu, R.N. (1979). Physiology of hydration-dehydration treatment in the maintenance of seed viability in wheat (*Triticum Aestivum* L.) *Indian Journal of Experimental Biology*, **17**, 768-771.

Rudrapal, A.B. and Basu, R.N. (1980). Iodine treatment of mungbean seeds for the maintenance of vigour and viability. *Current Science*, **49**, 319-320.

Rudrapal, A.B. and Basu, R.N. (1981). Use of chlorine and bromine in controlling mustard seed deterioration. *Seed Research*, **9**, 188-191.

Rudrapal, A.B. and Basu, R.N. (1982). Lipid peroxidation and membrane damage in deteriorating wheat and mustard seeds. *Indian Journal of Experimental Biology*, **20**, 465-470.

Rudrapal, D, and Nakamura, S. (1988b). The effects of hydration-dehydration pretreatments on eggplant and radish seed viability and vigour. *Seed Science and Technology*, **16**, 123-130.

Rudrapal, D. and Basu, R.N. (2002). Seed invigoration of mustard (*Brassica juncea* L. Czern & Coss) as influenced by seed vigour status at the time of treatment. *Indian Agriculturist*, **46**(1&2), 1-9.

Rudrapal, D. and Basu, R.N. (2004). The role of initial seed vigour status in the expression of invigoration effect of wet and dry treatments in French bean (*Phaseolus vulgaris* L.). *Seed Science and Technology*, **32**, 205-212.

Rudrapal, D. and Nakamura, S. (1988a). Use of halogens in controlling eggplant and radish seed deterioration. *Seed Science and Technology*, **16**, 115-121.

Saha, R. and Basu, R.N. (1981). Maintenance of soybean seed viability by hydration-dehydration treatments. *Indian Agriculturist,* **25**, 275-278.

Saha, R. and Basu, R.N. (1982). Preconditioning soybean seed for viability maintenance by soaking-drying treatment, *Seed Research*, **10**, 193-197.

Saha, R. and Basu, R.N. (1984). Invigoration of soybean seed for alleviation of soaking injury and ageing damage on germinability. *Seed Science and Technology*, **12**, 613-622.

Saha, R. and Basu, R.N. (1985). Effect of culture filtrate of *aspergillus flavus* on the vigour and viability of stored wheat seed. *Indian Journal of Mycology Research,* **23**(1), 19-26.

Saha, R., Mandal, A.K.and Basu, R.N. (1990). Physiology of seed invigoration treatments in soybean (*Glycine max* L.). *Seed Science and Technology*, **18**, 269-276.

Sashidar, V.R., Mekhri, A.A. and Krishnasastry, K.S. (1977). Proline accumulation in relation to seed hardening in groundnut genotypes. *Indian Journal of Agricultural Sciences*, **47**, 595-598.

Schoorel, A.F. (1956). Report of the activities of the committee on seedling vigour. In: *Proceedings of International Seed Testing Association*, **21**, 282-86.

Schoorel, A.F. (1960). Report on the activities of the vigour test committee. In: *Proceedings of International Seed Testing Association*, **25**, 519-524.

Schutter, O.D. (2011). Agroecology and the right to food. UN Special Rapporteur on the Right to Food Annual Report of the Human Rights Council, March 8, 2011.

Schwember, A.R., and Bradford, K.J. (2005). Drying rates following priming affect temperature sensitivity of germination and longevity of lettuce seeds. *Horticulture Science*, **40**, 778-781.

Scialabba, A., and Tamburello, C. (2002). Microwave effects on germination and growth of raddish (*Raphanus sativus* L.) seedlings. *Acta Botanica Gallica,* **149**, 113-123.

Sebastiampillai, A.R. and Anandappa, T.I. (1979). The influence of moisture and temperature on the germinability and longevity of tea (*Camellia sinensis* L.) seeds. *Tea Quaterly.*, **48**, 8-20.

Sen, S. and Osborne, D.J. (1974). Germination of rye embroys following hydration-dehydration treatments: enhancement of protein and RNA synthesis and earlier induction of RNA replication. *Journal of Experimental Botany*, **25**, 1010-1019.

Sengupta, A. K., De, B. K., & Mandal, A. K. (2005). Pre-storage seed invigoration treatments for the maintenance of vigour, viability and field performance of high-vigour onion seed (*Allium cepa* L.). *Seed Science and Technology*, **33**(3), 753-760.

Sevino, G., Haigh, P.M. and De Leo, P. (1979). Effect of pre-sowing treatments upon seed vigour and viability during storage. *Seed Science and Technology*, **7**, 57-64.

Shakhov, A. (1974). Irradiated seedlings for better crops. *UNESCO Courier*, 9090-**173**, 38-39.

Shaukat, S., Farooq, M., Sissidui, M,, and Zaidi, S. (2013). Effect of enhanced UV-B radiation on germination, seedling growth and biochemical responses of *Vigna mungo* (L.) Hepper, *Pakistan Journal of. Botany*, **45**, 779-785.

Shekari, F., Pakmehr, A., Rastgoo, M., Vazayefi, M., and Zangani, E. (2010). Effects of seed priming with salicylic acid on some morphological traits of a cowpea cultivar (*Vigna unguiculata* L.) under water deficit of podding stage. *Mod Tech Agr*, **4**, 5-26.

Shine, M.B., Guruprasad, K.N., and Anand, A. (2011). Enhancement of germination, growth, and photosynthesis in soybean by pre-treatment of seeds with magnetic field. *Bioelectromagnetics*, **32**, 474-484.

Siddiqui, A., Dawar, S., Javed Zaki, M., and Hamid, N. (2011). Role of ultra violet (UV-C) radiation in the control of root infecting fungi on groundnut and mung bean. *Pakistan Journal of Botany*, **43**, 2221-2224.

Singh, C., Bhan, A.K. and Kaul, B.L. (1974). The role of some physical and chemical agents in the improvement of seed germination in *Datura innoxia*. *Seed Science and Technology*, **2**, 421-425.

Singh, H., Morrs, S, and Orton, T.J. (1985). Effects of osmotic pretreatment and storage in germination of celery seeds. *Seed Science and Technology*, **13**, 551-558.

Sliwinska, E., and Jendrzejczak, E. (2002). Sugar-beet seed quality and DNA synthesis in the embryo in relation to hydration-dehydration cycles. *Seed Science and Technology*, **30**, 597-608.

Smith, M.T. and Adamson, J.H. (1989). Volatile lipid peroxidation breakdown products and viability in seeds of lettuce (*Lactuca sativa* L.). *South African Journal of Science*, **85**, 63-64.

Smith, P.T. and Cobb, B.G. (1991a). Physiological and enzymatic activity of pepper seeds (*Capsicum annuum*) during priming. *Physiologia Plantarum*, **82**, 433-439.

Smith, P.T. and Cobb, B.G. (1991b). Accelerated germination of pepper seed by priming with salt solutions and water. *Horticulture Science*,**26**, 417-419.

Spencer, G.F., Earle, F.A., Wolff, I.A. and Tullent, W.H. (1973). Oxygenation of unsaturated fatty acids during storage. *Chemistry and Physics of Lipids*, **10**, 191-202.

Stewart, R.R.C. and Bewley, J.D. (1980). Lipid peroxidation associated with accelerated ageing of soybean axes. *Plant Physiology*, **65**, 245-248.

Sung, J. M. (1996). Lipid peroxidation and peroxide-scavenging in soybean seeds during ageing. *Physiologia plantarum*, **97**(1), 85-89.

Sur (Mandal), K. and Basu, R.N. (1985). Effect of irradiation on the deteriorative senescence of embryo and endosperm of rice (*Oryza Sativa* L.) seed. *Indian Journal of Plant Physiology*, **28**(4), 303-309.

Sur (Mandal), K. and Basu, R.N. (1985). Influence of hydration-dehydration treatments of rice seed on the growth of embryo-endosperm transplants. *Current Science*, **54**(17), 862-864.

Sur, K. and Basu, R.N. (1990a). Reduced volatile production in wheat by seed invigoration treatments. *Current Science*, **59**, 799-800.

Sur, K. and Basu, R.N. (1986). Control of age and irradiation-induced seed deterioration in rice (*Oryza sativa* L.) by hydration-dehydration treatments. *Seed Research*, **14**, 197-205.

Sur, K. and Basu, R.N. (1990b). Vigour rating of wheat seeds. *Seed Science and Technology*, **18**, 661-671.

Sur, K. and Basu, R.N. (1993). Physiology of age- and gama- irradiation induced seed deterioration in rice (*oryza sativa*, L.). *Seed Research*, **1**, 255-261.

Sur, K., Lahiri, A.K. and Basu, R.N. (1987). Improvement of germinability of some forest tree seeds by acid scarification and hydration-dehydration treatments. *Indian Agriculturist*, **31**(2), 115-122.

Sur, K., Lahiri, A.K. and Basu, R.N. (1988). Hydration-dehydration treatments for improved seed storability of bamboo (*Dendrocalamus strictus* L.). *Indian Forester*, **14**(9), 560-563.

Suzuki, H., Obayashi, S., Yamagishi, J. and Inanaga, S. (1990). Effect of pH of tertiary phosphate solutions on radicle protrusion during priming of carrot seeds. *Journal of the Japanese Society for Horticultural Science*, **59**(3), pp.589-595.

Takayanagi, K. and Murakami, K. (1968). Rapid germinability test with exudates from seeds. *Nature*, London, **218**, 493-494.

Talei, D., Valdiani, A., Maziah, M, and Mohsenkhah, M. (2013). Germination response of MR 219 rice variety to different exposure times and periods of 2450 MHz microwave frequency. *The Scientific World Journal* 408026.

Tanaka, Y, and Umeda, K. (1974). Effects of γ-irradiation on the germination and metabolism of organic acids in raddish seed. *Journal of Food Science and Technology*, Tokyo, **21**, 538-544 (*Horticulture Abstracts*, **45**, 7350).

Tarquis, A.M., and Bradford, K.J. (1992). Prehydration and priming treatments that advance germination also increase the rate of deterioration of lettuce seed. *Journal of Experimental Botany*, **43**, 307-317.

Taylor, A.G., Allen, P.S., Bennet, M.A., Bradford, K.J., Burris, J.S., and Misra, M.K. (1998). Seed enhancements. *Seed Science Research*, **8**, 245-256.

Taylor, A.G., Huang, X.L. and Hill, H.J. (1988b). Sinapine leakage from non-viable cabbage seeds. *Journal of Experimental Botany*, **39**, 1443-1438.

Taylor, A.G., Klein, D.B. and Whitlow, T.H. (1988a). SMP: Solid matrix priming of seeds. *Scientia Horticulturae*, **37**, 1-11.

Toole, V.K. and Toole, E.H. (1953). Seed dormancy in relation to seed longevity. In: *Proceedings of International Seed Testing Association*, **18**, 325-328.

Torne, S.G. and Desai, N.P.R. (1975). Effect of ionizing radiation on seed germination of *Passiflora* species. *Current Science*, **44**, 112-113.

Trawatha, S.E., Steiner, J.J. and Bradford, K.J. (1990). Laboratory vigor tests to predict pepper seedling field emergence performance. *Crop Science*, **30**, 713-717.

Tudor, M, and Ioana, I. (1973). The effect of treatment of sunflower with ultrasound shown during plant development. *Analele Universitatiidin Craiova, Stiinte Agricole*, **4**, 143-157, (*Field Crop Abstracts*, **28**, 3926).

Uphoff, N. (2003). Higher yields with fewer external inputs: The system of rice intensification and potential contributions to agricultural sustainability. *International Journal of Agricultural Sustainability*, **1**, 38-50.

Van Staden, J., Darvey, J.E. and du Plessis, L.M. (1976). Lipid utilization in viable and nonviable *Protea compacta* embryos during germination. *Zeitschrift fur Pflanzenphysiologie*, **77**, 113-119.

Van Toai, T. T., McDonald, M. B. (Jr.) and Staby, G. L. (1986). Cultivar, fungicide seed treatment and storage environment interaction on carry over soybean seed quality. *Seed Science and Technology*, **14**, 301-312.

Vashisth, A., and Nagarajan, S. (2010). Effect on germination and early growth characteristics in sunflower (*Helianthus annuus*) seeds exposed to static magnetic field. *Journal of Plant Physiology*, **167**, 149-156.

Ventura, L., Dona, M., Macovei, A., Carbonera, D., Buttafava, A., Mondoni, A., Rossi, G., and Balestrazzi, A. (2012). Understanding the molecular pathways associated with seed vigour. *Plant Physiology and Biochemistry*, **60**, 196-206.

Verma, M.M. and Banerjee, S.K. (1983). Note on radiographic determination of seed quality in tea (*Camellia sinensis* L.). *Seed Research*, **11**, 228-231.

Vijaykumar, T., Raidu, D.V., Killi, J., Palli, M., Shah, P., kalavakonda, V. and Lakhey, S. (2009). Ecologically sound, economically viable community managed sustainable agriculture in Andhra Pradesh, India. The World Bank, Washington DC, pp.1-29.

Villiers, T. A. (1975). Genetic maintenance of seeds in imbibed storage. *Crop Genetic Resources for Today and Tomorrow*, IBP, **2**, 297-316.

Villiers, T.A. (1974). Seed ageing: Chromosome stability and extended viability of seeds stored fully imbibed. *Plant Physiology*, **53**, 873-878.

Villiers, T.A. and Edgcumbe, D.J. (1975). On the cause of seed deterioration in dry storage. *Seed Science and Technology*, **3**, 761-774.

Wang, W-Q., Liu, S-J., Song, S-Q., and Muller, I.M. (2015). Proteomics of seed development, desiccation tolerance, germination and vigour. *Plant Physiology and Biochemistry*, **86C**, 1-15.

Warren, J.E., and Bennett, M.A. (1997). Seed hydration using the drum priming system. *Horticulture Science,***32**, 1220-1221

Waterworth, W.M., Drury, G.E., Bray, C.M., and West, C.E. (2011). Repairing breaks in the plant genome: the importance of keeping it together. *New Phytologist*, **192**, 805-822.

Waterworth, W.M., Masnavi, G., Bhardwaj, R.M., Jiang, Q., Bray, C.M., and West, C.E. (2010). A plant DNA ligase is an important determinant of seed longevity. *Plant Journal*, **63**, 848-860.

Welbaum, G.E. and Bradford, K.J. (1991). Water relations of seed development and germination in muskmelon (*Cucumis melo* L.) VI. Influence of priming on germination responses to temperature and water potential during seed development. *Journal of Experimental Botany*, **42**, 393-399.

Wesley, R.A., Garner, W.E. and Garner, T.H. (1974). Microwave drying of cotton seed. Energy utilization and its effect on seed-mass temperature and seed germination, ARS US, *Agricultural Research Service*, South Reg., **37**, 13.

Whalley, W.R., Ober, E.S., and Jenkins, M. (2013). Measurement of the matric potential of soil water in the rhizosphere. *Journal of Experimental Botany*, **64**, 3951-3963.

Wilson, D. J. (Jr.) and McDonald, M.B. (Jr.) (1986b). The lipid peroxidation model of seed ageing. *Seed Science and Technology*, **14**, 269-300.

Wilson, D. J. (Jr.). (1995). Storage of orthodox seeds. In: *Seed Quality: Basic Mechanisms and Agricultural Implications* (ed. A. S. Basra), Food Products Press, NY. USA, pp. 173-208.

Wilson, D.O. and McDonald, M.B., Jr. (1986a). A convenient volatile aldehyde assay for measuring seed vigour. *Seed Science and Technology*, **14**, 259-268.

Wolff, S. and Sicard, A.M. (1961). Post-irradiation storage and the growth of barley seedlings. In: *Effects of Ionizing Radiation on Seeds*, pp. 171-179. International Atomic Energy Agency, Vienna.

Wonky-Appiah J.B. (1974). Effect of duration of heat treatment on germination of *dura oil* palm seed. *Ghana Journal of Agricultural Science*, **7**, 57-59.

Woodstock, L.W. and Taylorson, R.B. (1981). Ethanol and acetaldehyde in imbibing soybean seeds in relation to deterioration. *Plant Physiology*, **67**, 424-428.

Woodstock, L.W., Maxon, S., Faust, K. and Bass, L.N. (1983). Use of freeze-drying and acetone impregnation with natural and synthetic antioxidants to improve storability of onion, pepper and parsley seeds. *Journal of the American Society for Horticultural Science*, **108**, 692-696.

Wright, B., Rowse, H., and Whipps, J.M. (2003). Microbial population dynamics on seeds during drum and steeping. *Plant and Soil,*, **255**, 631-640.

Yacoubi, R., Job, C., Belghazi, M., Chaibi, W., and Job, D. (2013). Proteomic analysis of the enhancement of seed vigour on osmoprimed alfalfa seeds germinated under salinity stress. *Seed Science Research*, **23**, 99-110.

Yaldagard, M., Mortazavi, S.A., and Tabatabaie, F. (2008). Application of ultrasonic waves as a priming technique for accelerating and enhancing the germination of barley seed; optimization of method by the Taguchi Approach. *Journal of the Institute of Brewing*, **114**, 14-21.

Yang, D., Wang, N., Yan, X., Shi, J., Zhang, M., Wang, Z., and Yuan, H. (2014). Microencapsulation of seed-coating tebuconazole and its effects on physiology and biochemistry of maize seedlings. *Colloids and Surfaces B: Biointerfaces*, **114**, 241-246.

Yao, Z., Liu, L., Gao, F., Rampitsch, C., Reinecke, D.M., Ozga, J.A., and Ayele, B.T. (2012). Developmental and seed aging mediated regulation of antioxidative genes and differential expression of proteins during pre- and post-germinative phases in pea. *Journal of Plant Physiology*, **15**, 1477-1488.

Zolyneak, C.C. and Lupascu, G. (1972). Combined treatments of plant material with ionizing radiation and electromagnetic fields. 3. Plant material of different sorts under some combined treatments with γ-rays and electromagnetic fields, *Analele Stiinte Universitatii, Al, I, Cuza*, **18**, 213-217, (*Field Crop Abstracts*, **28**, 3113).